Lecture Notes in Mathematics 1358

Springer
Berlin
Heidelberg
New York
Barcelona
Hong Kong
London
Milan
Paris
Singapore
Tokyo

David Mumford

The Red Book of Varieties and Schemes

Second, Expanded Edition

Includes the Michigan Lectures (1974) on Curves and their Jacobians

Springer

Authors

David B. Mumford
Division of Applied Mathematics, Box F
Brown University
Providence, RI 02912, USA
E-mail: David_Mumford@brown.edu

Enrico Arbarello
Scuola Normale Superiore
7, Piazza dei Cavalieri
56126 Pisa, Italy
E-mail: ea@bibsns.sns.it

Cataloging-in-Publication Data applied for

Die Deutsche Bibliothek - CIP-Einheitsaufnahme

Mumford, David:
The red book of varieties and schemes : includes the Michigan lectures (1974) on curves and their Jacobians / David Mumford. - 2., expanded ed. - Berlin ; Heidelberg ; New York ; Barcelona ; Hong Kong ; London ; Milan ; Paris ; Singapore ; Tokyo : Springer, 1999
(Lecture notes in mathematics ; 1358)
ISBN 3-540-63293-X

Mathematics Subject Classification (1991): 14-01, 14A15, 14H10, 14H42, 14H45

ISSN 0075-8434
ISBN 3-540-63293-X Springer-Verlag Berlin Heidelberg New York

Typesetting: I. Jebram using a Springer-TEX macro-package
SPIN: 10979419 41/3111 - 5 4 3 2 1 – Printed on acid free paper

Preface to the Second Edition

At the same time that the first edition of the Springer Lecture Notes version of the "Red Book of Varieties and Schemes" was sold out, the supply of my lecture notes "Curves and their Jacobians", published by the University of Michigan Press, was also exhausted and the copyright returned to me. These two sets of notes have similar purposes. At one time, I had had ambitious plans to write a multi-volume introduction to algebraic geometry and specifically to moduli problems, this being the part of the field which I knew best. Before writing these, I gave lectures on various occasions to explain the ideas informally with the idea that these might be a first draft for parts of a longer more polished version. What I found, however, was that these first drafts, written informally, with pictures and diagrams, printed in a typewriter font and, above all, being short, were more appealing to many readers than a full-blown Grundlehren-style book. Therefore, Springer-Verlag and I made the plan to keep both sets of notes in print a bit longer, reprinting them together, with the original "Red Book" as part one and "Curves and their Jacobians" as part two.

The "Red Book" originated in mimeographed lectures delivered at Harvard in the 60's, whose primary goal was to show that the language of schemes was fundamentally geometric and expresses the intuitions of algebraic geometry clearly. [1] "Curves and their Jacobians", on the other hand, is a write-up of four lectures given at the University of Michigan in 1974, whose goal was to give a bird's eye view of curves, their moduli spaces, their jacobians and how these relate to each other. Although the styles differ because the "Red Book" was addressed to students and "Curves and their Jacobians" to faculty as well, both try to describe ideas as informally and simply as possible.

Since my research now is in stochastic methods in signal processing and image analysis (which I like to call "Pattern Theory" following Grenander), I have not kept up with the literature on the moduli of curves. The last section of the original notes entitled "Guide to the Literature and References" is quite out of date and it is impossible for me to bring it up to date, although I have added references to Harris' work with me and Eisenbud on the Kodaira dimension of the Moduli Space.

---

[1] (Added in proof) Rereading Chapter 3, I realize that I particularly wished to show that Zariski's work, on which I had grown up, could be recast, with no loss of its force, in the language of schemes.

For the Schottky problem, there has been a great deal of progress and Professor Enrico Arbarello has generously contributed a new section "Supplementary Bibliography on the Schottky Problem" to give pointers to this work. Another useful reference for recent work is the book "The Moduli Space of Curves", edited by Dijkgraaf, Faber and van der Geer, Birkhäuser-Boston.

Cambridge, Mass.
May, 1999.

David Mumford

Preface to the First Edition

These notes originated in several classes that I taught in the mid 60's to introduce graduate students to algebraic geometry. I had intended to write a book, entitled "Introduction to Algebraic Geometry", based on these courses and, as a first step, began writing class notes. The class notes first grew into the present three chapters. As there was a demand for them, the Harvard mathematics department typed them up and distributed them for a while (this being in the dark ages before Springer Lecture Notes came to fill this need). They were called "*Introduction to Algebraic Geometry: preliminary version of the first 3 chapters*" and were bound in red. The intent was to write a much more inclusive book, but as the years progressed, my ideas of what to include in this book changed. The book became two volumes, and eventually, with almost no overlap with these notes, the first volume appeared in 1976, entitled "*Algebraic Geometry I: complex projective varieties*". The present plan is to publish shortly the second volume, entitled "*Algebraic Geometry II: schemes and cohomology*", in collaboration with David Eisenbud and Joe Harris.

David Gieseker and several others have, however, convinced me to let Springer Lecture Notes reprint the original notes, long out of print, on the grounds that they serve a quite distinct purpose. Whereas the longer book "*Algebraic Geometry*" is a systematic and fairly comprehensive exposition of the basic results in the field, these old notes had been intended only to explain in a quick and informal way what varieties and schemes are, and give a few key examples illustrating their simplest properties. The hope was to make the basic objects of algebraic geometry as familiar to the reader as the basic objects of differential geometry and topology: to make a variety as familiar as a manifold or a simplicial complex. This volume is a reprint of the old notes without change, except that the title has been changed to clarify their aim.

The weakness of these notes is what had originally driven me to undertake the bigger project: there is no real *theorem* in them! I felt it was hard to convince people that algebraic geometry was a great and glorious field unless you offered them a theorem for their money, and that takes a much longer book. But for a puzzled non-algebraic geometer who wishes to find the facts needed to make sense of some algebro-geometric statement that they want to apply, these notes may be a convenient way to learn quickly the basic definitions. In twenty years of giving colloquium talks about algebraic geometry to audiences of mostly non-algebraic geometers, I have learned only too well that algebraic geometry is not so easily accessible, nor are its basic definitions universally known.

It may be of some interest to recall how hard it was for algebraic geometers, even knowing the phenomena of the field very well, to find a satisfactory language in which to communicate to each other. At the time these notes were written, the field was just emerging from a twenty-year period in which every researcher used his own definitions and terminology, in which the "foundations" of the subject had been described in at least half a dozen different mathematical "languages". Classical style researchers wrote in the informal geometric style of the Italian school, Weil had introduced the concept of *specialization* and made this the cornerstone of his language and Zariski developed a hybrid of algebra and geometry with valuations, universal domains and generic points relative to various fields k playing important roles. But there was a general realization that not all the key phenomena could be clearly expressed and a frustration at sacrificing the suggestive geometric terminology of the previous generation.

Then Grothendieck came along and turned a confused world of researchers upside down, overwhelming them with the new terminology of schemes as well as with a huge production of new and very exciting results. These notes attempted to show something that was still very controversial at that time: that schemes really were the most natural language for algebraic geometry and that you did not need to sacrifice geometric intuition when you spoke "scheme". I think this thesis is now widely accepted within the community of algebraic geometry, and I hope that eventually schemes will take their place alongside concepts like Banach spaces and cohomology, i.e. as concepts which were once esoteric and abstruse, but became later an accepted part of the kit of the working mathematician. Grothendieck being sixty this year, it is a great pleasure to dedicate these notes to him and to send him the message that his ideas remain the framework on which subsequent generations will build.

Cambridge, Mass.
Feb. 21, 1988.

Table of Contents

I. Varieties

The basic object of study in algebraic geometry is an arbitrary prescheme. However, among all preschemes, the classical ones known as varieties are by far the most accessible to intuition. Moreover, in dealing with varieties one can carry over without any great difficulty the elementary methods and results of the other geometric categories, i.e., of topological spaces, differentiable manifolds or of analytic spaces. Finally, in any study of general preschemes, the varieties are bound, for many reasons which I will not discuss here, to play a unique and central role. Therefore it is useful and helpful to have a basic idea of what a variety is before plunging into the general theory of preschemes. We will fix throughout an algebraically closed ground field k which will never vary. We shall restrict ourselves to the purely geometric operations on varieties in keeping with the aim of establishing an intuitive and geometric background: thus we will not discuss specialization, nor will we use generic points. This set-up is the one pioneered by Serre in his famous paper "Faisceaux algébriques cohérents". There is no doubt that it is completely adequate for the discussion of nearly all purely geometric questions in algebraic geometry.

§1. Some algebra

We want to study the locus V of roots of a finite set of polynomials $f_i(X_1, \ldots, X_n)$ in k^n (k being an algebraically closed field). However, the basic tool in this study is the ring of functions from V to k obtained by restricting polynomials from k^n to V. And we cannot get very far without knowing something about the algebra of such a ring. The purpose of this section is to prove 2 basic theorems from commutative algebra that are key tools in analyzing these rings, and hence also the loci such as V. We include these results because of their geometric meaning, which will emerge gradually in this chapter (cf. §7). On the other hand, we assume known the following topics in algebra:

1) The essentials of field theory (Galois theory, separability, transcendence degree).
2) Localization of a ring, the behaviour of ideals in localization, the concept of a local ring.
3) Noetherian rings, and the decomposition theorem of ideals in these rings.
4) The concept of integral dependence, (cf., for example, Zariski-Samuel, vol. 1).

The first theorem is:

Noether's Normalization Lemma. *Let R be an integral domain, finitely generated over a field k. If R has transcendence degree n over k, then there exist elements $x_1, \ldots, x_n \in R$, algebraically independent over k, such that R is integrally dependent on the subring $k[x_1, \ldots, x_n]$ generated by the x's.*

Proof (Nagata). Since R is finitely generated over k, we can write R as a quotient:

$$R = k\,[Y_1, \ldots, Y_m]\,/P\ ,$$

for some prime ideal P. If $m = n$, then the images $y_1, \ldots, y_m$ of the Y's in R must be algebraically independent themselves. Then $P = (0)$, and if we let $x_i = y_i$, the lemma follows. If $m > n$, we prove the theorem by induction on m. It will suffice to find a subring S in R generated by $m - 1$ elements and such that R is integrally dependent on S. For, by induction, we know that S has a subring $k\,[x_1, \ldots, x_n]$ generated by n independent elements over which it is integrally dependent; by the transitivity of integral dependence, R is also integrally dependent on $k\,[x_1, \ldots, x_n]$ and the lemma is true for R.

Now the m generators $y_1, \ldots, y_m$ of R cannot be algebraically independent over k since $m > n$. Let

$$f\,(y_1, \ldots, y_m) = 0$$

by some non-zero algebraic relation among them (i.e., $f\,(y_1, \ldots, y_m)$ is a non-zero polynomial in P). Let $r_1, \ldots, r_m$ be positive integers, and let

$$z_2 = y_2 - y_1^{r_2}, \qquad z_3 = y_3 - y_1^{r_3}, \ldots, z_m = y_m - y_1^{r_m}\ .$$

Then

$$f\,(y_1, z_2 + y_1^{r_2}, \ldots, z_m + y_1^{r_m}) = 0\ ,$$

i.e., $y_1, z_2, \ldots, z_m$ are roots of the polynomial $f\,(Y_1, Z_2 + Y_1^{r_2}, \ldots, Z_m + Y_1^{r_m})$.

Each term $a \cdot \prod_{i=1}^{m} y_i^{b_i}$ in f gives rise to various terms in this new polynomial, including one monomial term

$$a \cdot y_1^{b_1 + r_2 b_2 + \ldots + r_m b_m}\ .$$

A moment's reflection will convince the reader that if we pick the r_i's to be large enough, and increasing rapidly enough:

$$0 \ll r_2 \ll r_3 \ll \ldots \ll r_m,$$

then these new terms $a \cdot Y_1^{b_1 + \ldots + r_m b_m}$ will all have distinct degrees, and one of them will emerge as the term of highest order in this new polynomial. Therefore,

$$f\,(Y_1, Z_2 + Y_1^{r_2}, \ldots, Z_m + Y_1^{r_m}) = b \cdot Y_1^N + \ \ [\text{ terms of degree } < N]\ ,$$

$(b \neq 0)$. This implies that the equation $f\,(y_1, z_2 + y_1^{r_2}, \ldots, z_m + y_1^{r_m}) = 0$ is an equation of integral dependence for y_1 over the ring $k[z_2, \ldots, z_m]$. Thus y_1 is integrally dependent on $k[z_2, \ldots, z_m]$, so $y_2, \ldots, y_m$ are too since $y_i = z_i + y_1^{r_i}$ $(i = 2, \ldots, m)$. Therefore the whole ring R is integrally dependent on the subring $S = k\,[z_2, \ldots, z_m]$. By induction, this proves the lemma. □

The second important theorem is:

Going-up theorem of Cohen-Seidenberg. *Let R be a ring (commutative as always) and $S \subset R$ a subring such that R is integrally dependent on S. For all prime ideals $P \subset S$, there exist prime ideals $P' \subset R$ such that $P' \cap S = P$.*

Proof. Let M be the multiplicative system $S - P$. Then we may as well replace R and S by their localizations R_M and S_M with respect to M. For S_M is still a subring of R_M, and R_M is still integrally dependent on S_M. In fact, we get a diagram:

$$\begin{array}{ccc} R & \xrightarrow{\;j\;} & R_M \\ \cup & & \cup \\ S & \xrightarrow[\;i\;]{} & S_M \end{array} .$$

Moreover S_M is a local ring, with maximal ideal $P_M = i(P) \cdot S_M$ and $P = i^{-1}(P_M)$. If $P^* \subset R_M$ is a prime ideal of R_M such that $P^* \cap S_M = P_M$, then $j^{-1}(P^*)$ is a prime ideal in R such that

$$j^{-1}(P^*) \cap S = i^{-1}(P^* \cap S_M) = i^{-1}(P_M) = P.$$

Therefore, it suffices to prove the theorem for R_M and S_M.

Therefore we may assume that S is a local ring and P is its unique maximal ideal. In this case, for all ideals $A \subset R$, $A \cap S \subseteq P$. I claim that for *all* maximal ideals $P' \subset S$, $P' \cap S$ equals P. Since maximal ideals are prime, this will prove the theorem. Take some maximal ideal P'. Then consider the pair of quotient rings:

$$\begin{array}{ccc} R & \longrightarrow & R/P' \\ \cup & & \cup \\ S & \longrightarrow & S/S \cap P' \end{array} .$$

Since P' is maximal, R/P' is a field. If we can show that the subring $S/S \cap P'$ is a field too, then $S \cap P'$ must be a maximal ideal in S, so $S \cap P'$ must equal P and the theorem follows. Therefore, we have reduced the question to:

Lemma. *Let R be a field, and $S \subset R$ a subring such that R is integrally dependent on S. Then S is a field.*

Note that this is a special case of the theorem: For if S were not a field, it would have non-zero maximal ideals and these could not be of the form $P' \cap S$ since R has no non-zero ideals at all.

Proof of lemma. Let $a \in S$, $a \neq 0$. Since R is a field, $1/a \in R$. By assumption, $1/a$ is integral over S, so it satisfies an equation

$$X^n + b_1 X^{n-1} + \ldots + b_n = 0,$$

$b_i \in S$. But this means that

$$\frac{1}{a^n} + \frac{b_1}{a^{n-1}} + \ldots + b_n = 0.$$

Multiply this equation by a^{n-1} and we find

$$\frac{1}{a} = -b_1 - ab_2 - \ldots - a^{n-1}b_n \in S.$$

Therefore S is a field. □

Using both of these results, we can now prove:

Weak Nullstellensatz. *Let k be an algebraically closed field. Then the maximal ideals in the ring $k[X_1, \ldots, X_n]$ are the ideals*

$$(X_1 - a_1, X_2 - a_2, \ldots, X_n - a_n),$$

where $a_1, \ldots, a_n \in k$.

Proof. Since the ideal $(X_1 - a_1, \ldots, X_n - a_n)$ is the kernel of the surjection:

$$\begin{aligned} k[X_1, \ldots, X_n] &\longrightarrow k \\ f[X_1, \ldots, X_n] &\longmapsto f(a_1, \ldots, a_n), \end{aligned}$$

it follows that $k[X_1, \ldots, X_n] / (X_1 - a_1, \ldots, X_n - a_n) = k$, hence the ideal $(X_1 - a_1, \ldots, X_n - a_n)$ is maximal. Conversely, let $M \subset k[X_1, \ldots, X_n]$ be a maximal ideal. Let $R = k[X_1, \ldots, X_n]/M$. R is a field since M is maximal, and R is also finitely generated over k as a ring. Let r be the transcendence degree of R over k.

The crux of the proof consists in showing that $r = 0$: By the normalization lemma, find a subring $S \subset R$ of the form $k[y_1, \ldots, y_r]$ such that R is integral over S. Since the y_i's are algebraically independent, S is a polynomial ring in r variables. By the going-up theorem – in fact, by the special case given in the lemma – S must be a field too. But a polynomial ring in r variables is a field only when $r = 0$.

Therefore R is an algebraic extension field of k. Since k is algebraically closed, R must equal k. In other words, the subset k of $k[X_1, \ldots, X_n]$ goes *onto* $k[X_1, \ldots, X_n]/M$. Therefore

$$k + M = k[X_1, \ldots, X_n].$$

In particular, each variable X_i is of the form $a_i + m_i$, with $a_i \in k$ and $m_i \in M$. Therefore, $X_i - a_i \in M$ and M contains the ideal $(X_1 - a_1, \ldots, X_n - a_n)$. But the latter is maximal already, so $M = (X_1 - a_1, \ldots, X_n - a_n)$. □

The great importance of this result is that it gives us a way to translate affine space k^n into pure algebra. We have a bijection between k^n, on the one hand, and the set of maximal ideals in $k[X_1, \ldots, X_n]$ on the other hand. This is the origin of the connection between algebra and geometry that gives rise to our whole subject.

§2. Irreducible algebraic sets

For the rest of this chapter k will denote a fixed algebraically closed field, known as the *ground field.*

Definition 1. A *closed algebraic subset* of k^n is a set consisting of all roots of a finite collection of polynomial equations: i.e.,

$$\{(x_1, \ldots, x_n) \mid f_1(x_1, \ldots, x_n) = \ldots = f_m(x_1, \ldots, x_n) = 0\} .$$

It is clear that the above set depends only on the ideal $A = (f_1, \ldots, f_m)$ generated by the f_i's in $[X_1, \ldots, X_n]$ and not on the actual polynomials f_i. Therefore, if A is any ideal in $k[X_1, \ldots, X_n]$, we define

$$V(A) = \{x \in k^n \mid f(x) = 0 \quad \text{for all } f \in A\} .$$

Since $k[X_1, \ldots, X_m]$ is a noetherian ring, the subsets of k^n of the form $V(A)$ are exactly the closed algebraic sets. On the other hand, if Σ is a closed algebraic set, we define

$$I(\Sigma) = \{f \in k[X_1, \ldots, X_n] \mid f(x) = 0 \quad \text{for all } x \in \Sigma\} .$$

Clearly $I(\Sigma)$ is an ideal such that $\Sigma = V(I(\Sigma))$. The key result is:

Theorem 1 (Hilbert's Nullstellensatz).

$$I(V(A)) = \sqrt{A} .$$

Proof. It is clear that $\sqrt{A} \subset I(V(A))$. The problem is to show the other inclusion. Put concretely this means the following:

> Let $A = (f_1, \ldots, f_m)$. If $g \in k[X_1, \ldots, X_n]$ satisfies:
>
> $$\{f_1(a_1, \ldots, a_n) = \ldots = f_m(a_1, \ldots, a_n) = 0\} \Longrightarrow g(a_1, \ldots, a_n) = 0$$
>
> then there is an integer ℓ and polynomials $h_1, \ldots, h_m$ such that
>
> $$g^\ell(X) = \sum_{i=1}^m h_i(X) \cdot f_i(X).$$

To prove this, introduce the ideal

$$B = A \cdot k[X_1, \ldots, X_n, X_{n+1}] + (1 - g \cdot X_{n+1})$$

in $k[X_1, \ldots, X_{n+1}]$. There are 2 possibilities: either B is a proper ideal, or $B = k[X_1, \ldots, X_{n+1}]$. In the first case, let M be a maximal ideal in $k[X_1, \ldots, X_{n+1}]$ containing B. By the weak Nullstellensatz of §1,

$$M = (X_1 - a_1, \ldots, X_n - a_n, X_{n+1} - a_{n+1})$$

for some elements $a_i \in k$. Since M is the kernel of the homomorphism:

$$\begin{aligned} k[X_1, \ldots, X_n, X_{n+1}] &\longrightarrow k \\ f &\longmapsto f(a_1, \ldots, a_{n+1}), \end{aligned}$$

$B \subset M$ means that:

$$i) \qquad f_1(a_1, \ldots, a_n) = \ldots = f_m(a_1, \ldots, a_n) = 0$$

and

$$ii) \qquad 1 = g(a_1, \ldots, a_n) \cdot a_{n+1}.$$

But by our assumptionon g, (i) implies that $g(a_1, \ldots, a_n) = 0$, and this contradicts (ii). We can only conclude that the ideal B would not have been a proper ideal.

But then $1 \in B$. This means that there are polynomials $h_1, \ldots, h_m, h_{m+1} \in k[X_1, \ldots, X_{n+1}]$ such that:

$$\begin{aligned} 1 = &\sum_{i=1}^{m} h_i(X_1, \ldots, X_{m+1}) \cdot f_i(X_1, \ldots, X_n) \\ &+ (1 - g(X_1, \ldots, X_n) \cdot X_{n+1}) \cdot h_{m+1}(X_1, \ldots, X_{n+1}). \end{aligned}$$

Substituting g^{-1} for X_{n+1} in this formula, we get:

$$1 = \sum_{i=1}^{m} h_i(X_1, \ldots, X_n, 1/g) \cdot f_i(X_1, \ldots, X_n).$$

Clearing denominators, this gives:

$$g^{\ell}(X_1, \ldots, X_n) = \sum_{i=1}^{m} h_i^*(X_1, \ldots, X_n) \cdot f_i(X_1, \ldots, X_n)$$

for some new polynomials $h_i^* \in k[X_1, \ldots, X_n]$, i.e., $g \in \sqrt{A}$. □

Corollary. *V and I set up a bijection between the set of closed algebraic subsets of k^n and the set of ideals $A \subset k[X_1, \ldots, X_n]$ such that $A = \sqrt{A}$.*

This correspondence between algebraic sets and ideals is compatible with the lattice structures:

i) $A \subset B \Longrightarrow V(A) \supset V(B)$
ii) $\Sigma_1 \subset \Sigma_2 \Longrightarrow I(\Sigma_1) \supset I(\Sigma_2)$
iii) $V\left(\sum_\alpha A_\alpha\right) = \bigcap_\alpha V(A_\alpha)$

iv) $V(A \cap B) = V(A) \cup V(B)$

where A, B, A_α are ideals, Σ_1, Σ_2 closed algebraic sets.

Proof. All are obvious except possibly (iv). But by (i), $V(A \cap B) \supset V(A) \cup V(B)$. Conversely, if $c \notin V(A) \cup V(B)$, then there exist polynomials $f \in A$ and $g \in B$ such that $f(x) \neq 0$, $g(x) \neq 0$. But then $f \cdot g \in A \cap B$ and $(f \cdot g)(x) \neq 0$, hence $x \notin V(A \cap B)$. □

Definition 2. A closed algebraic set is *irreducible* if it is not the union of two strictly smaller closed algebraic sets. (We shall omit "closed" in referring to these sets).

Recall that by the noetherian decomposition theorem, if $A \subset k[X_1, \ldots, X_n]$ is an ideal such that $A = \sqrt{A}$, then A can be written in exactly one way as an intersection of a finite set of *prime* ideals, none of which contains any other. And a prime ideal is not the intersection of any two strictly bigger ideals. Therefore:

Proposition 2. *In the bijection of the Corollary to Theorem 1, the irreducible algebraic sets correspond exactly to the prime ideals of* $k[X_1, \ldots, X_n]$. *Moreover, every closed algebraic set* Σ *can be written in exactly one way as:*

$$\Sigma = \Sigma_1 \cup \ldots \cup \Sigma_k$$

where the Σ_i *are irreducible sets and* $\Sigma_i \not\subset \Sigma_j$ *if* $i \neq j$.

Definition 3. The Σ_i of Proposition 2 will be called the *components* of Σ.

In the early 19th century it was realized that for many reasons it was inadequate and misleading to consider only the above "affine" algebraic sets. Among others, Poncelet realized that an immense simplification could be introduced in many questions by considering "projective" algebraic sets (cf. Felix Klein, *Die Entwicklung der Mathematik*, part I, pp. 80–82). Even to this day, there is no doubt that projective algebraic sets play a central role in algebro-geometric questions; therefore we shall define them as soon as possible.

Recall that, by definition, $\mathbb{P}_n(k)$ is the set of $(n+1)$-tuples $(x_0, \ldots, x_n) \in k^{n+1}$ such that some $x_i \neq 0$, modulo the equivalence relation

$$(x_0, \ldots, x_n) \sim (\alpha x_0, \ldots, \alpha x_n), \qquad \alpha \in k^*,$$

(where k^* is the multiplicative group of non-zero elements of k). Then an $(n+1)$-tuple $(x_0, x_1, \ldots, x_n)$ is called a set of *homogeneous coordinates* for the point associated to it. $\mathbb{P}_n(k)$ can be covered by $n+1$ subsets $U_0, U_1, \ldots, U_n$, where

$$U_i = \left\{ \begin{array}{l} \text{points represented by homogeneous} \\ \text{coordinates} \quad (x_0, x_1, \ldots, x_n) \text{ with } x_i \neq 0 \end{array} \right\}.$$

Each U_i is naturally isomorphic to k^n under the map

$$\begin{array}{rcl} U_i & \longrightarrow & k^n \\ (x_0, x_1, \ldots, x_n) & \longmapsto & \left(\frac{x_0}{x_i}, \frac{x_1}{x_i}, \ldots, \frac{x_n}{x_i}\right), \quad \left(\frac{x_i}{x_i} \text{ omitted}\right). \end{array}$$

The original motivation for introducing $\mathbb{P}_n(k)$ was to add to the affine space $k^n \cong U_0$ the extra "points at infinity" $\mathbb{P}_n(k) - U_0$ so as to bring out into the open the mysterious things that went on at infinity.

Recall that to all subvectorspaces $W \subset k^{n+1}$ one associates the set of points $P \in \mathbb{P}_n(k)$ with homogeneous coordinates in W: the sets so obtained are called the linear subspaces L of $\mathbb{P}_n(k)$. If W has codimension 1, then we get a hyperplane. In particular, the points "at infinity" with respect to the affine piece U_i form the hyperplane associated to the subvectorspace $x_i = 0$. Moreover, by introducing a basis into W, the linear subspace L associated to W is naturally isomorphic to $\mathbb{P}_r(k)$, $(r = \dim W - 1)$. The linear subspaces are the simplest examples of projective algebraic sets:

Definition 4. A *closed algebraic set* in $\mathbb{P}_n(k)$ is a set consisting of all roots of a finite collection of homogeneous polynomials $f_i \in k[X_0, \ldots, X_n]$, $1 \le i \le m$. This makes sense because if f is homogeneous, and $(x_0, \ldots, x_n)$, $(\alpha x_0, \ldots, \alpha x_n)$ are 2 sets of homogeneous coordinates of the same point, then

$$f(x_0, \ldots, x_n) = 0 \Longleftrightarrow f(\alpha x_0, \ldots, \alpha x_n) = 0.$$

We can now give a projective analog of the V, I correspondence used in the affine case. We shall, of course, now use only homogeneous ideals $A \subset k[X_0, \ldots, X_n]$: i.e., ideals which, when they contain a polynomial f, also contain the homogeneous components of f. Equivalently, homogeneous ideals are the ideals generated by a finite set of homogeneous polynomials. If A is a homogeneous ideal, define

$$V(A) = \left\{ P \in \mathbb{P}_n(k) \;\middle|\; \begin{array}{l} \text{If } (x_0, \ldots, x_n) \text{ are homogeneous coordinates} \\ \text{of } P, \text{ then } f(x) = 0, \text{ all } f \in A \end{array} \right\}.$$

If $\Sigma \subset \mathbb{P}_n(k)$ is a closed algebraic set, then define

$$I(\Sigma) = \left\{ \begin{array}{l} \text{ideal generated by all homogeneous polynomials} \\ \text{that vanish identically on } \Sigma \end{array} \right\}.$$

Theorem 3. *V and I set up a bijection between the set of closed algebraic subsets of $\mathbb{P}_n(k)$, and the set of all homogeneous ideals $A \subset k[X_0, \ldots, X_n]$, such that $A = \sqrt{A}$ except for the one ideal $A = (X_0, \ldots, X_n)$.*

Proof. It is clear that if Σ is a closed algebraic set, then $V(I(\Sigma)) = \Sigma$. Therefore, in any case, V and I set up a bijection between closed algebraic subsets of $\mathbb{P}_n(k)$ and those homogeneous ideals A such that:

$$(*) \qquad A = I(V(A)).$$

These ideals certainly equal their own radical. Moreover, the empty set is $V((X_0,\ldots,X_n))$, hence $1 \in I(V((X_0,\ldots,X_n)))$; so $(X_0,\ldots,X_n)$ does not satisfy (*) and must be excluded. Finally, let A be any other homogeneous ideal which equals its own radical. Let $V^*(A)$ be the closed algebraic set corresponding to A in the *affine* space k^{n+1} with coordinates $X_0,\ldots,X_n$. Then $V^*(A)$ is invariant under the substitutions

$$(X_0,\ldots,X_n) \longrightarrow (\alpha X_0,\ldots,\alpha X_n) ,$$

all $\alpha \in k$. Therefore, either

1) $V^*(A)$ is empty,
2) $V^*(A)$ equals the origin only, or
3) $V^*(A)$ is a union of lines through the origin: i.e., it is the cone over the subset $V(A)$ in $\mathbb{P}_n(k)$.

Moreover, by the affine Nullstellensatz, we know that

$$(**) \qquad A = I(V^*(A)).$$

In case 1), (**) implies that $A = k[X_0,\ldots,X_n]$, hence $I(V(A))$ – which always contains A – must equal A since there is no bigger ideal. In case 2), (**) implies that $A = (X_0,\ldots,X_n)$ which we have excluded. In case 3), if f is a homogeneous polynomial, then f vanishes on $V(A)$ if and only if f vanishes on $V^*(A)$. Therefore by (**), if f vanishes on $V(A)$, then $f \in A$, i.e.,

$$A \supset I(V(A)).$$

Since the other inclusion is obvious, Theorem 3 is proven. □

The same lattice-theoretic identities hold as in the affine case. Moreover, we define *irreducible* algebraic sets exactly as in the affine case. And we obtain the analog of Proposition 2:

Proposition 4. *In the bijection of Theorem 3, the irreducible algebraic sets correspond exactly to the homogeneous prime ideals ($(X_0,\ldots,X_n)$ being accepted). Moreover, every closed algebraic set Σ in $\mathbb{P}_n(k)$ can be written in excatly one way as:*

$$\Sigma = \Sigma_1 \cup \ldots \cup \Sigma_k ,$$

where the Σ_i are irreducible algebraic sets and $\Sigma_i \not\subset \Sigma_j$ if $i \neq j$.

Problem. Let $\Sigma \subset \mathbb{P}_n(k)$ be a closed algebraic set, and let H be the hyperplane $X_0 = 0$. Identify $\mathbb{P}_n - H$ with k^n in the usual way. Prove that $\Sigma \cap (\mathbb{P}_n - H)$ is a closed algebraic subset of k^n and show that the ideal of $\Sigma \cap (\mathbb{P}_n - H)$ is derived from the ideal of Σ in a very natural way. For details on the relationship between the affine and projective set-up and on everything discussed so far, read Zariski-Samuel, vol. 2, Ch. 7, §§3,4,5 and 6.

Example A. Hypersurfaces. Let $f(X_0, \ldots, X_n)$ be an irreducible homogeneous polynomial. Then the principal ideal (f) is prime, so $f = 0$ defines an irreducible algebraic set in $\mathbb{P}_n(k)$ called a hypersurface (e.g. plane curve, surface in 3-space, etc.).

Example B. The twisted cubic in $\mathbb{P}_3(k)$. This example is given to show the existence of nontrivial examples: Start with the ideal:

$$A_0 = (xz - y^2, yw - z^2) \subset k[x, y, z, w].$$

$V(A_0)$ is just the intersection in $\mathbb{P}_3(k)$ of the 2 quadrics $xz = y^2$ and $yw = z^2$. Look in the affine space with coordinates

$$X = x/w, \qquad Y = y/w, \qquad Z = z/w$$

(the complement of $w = 0$). In here, $V(A_0)$ is the intersection of the ordinary cone $XZ = Y^2$, and of the cylinder over the parabola $Y = Z^2$. This intersection falls into 2 pieces: the line $Y = Z = 0$, and the twisted cubic itself. Correspondingly, the ideal A_0 is an intersection of the ideal of the line and of the twisted cubic:

$$A_0 = (y, z) \cap \underbrace{(xz - y^2,\ yw - z^2,\ xw - yz)}_{A}.$$

The twisted cubic is, by definition, $V(A)$. [To check that A is prime, the simplest method is to verify that A is the kernel of the homomorphism ϕ:

$$k[x, y, z, w] \xrightarrow{\phi} k[s, t]$$

$$\begin{aligned} \phi(x) &= s^3 \\ \phi(y) &= s^2 t \\ \phi(z) &= st^2 \\ \phi(w) &= t^3 \ ]. \end{aligned}$$

In practice, it may be difficult to tell whether a given ideal is prime or whether a given algebraic set is irreducible. It is relatively easy for principal ideals, i.e., for hypersurfaces, but harder for algebraic sets of higher codimension. A good deal of effort used to be devoted to compiling lists of *all* types of irreducible algebraic sets of given dimension and "degree" when these were small numbers. In Semple and Roth, *Algebraic Geometry*, one can find the equivalent of such lists. A study of these will give one a fair feeling for the menagerie of algebraic sets that live in $\mathbb{P}_3$, $\mathbb{P}_4$ or $\mathbb{P}_5$ for example. As for the general theory, it is far from definitive however.

§3. Definition of a morphism

We will certainly want to know when 2 algebraic sets are to be considered isomorphic. More generally, we will need to define not just the *set* of all algebraic sets, but the *category* of algebraic sets (for simplicity, in Chapter 1, we will stick to the irreducible ones).

Example C. Look at

a) k, the affine line,
b) $y = x^2$ in k^2, the parabola.

Projecting the parabola onto the x-axis should surely be an isomorphism between these algebraic sets:

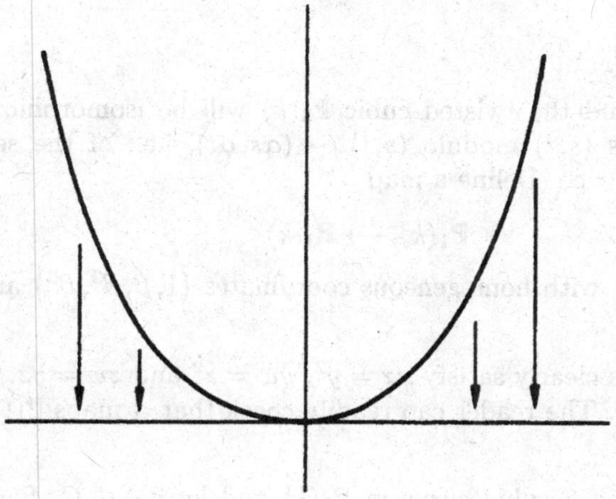

More generally, if $V \subset k^n$ is an irreducible algebraic set, and if $f \in k[X_1, \ldots, X_n]$, then the set of points:

$$V^* = \{(x_1, \ldots, x_n, f(x_1, \ldots, x_n)) \mid (x_1, \ldots, x_n) \in V\} \subset k^{n+1}$$

is an irreducible algebraic set. And the projection

$$(x_1, x_2, \ldots, x_{n+1}) \longmapsto (x_1, x_2, \ldots, x_n)$$

should define an isomorphism from V^* to V.

Example D. An irreducible conic $C \subset \mathbb{P}_2(k)$ will turn out to be *isomorphic* to the projective line $\mathbb{P}_1(k)$ under the following map: fix a point $P_0 \in C$. Identify $\mathbb{P}_1(k)$ with the set of all lines through P_0 in the classical way. Then define a map

$$\mathbb{P}_1(k) \xrightarrow{\alpha} C$$

by letting $\alpha(\ell)$ for all lines ℓ through P_0 be the second point in which ℓ meets C, besides P_0. Also, if ℓ is the tangent line to C at P_0, define $\alpha(\ell)$ to be P_0 itself (since P_0 is a "double" intersection of C and this tangent line).

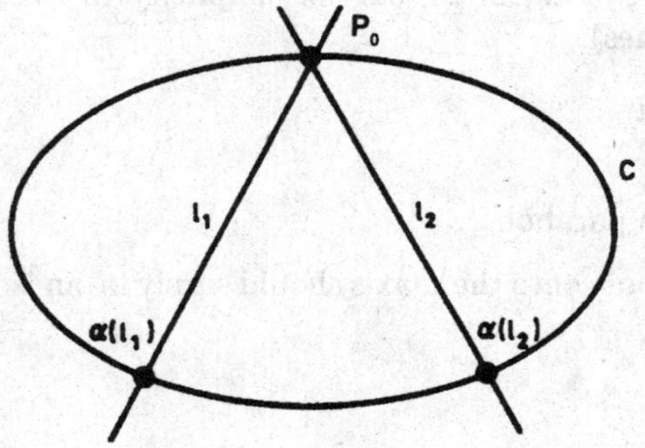

Example E. $\mathbb{P}_1(k)$ and the twisted cubic $\mathbb{P}_3(k)$ will be isomorphic. For $\mathbb{P}_1(k)$ consists of the pairs (s,t) modulo $(s,t) \sim (\alpha s, \alpha t)$, i.e., of the set of ratios $\beta = s/t$ including $\beta = \infty$. Define a map

$$\mathbb{P}_1(k) \xrightarrow{\alpha} \mathbb{P}_3(k)$$

by $\alpha(\beta) =$ the point with homogeneous coordinates $(1, \beta, \beta^2, \beta^3)$ and $\alpha(\infty) = (0,0,0,1)$.

The image points clearly satisfy $xz = y^2$, $yw = z^2$ and $zw = yz$, so they are on the twisted cubic. The reader can readily check that α maps $\mathbb{P}_1(k)$ *onto* the twisted cubic.

Example F. Let C be a cubic curve in $\mathbb{P}_2(k)$ and let $P_0 \in C$. For any point $P \in C$, let ℓ be the line through P and P_0 and let $\alpha(P)$ be the third point in which ℓ meets C. Although this may not seem as obvious as the previous examples, α will be an automorphism of C of order 2.

We shall use this example later to work out our definition in a nontrivial case.

Now turn to the problem of actually defining morphisms, and hence isomorphisms, of irreducible algebraic sets. First consider the case of 2 irreducible *affine* algebraic sets.

Definition 1. Let $\Sigma_1 \subset k^{n_1}$ and $\Sigma_2 \subset k^{n_2}$ be two irreducible algebraic sets. A map

$$\alpha : \Sigma_1 \longrightarrow \Sigma_2$$

will be called a *morphism* if there exist n_2 polynomials $f_1, \ldots, f_{n_2}$ in the variables $X_1, \ldots, X_{n_1}$ such that

$$(*) \qquad \alpha(x) = \left(f_1\left(x_1, \ldots, x_{n_1}\right), \ldots, f_{n_2}\left(x_1, \ldots, x_{n_1}\right)\right)$$

for all points $x = (x_1, \ldots, x_{n_1}) \in \Sigma_1$.

Note one feature of this definition: it implies that every morphism α from Σ_1 to Σ_2 is the restriction of a morphism α' from k^{n_1} to k^{n_2}. This may look odd at first, but it turns out to be reasonable – cf. §4. Note also that with this definition the map in Example C above is an isomorphism, i.e., both it and its inverse are morphisms.

To analyze the definition further, suppose

$$\begin{aligned} P_1 &\subset k[X_1,\ldots,X_{n_1}] \\ P_2 &\subset k[X_1,\ldots,X_{n_2}] \end{aligned}$$

are the prime ideals $I(\Sigma_1)$ and $I(\Sigma_2)$ respectively. Set

$$R_i = k[X_1,\ldots,X_{n_i}]/P_i, \qquad i = 1,2 \ .$$

Then R_1 (resp. R_2) is just the ring of k-valued functions on Σ_1 (resp. Σ_2) obtained by restricting the ring of polynomial functions on the ambient affine space. Suppose $g \in R_2$. Regarding g as a function on Σ_2, the definition of morphism implies that the function $g \cdot \alpha$ on Σ_1 is in R_1 – in fact

$$(g \cdot \alpha)(X_1,\ldots,X_{n_1}) = g\left(f_1(X_1,\ldots,X_{n_1}),\ldots,f_{n_2}(X_1,\ldots,X_{n_1})\right).$$

Therefore α induces a k-homomorphism:

$$\alpha^* : R_2 \longrightarrow R_1 .$$

Moreover, note that α is determined by α^*. This is so because the polynomials $f_1,\ldots,f_{n_2}$ can be recovered – up to an element of P_1 – as $\alpha^*(X_1),\ldots,\alpha^*(X_{n_2})$; and the point $\alpha(x)$, for $x \in \Sigma_1$, is determined via $f_1,\ldots,f_{n_2}$ *modulo* P_1 by equation $(*)$. Even more is true. Suppose you start with an arbitrary k-homomorphism

$$\lambda : R_2 \longrightarrow R_1 .$$

Let f_i be a polynomial in $k[X_1,\ldots,X_{n_1}]$ whose image modulo P_1 equals $\lambda(X_i)$, for all $1 \le i \le n_2$. Then define a map

$$\alpha' : k^{n_1} \longrightarrow k^{n_2}$$

by

$$\alpha'(x_1,\ldots,x_{n_1}) = \left(f_1(x_1,\ldots,x_{n_1}),\ldots,f_{n_2}(x_1,\ldots,x_{n_1})\right).$$

If $x = (x_1,\ldots,x_{n_1}) \in \Sigma_1$, then actually $\alpha'(x)$ will be in Σ_2: for if $g \in P_2$, then

$$g(\alpha'(x)) = g(f_1(x),\ldots,f_{n_2}(x)) .$$

But $g(f_1,\ldots,f_{n_2}) \equiv g(\lambda(X_1),\ldots,\lambda(X_{n_2}))$ modulo P_1

$$\begin{aligned} &\equiv \lambda(g) \text{ modulo } P_1 \\ &\equiv 0 \quad \text{ modulo } P_1 \ . \end{aligned}$$

Therefore, $g(\alpha'(x)) = 0$ and $\alpha'(x) \in \Sigma_2$.

We can summarize this discussion in the following:

Definition 2. Let $\Sigma \subset k^n$ be an irreducible algebraic set. Then the *affine ring* $\Gamma(\Sigma)$ is the ring of k-valued functions on Σ given by polynomials in the coordinates, i.e.,

$$k[X_1, \ldots, X_n]/I(\Sigma).$$

Proposition 1. *If Σ_1, Σ_2 are two irreducible algebraic sets, then the set of morphisms from Σ_1 to Σ_2 and the set of k-homomorphisms from $\Gamma(\Sigma_2)$ to $\Gamma(\Sigma_1)$ are canonically isomorphic:*

$$\hom(\Sigma_1, \Sigma_2) \cong \hom_k(\Gamma(\Sigma_2), \Gamma(\Sigma_1)).$$

Corollary. *If Σ is an irreducible algebraic set, then $\Gamma(\Sigma)$ is canonically isomorphic to the set of morphisms from Σ to k.*

Proof. Note that $\Gamma(k)$ is just $k[X]$.

Using $\Gamma(\Sigma)$, we can define a subset $V(A)$ of Σ for any ideal $A \subset \Gamma(\Sigma)$ as

$$\{x \in \Sigma \mid f(x) = 0 \text{ for all } f \in A\}.$$

Moreover, the Nullstellensatz for k^n implies immediately the Nullstellensatz for Σ:

$$\{f \in \Gamma(\Sigma) \mid f(x) = 0 \text{ for all } x \in V(A)\} = \sqrt{A}.$$

Even more than Proposition 1 is true:

Proposition 2. *The assignment*

$$\Sigma \longmapsto \Gamma(\Sigma)$$

extends to a contravariant functor Γ:

$$\left\{\begin{array}{l}\textit{Category of irreducible}\\ \textit{algebraic sets + morphisms}\end{array}\right\} \longrightarrow \left\{\begin{array}{l}\textit{Category of finitely generated integral}\\ \textit{domains over } k + k\textit{-homomorphisms}\end{array}\right\}$$

which is an equivalence of categories.

Proof. Prop. 1 asserts that Γ is a fully faithful functor. The other fact to check is that every finitely generated integral domain R over k occurs as $\Gamma(\Sigma)$. But every such domain can be represented as:

$$R \cong k[X_1, \ldots, X_n]/(f_1, \ldots, f_m),$$

hence as $\Gamma(\Sigma)$ where Σ is the locus of zeroes of $f_1, \ldots, f_m$ in k^n. □

Because of the usefulness of continuity in topology and other parts of geometry, another natural question is whether there is a natural topology on irreducible algebraic sets in which all morphisms are continuous. We will certainly want points of k to be closed, so their inverse images by morphisms must be closed. If we take the weakest topology satisfying this condition, we get the following:

Definition 3. A *closed set* in k^n is to be a closed algebraic set $V(A)$. By the results of §2, these define a topology in k^n, called the *Zariski topology.* An irreducible algebraic set $\Sigma \subset k^n$ is given the induced topology, again called the Zariski topology. It is clear that the closed sets of Σ are exactly the sets $V(A)$, where A is an ideal in $\Gamma(\Sigma)$.

It is easy to check that all morphisms are continuous in the Zariski topology. A basis for the open sets in the Zariski topology on Σ is given by the open sets:

$$\Sigma_f = \{x \in \Sigma \mid f(x) \neq 0\}$$

for elements $f \in \Gamma(\Sigma)$. In fact, $\Sigma_f = \Sigma - V((f))$, hence Σ_f is open. And if $U = \Sigma - V(A)$ is an arbitrary open set, then

$$U = \bigcup_{f \in A} \Sigma_f.$$

One should notice that the Zariski topology is very weak. On k itself, for instance, it is just the topology of finite sets, the weakest T_1 topology (since any ideal A in $k[X]$ is principal – $A = (f)$ – therefore $V(A)$ is just the finite set of roots of f). It follows that any bijection $\alpha : k \to k$ is continuous, so not all continuous maps are morphisms. In any case this is a very unclassical type of topological space.

Definition 4. A topological space X is *noetherian* if its closed sets satisfy the descending chain condition (d.c.c.). It is equivalent to require that all open sets be quasi-compact (= having the Heine-Borel covering property, but not necessarily T2).

Now since ideals in $k[X_1, \ldots, X_2]$ satisfy the a.c.c., it follows that closed sets satisfy the d.c.c. – so the Zariski topology is noetherian.

Our simple definition of morphisms for affine algebraic sets does not work for projective algebraic sets. The trouble is that it automatically implied that the morphism will extend to a morphism of the ambient affine space. There is no analogous fact in the projective case. Look at the case of Example D. Let Σ_1 be the conic with homogeneous equation

$$(*) \qquad xz = y^2$$

in $\mathbb{P}_2(k)$. Let $\Sigma_2 = \mathbb{P}_1(k)$. Let $P_0 \in \Sigma_1$ be the point $(0,0,1)$. To every point $Q \in \mathbb{P}_2(k) - \{P_0\}$, we can associate the line P_0Q, and by identifying the pencil of lines through P_0 with $\mathbb{P}_1(k)$ we get a point of $\mathbb{P}_1(k)$. In terms of coordinates, this can be expressed by the map:

$$(a, b, c) \longmapsto (a, b)$$

as long as (a, b, c) are not homogeneous coordinates for P_0, i.e., a or b is not zero. Let (s, t) be homogeneous coordinates in $\mathbb{P}_1(k)$. Then the map from Σ_1 to Σ_2 should be defined by:

$$(A) = \begin{cases} s &= x \\ t &= y \, . \end{cases}$$

Unfortunately, this is undefined at P_0 itself. But consider the second map defined by:

$$(B) = \begin{cases} s &= y \\ t &= z \, . \end{cases}$$

This is defined except at the point $P_1 \in \Sigma_1$ with coordinates $(1,0,0)$; moreover, at points on Σ_1 not equal to P_1 or P_2, the ratios $(x:y)$ and $(y:z)$ are equal in view of the equation $(*)$. Therefore A and B together define everywhere a map from Σ_1 to Σ_2. On the other hand, it will turn out that there are no surjective morphisms at all from $\mathbb{P}_2(k)$ to $\mathbb{P}_1(k)$ (cf. §7).

Thus defining morphisms between projective sets is more subtle. We find that we must define morphisms locally and patch them together. But the problem arises: on which local places. We could use the affine algebraic sets

$$\Sigma - \Sigma \cap H$$

where $H \subset \mathbb{P}_n(k)$ is a hyperplane. But in general these will not be small enough. We shall need arbitrarily small open sets in the Zariski-topology:

Definition 5. A *closed set* in $\mathbb{P}_n(k)$ is to be a closed algebraic set $V(A)$. By the results of §2, these define a topology in $\mathbb{P}_n(k)$, the *Zariski-topology*. Irreducible algebraic sets themselves are again given the induced topology. As in the affine case, a basis for the open sets is given by:

$$[\mathbb{P}_n(k)]_f = \{x \in \mathbb{P}_n(k) \mid f(x) \neq 0\}$$

where f is a *homogeneous* polynomial. Moreover, it is clear that the Zariski topology on $\mathbb{P}_n(k)$ is noetherian.

Problem. Check that $\mathbb{P}_n(k)_{X_0}$ is *homeomorphic* to k^n under the usual map

$$(x_0, \ldots, x_n) \longmapsto (x_1/x_0, x_2/x_0, \ldots, x_n/x_0) \, .$$

Finally, to define morphisms locally, we will need to attach affine coordinate rings to a lot of the Zariski-open sets U and give a definition of affine morphism in terms of local properties. Clearly, we should begin by constructing the apparatus used for defining things locally.

§4. Sheaves and affine varieties

Definition 1. Let X be a topological space. A *presheaf* F on X consists of

i) for all open $U \subset X$, a set $F(U)$
ii) for all pairs of open sets $U_1 \subset U_2$, a map ("restriction")

$$\mathrm{res}_{U_2 U_1} : F(U_2) \longrightarrow F(U_1)$$

such that the following axioms are satisfied:

a) $\mathrm{res}_{U,U} = \mathrm{id}_{F(U)}$ for all U.
b) If $U_1 \subset U_2 \subset U_3$, then

$$\begin{array}{ccc} F(U_3) & & \\ \downarrow {\scriptstyle \mathrm{res}_{U_3 U_1}} & \searrow^{\mathrm{res}_{U_3 U_2}} & \\ & & F(U_2) \\ \downarrow & \swarrow_{\mathrm{res}_{U_2 U_1}} & \\ F(U_1) & & \end{array}$$

commutes.

Definition 2. If F_1, F_2 are presheaves on X, a *map* $\varphi : F_1 \to F_2$ is a collection of maps $\varphi(U) : F_1(U) \to F_2(U)$ for each open U such that if $U \subset V$,

$$\begin{array}{ccc} F_1(V) & \xrightarrow{\varphi(V)} & F_2(V) \\ {\scriptstyle \mathrm{res}_{V,U}} \downarrow & & \downarrow {\scriptstyle \mathrm{res}_{V,U}} \\ F_1(U) & \xrightarrow[\varphi(U)]{} & F_2(U) \end{array}$$

commutes.

Definition 3. A presheaf F is a *sheaf* if for every collection $\{U_i\}$ of open sets in X with $U = \cup U_i$, the diagram

$$F(U) \longrightarrow \prod F(U_i) \overset{\longrightarrow}{\longrightarrow} \prod_{i,j} F(U_i \cap U_j)$$

is exact, i.e., the map

$$\prod \mathrm{res}_{U,U_i} : F(U) \longrightarrow \prod F(U_i)$$

is injective, and its image is the set on which

$$\prod \mathrm{res}_{U_i, U_i \cap U_j} : \prod_i F(U_i) \longrightarrow \prod_{i,j} F(U_i \cap U_j)$$

and

$$\prod \mathrm{res}_{U_j, U_i \cap U_j} : \prod_j F(U_j) \longrightarrow \prod_{i,j} F(U_i \cap U_j)$$

agree.

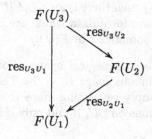

When we pull this high-flown terminology down to earth, it says this.

1) If $x_1, x_2 \in F(U)$ and for all i, $\mathrm{res}_{U,U_i} x_1 = \mathrm{res}_{U,U_i} x_2$, then $x_1 = x_2$. (That is, elements are uniquely determined by local data.)
2) If we have a collection of elements $x_i \in F(U_i)$ such that $\mathrm{res}_{U_i, U_i \cap U_j} x_j = \mathrm{res}_{U_j, U_i \cap U_j} x_j$ for all i and j then there is an $x \in F(U)$ such that $\mathrm{res}_{U,U_i} x = x_i$ for all i. (That is, if we have local data which are compatible, they actually "patch together" to form something in $F(U)$.)

Example G. Let X and Y be topological spaces. For all open sets $U \subset X$, let $F(U)$ be the set of continuous maps $U \to Y$. This is a presheaf with the restriction maps given by simply restricting maps to smaller sets; it is a sheaf because a function is continuous on $\cup U_i$ if and only if its restrictions to each U_i are continuous.

Example H. X and Y differentiable manifolds. $F(U)$ = differentiable maps $U \to Y$. This again is a sheaf because differentiability is a local condition.

Example I. X, Y topological spaces, $G(U)$ = continuous functions $U \to Y$ which have relatively compact image. This is a subpresheaf of the first example, but clearly need not be a sheaf.

Example J. X a topological space, $F(U)$ = the vector space of locally constant real-valued functions on U, modulo the constant functions on U. This is clearly a presheaf. But every $s \in F(U)$ goes to zero in $\prod F(U_i)$ for some covering $\{U_i\}$, while if U is not connected, $F(U) \neq (0)$. Therefore it is not a sheaf.

Sheaves are almost standard nowadays, and we will not develop their properties in detail. Recall two important ideas:

(1) *Stalks.* Let F be a sheaf on X, $x \in X$. The collection of $F(U)$, U open containing x, is an inverse system and we can form

$$F_x = \varinjlim_{x \in U} F(U) ,$$

called the *stalk of F at x.*

Example. Let $F(U)$ = continuous functions $U \to \mathbb{R}$. Then F_x is the set of *germs of continuous functions at x.* It is $\cup_{x \in U} F(U)$ modulo an equivalence relation: $f_1 \sim f_2$ if f_1 and f_2 agree in a neighbourhood of x.

(2) *Sheafification of a presheaf.* Let F_0 be a presheaf on X. Then there is a sheaf F and a map $f : F_0 \to F$ such that if $g : F_0 \to F'$ is any map with F' a sheaf, there is a unique map $h : F \to F'$ such that

$$\begin{array}{ccc} & F & \\ {\scriptstyle f}\nearrow & & \searrow{\scriptstyle h} \\ F_0 & \xrightarrow{\quad g \quad} & F' \end{array}$$

commutes.

(F is "the best possible sheaf you can get from F_0". It is easy to imagine how to get it: first identify things which have the same restrictions, and then add in all the things which can be patched together.) Thus in Example I above, if X is locally compact, the sheafification of this presheaf is the sheaf of all continuous functions in functions on X; and in example J, the sheafification of this presheaf is (0).

Notation. We may write $\Gamma(U, F)$ for $F(U)$, and call it the set of ***sections of** F* ***over*** U. $\Gamma(X, F)$ is the set of ***global sections*** of F. In other contexts we may denote $F(X)$ by $H^0(X, F)$ and call it the ***zeroth cohomology group.*** (In those contexts it will be a group, and there will be higher cohomology groups.)

Suppose that for all U, $F(U)$ is a group [ring, etc.] and that all the restriction maps are group [ring, etc.] homomorphisms. Then F is called a ***sheaf of groups [rings, etc.]***. In this case F_x is a group [ring, etc.], and so on.

Example K. For any topological space X, let $F_{\text{cont},X}(U) =$ continuous functions $U \to \mathbb{R}$. Then $F_{\text{cont},X}(U)$ is a sheaf of rings.

Note that if $g : X \to Y$ is a continuous function, the operation $f \longmapsto f \cdot g$ gives us the following maps: for every open $U \subset Y$ a map $F_{\text{cont},Y}(U) \to F_{\text{cont},X}\left(g^{-1}U\right)$ such that

$$\begin{array}{ccc} F_Y(U) & \longrightarrow & F_X(g^{-1}U) \\ \Big\downarrow{\scriptstyle \text{res}} & & \Big\downarrow{\scriptstyle \text{res}} \\ F_Y(V) & \longrightarrow & F_X(g^{-1}V) \end{array}$$

commutes for all open sets $V \subset U$. This set-up is called a morphism of the pair (X, F_X) to the pair (Y, F_Y).

Example L. Suppose that X and Y are differentiable manifolds, and that $F_{\text{diff},X}$ and $F_{\text{diff},Y}$ are the subsheaves of $F_{\text{cont},X}$ and $F_{\text{cont},Y}$ of differentiable functions. Let $g : X \to Y$ be a continuous map. Then g is differentiable if and only if for all open sets $U \subset Y$, $f \in F_{\text{diff},Y}(U) \Longrightarrow f \cdot g \in F_{\text{diff},X}\left(g^{-1}U\right)$.

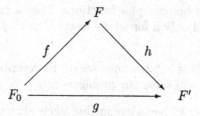

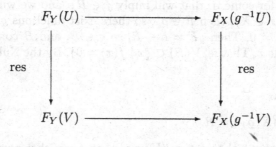

Example M. Similarly, say X, Y are complex analytic manifolds. Let $F_{\mathrm{an},X}$ and $F_{\mathrm{an},Y}$ be the sheaves of holomorphic functions. Then a continuous map $g : X \to Y$ is holomorphic if and only if for all open sets $U \subset Y$, $f \in F_{\mathrm{an},Y}(U) \Longrightarrow f \cdot g \in F_{\mathrm{an},X}\left(g^{-1}U\right)$.

Thus the idea of using a "structure sheaf" to describe an object is useful in many contexts, and it will solve our problems too.

Definition 4. Let $X \subset k^n$ be an irreducible algebraic set, R its affine coordinate ring. Since X is irreducible, $I(X)$ is prime and R is an integral domain. Let K be its field of fractions. Recall that R has been identified with a ring of functions on X. For $x \in X$, let $m_x = \{f \in R \mid f(x) = 0\}$. This is a maximal ideal, the kernel of the homomorphism $R \to k$ given by $f \mapsto f(x)$. Let $\underline{o}_x = R_{m_x}$. We have then $\underline{o}_x = \{f/g \mid f, g \in R, g(x) \neq 0\} \subset K$. Now for U open in X, let

$$\underline{o}_X(U) = \bigcap_{X \in U} \underline{o}_x \quad .$$

All the $\underline{o}_X(U)$ are subrings of K. If $V \subset U$, then $\underline{o}_X(U) \subset \underline{o}_X(V)$; if we take the inclusion as the restriction map, this defines a sheaf $\underline{o}_X$.

The elements of $\underline{o}_X(U)$ can be viewed as functions on U. Say $F \in \underline{o}_X(U)$, and $x \in U$. Then $F \in \underline{o}_x$, so we can write $F = f/g$ with $g(x) \neq 0$. We then define $F(x) = f(x)/g(x)$. Clearly $F(x) = 0$ for all $x \in U$ implies $F = 0$, so we can identify $\underline{o}_X(U)$ with the associated ring of functions on U.

Proposition 1. *Let X be an irreducible algebraic set and let $R = \Gamma(\Sigma)$. Let $f \in R$, and $X_f = \{x \in X \mid f(x) \neq 0\}$. Then $\underline{o}_X(X_f) = R_f$.*

Proof. If $g/f^n \in R_f$, then $g/f^n \in \underline{o}_x$ for all $x \in X_f$, since by definition $f(x) \neq 0$. Thus $R_f \subset \underline{o}_X(X_f)$.

Now suppose $f \in \underline{o}_X(X_f) \subset K$. Let $B = \{g \in R \mid g \cdot F \in R\}$. If we can prove $f^n \in B$, for some n, that will imply $f \in R_f$, and we will be through. By assumption, if $x \in X_f$, then $F \in \underline{o}_x$, so there exist functions $g, h \in R$ such that $F = h/g$, $g(x) \neq 0$. Then $gF = h \in R$, so $g \in B$, and B contains an element not vanishing at x. That is, $V(B) \subset \{x \mid f(x) = 0\}$. By the Nullstellensatz, then $f \in \sqrt{B}$. □

In particular,

Corollary. $\Gamma(X, \underline{o}_X) = R$.

Remarks. I) Assume that $f \in \underline{o}_X(U)$ and that f vanishes nowhere on U. Then $1/f \in \underline{o}_X(U)$.

Proof. Obvious, since $f(x) \neq 0 \Longrightarrow 1/f \in \underline{o}_x$.

II) The stalk of $\underline{o}_X$ at x is $\underline{o}_x$.

Proof. Since the sets X_f are a basis of the Zariski topology of X, we have

$$\varinjlim_{x\in U} \varrho_X(U) = \varinjlim_{x\in X_f} \varrho_X(X_f) = \varinjlim_{f(x)\neq 0} R_f .$$

Since all restriction maps in our sheaf are injective, this is just $\bigcup_{f(x)\neq 0} R_f$, which is clearly ϱ_x. □

III) The field K can also be recovered from the sheaf ϱ_X. Recall that X is irreducible, i.e., not the union of two proper closed subsets. Equivalently, the intersection of any two nonempty open sets is nonempty. But this means that we actually have an inverse system of *all* open sets, just like our previous inverse systems of open sets containing a given point x; in this way we can define a *generic stalk* of any sheaf F on X. In particular, it is evident that K is the generic stalk of the structure sheaf ϱ_X.

IV) If $h \in \varrho_X(U)$ for some open $U \subset X$, then it need not be true that $h = f/g$ with $f, g \in R$ and g vanishing nowhere on U. For example, let $X \subset k^4$ be $V(xw - yz)$, and let $U = X_y \cup X_w$. The following function $h \in \varrho_X(U)$ is not equal to f/g, $g \neq 0$ in U: $h = x/y$ on X_y, and $h = z/w$ on X_w^*[2]. The proposition shows that this is true however if U has the form X_g.

Proposition 2. *Let $X \subset k^n$, $Y \subset k^m$ be irreducible algebraic sets, and let $f : X \to Y$ be a continuous map. The following conditions are equivalent:*

i) *f is a morphism*
ii) *for all $g \in \Gamma(Y, \varrho_Y)$, $g \cdot f \in \Gamma(X, \varrho_X)$*
iii) *for all open $U \subset Y$, and $g \in \Gamma(U, \varrho_Y) \Longrightarrow g \cdot f \in \Gamma(f^{-1}U, \varrho_X)$*
iv) *for all $x \in X$, and $g \in \varrho_{f(x)} \Longrightarrow g \cdot f \in \varrho_x$.*

Proof. Trivially iii) $\Longrightarrow$ ii), and iv) $\Longrightarrow$ iii) by the definition of ϱ_X. i) $\Longleftrightarrow$ ii) is essentially proved in Proposition 1, §3. We assume ii), then, and prove iv). Let $g \in \varrho_{f(x)}$. We write $g = a/b$, $a, b \in \Gamma(Y, \varrho_Y)$, $b(f(x)) \neq 0$. By ii), $a \cdot f, b \cdot f \in \Gamma(X, \varrho_X)$; hence $g \cdot f = a \cdot f / b \cdot f \in \varrho_x$, since we have $b \cdot f(x) \neq 0$. □

This shows, among other things, that our sheaf gives us all the information we need for defining morphisms. We are ready, then, to cut loose from the ambient spaces and define:

[2] The proof that this h is not of the form f/g, $g \neq 0$ in U, requires a later result but it goes like this: Assume $h = f/g$. Let $Z = V(y, w)$: then Z is a plane in X and $U = X - Z$. By assumption $V(g) \cap X \subset Z$. Since all components of $V(g) \cap X$ have dimension 2 (cf. §7), and since Z is irreducible, either $V(g) \cap X = z$ or $V(g) \cap X$ is empty. If $V(g) \cap X = \emptyset$, then $h = f/g \in \varrho_X(X)$ which is absurd (since $x = y \cdot h$ and at some points of $X, x = 1$ and $y = 0$). Now let $Z' = V(x, z)$. Then

$$\{(0,0,0,0)\} = Z \cap Z' = V(g) \cap Z' .$$

In other words, g would be a polynomial function on the plane Z' that vanishes only at the origin. This is impossible too.

Definition 5. An *affine variety* is a topological space X plus a sheaf of k-valued functions $\underline{o}_X$ on X which is isomorphic to an irreducible algebraic subset of some k^n plus the sheaf just defined.

Definition 6. The affine variety $(k^n, \underline{o}_{k^n})$ is $\mathbb{A}^n$, *affine n-space.*

Definition 5 and Proposition 2 set up the category of affine varieties in precise analogy with the category of topological spaces, differentiable manifolds, and analytic spaces. There are, however, some very categorical differences between these examples. Consider the following statement:

Bijective morphisms are isomorphisms.

This is correct, for example, in the category of compact topological spaces, of Banach spaces, and of complex analytic *manifolds.* On the other hand, it is false for differentiable manifolds – consider the map:

$$\mathbb{R} \xrightarrow{f} \mathbb{R}$$

where $f(x) = x^3$.

The statement is also false in the category of affine varieties: a bijection $f : X_1 \to X_2$ of varieties may well correspond to an isomorphism of the ring of X_2 with a proper subring of the ring of X_1. Here are 3 key examples to bear in mind.

Example N. Let $\mathrm{char}(k) = p \neq 0$. Define the morphism

$$\mathbb{A}^1 \xrightarrow{f} \mathbb{A}^1$$

by $f(t) = t^p$. This is bijective. On the ring level, this corresponds to the inclusion map in the pair of rings:

$$k[X] \hookleftarrow k[X^p].$$

f is not an isomorphism since these rings are not equal.

Example O. Let k be any algebraically closed field. Define the morphism

$$\mathbb{A}^1 \xrightarrow{f} \mathbb{A}^2$$

by $f(t) = (t^2, t^3)$. The image of this morphism is the irreducible closed curve

$$C : X^3 = Y^2.$$

The morphism f from $\mathbb{A}^1$ to C is a bijection which corresponds to the inclusion map in the pair of rings:

$$k[T] \hookleftarrow k\left[T^2, T^3\right].$$

These rings are not equal, so f is not an isomorphism.

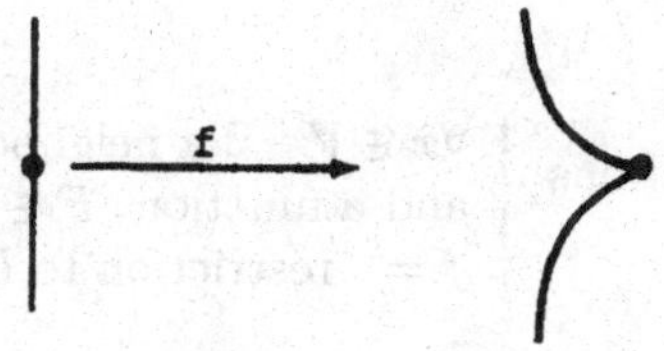

Example P. Define $\mathbb{A}^1 \xrightarrow{f} \mathbb{A}^2$ by $X = t^2 - 1$, $Y = t\left(t^2 - 1\right)$. It is not hard to check that the image of this morphism is the curve D:

$$(1) \qquad Y^2 = X^2(X+1)\,.$$

(Simply note that one can solve for the coordinate t of the point in $\mathbb{A}^1$ by the equation $t = Y/X$. Then substitute this into $X = t^2 - 1$.) Also, f is bijective between $\mathbb{A}^1$ and D except that both the points $t = -1$ and $t = 1$ are mapped to the origin. Let $X_1 = \mathbb{A}^1 - \{1\}$, an affine variety with coordinate ring $k\left[T, (T-1)^{-1}\right]$ (cf. Proposition 4 below). Then f restricts to a bijection f' from X_1 to D. This morphism corresponds to the inclusion in the pair of rings:

$$k\left[T, (T-1)^{-1}\right] \hookleftarrow k\left[T^2 - 1, T(T^2 - 1)\right].$$

Since these rings are unequal, f' is not an isomorphism.

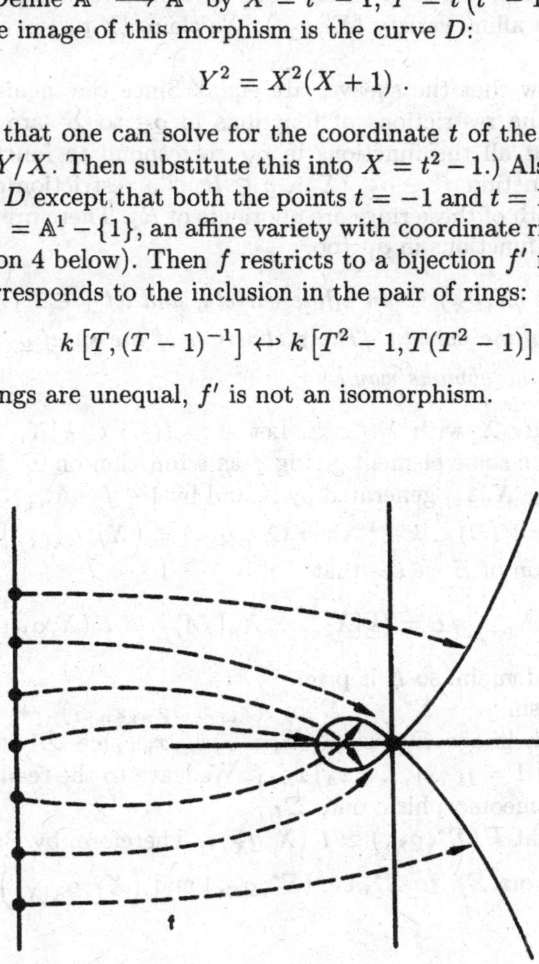

The last topic we will take up in this section is the induced variety structure on open and closed subsets of affine varieties.

Let Y be an irreducible closed subset of an affine variety $(X, \underline{o}_X)$. [Irreducible, now, in the sense given by the topology on X.] Define an induced sheaf $\underline{o}_X$ of functions on Y as follows:

If V is open in Y,

$$\underline{o}_Y(V) = \left\{ \begin{array}{l|l} k\text{-valued functions} & \forall x \in V,\ \exists \text{ a neighbourhood } U \text{ of } x \text{ in } X \\ f \text{ on } V & \text{and a function } F \in \underline{o}_X(U) \text{ such that} \\ & f = \text{ restriction to } U \cap V \text{ of } F. \end{array} \right\}.$$

Proposition 3. $(Y, \underline{o}_Y)$ *is an affine variety.*

Proof. Say X is isomorphic to $(\Sigma, \underline{o}_\Sigma)$ in k^n. Let $R = k[X_1, \ldots, X_n]$, $A = I(\Sigma) \subset R$. Let Y correspond to $\Sigma' \subset \Sigma$. Then Σ' is an irreducible algebraic set in k^n, so we have an affine variety $(\Sigma', \underline{o}_{\Sigma'})$. We claim $(Y, \underline{o}_Y)$ is isomorphic to $(\Sigma', \underline{o}_{\Sigma'})$.

It suffices to show that the sheaves are equal. Since the inclusion of Σ' in Σ is a morphism, the restrictions of functions in $\underline{o}_\Sigma$ to Σ' are functions in $\underline{o}_{\Sigma'}$. This shows that all the functions in $\underline{o}_Y$ correspond to functions in $\underline{o}_{\Sigma'}$. Conversely, every function $f' \in \underline{o}_{\Sigma'}(\Sigma'_g)$, $g \in R$, is a restriction of a function $f \in \underline{o}_\Sigma(\Sigma_g)$ since both of these rings are quotients of R_g. Therefore all functions in $\underline{o}_\Sigma$ correspond to functions in $\underline{o}_Y$ too. □

Proposition 4. *Let* $(X, \underline{o}_X)$ *be an affine variety, and let* $f \in \Gamma(X, \underline{o}_X)$. *Then* $\left(X_f, \underline{o}_{X|X_f}\right)$ *is an affine variety. [The restriction of the sheaf* $\underline{o}_X$ *to the open set* X_f *is defined in the obvious way.]*

Proof. Say we identify X with $\Sigma \subset k^n$. Let $A = I(\Sigma) \subset k[X_1, \ldots, X_n]$. Let $f_1 \in k[X_1, \ldots, X_n]$ be some element giving f as a function on Σ. Let B be the ideal in $k[X_1, \ldots, X_n, X_{n+1}]$ generated by A and by $1 - f_1 \cdot X_{n+1}$. We claim B is prime and, if $\Sigma^* = V(B) \subset k^{n+1}$, then $(\Sigma^*, \underline{o}_{\Sigma^*}) \cong \left(X_f, \underline{o}_{X|X_f}\right)$.

From the definition of B we see that

$$k[X_1, \ldots, X_{n+1}]/B = (k[X_1, \ldots, X_n]/A)_{f_1} \cong \Gamma(X, \underline{o}_X)_f,$$

which is an integral domain, so B is prime.

Define a morphism $\alpha : \Sigma^* \to \Sigma$ by $(x_1, \ldots, x_n, x_{n+1}) \to (x_1, \ldots, x_n)$. It's an injection with image Σ_{f_1}, since $(x_1, \ldots, x_n, x_{n+1}) \in \Sigma^*$ if and only if $(x_1, \ldots, x_n) \in \Sigma$ and $1 = f_1(x_1, \ldots, x_n)\, x_{n+1}$. We leave to the reader the verification that it is a homeomorphism onto Σ_{f_1}.

We saw above that $\Gamma(\Sigma^*, \underline{o}_{\Sigma^*}) \simeq \Gamma(X, \underline{o}_X)_f$. Therefore, by Proposition 2, α^{-1} is a morphism from Σ_{f_1} to Σ^*, i.e. $(\Sigma^*, \underline{o}_{\Sigma^*})$ and $\left(X_f, \underline{o}_{X|X_f}\right)$ are isomorphic. □

What we have done to get X_f is to push the zeroes of f out to infinity. For example, suppose $X = \mathbb{A}^1$ and f is the coordinator X_1. Then $B = (1 - X_1 X_2)$, giving a hyperbola:

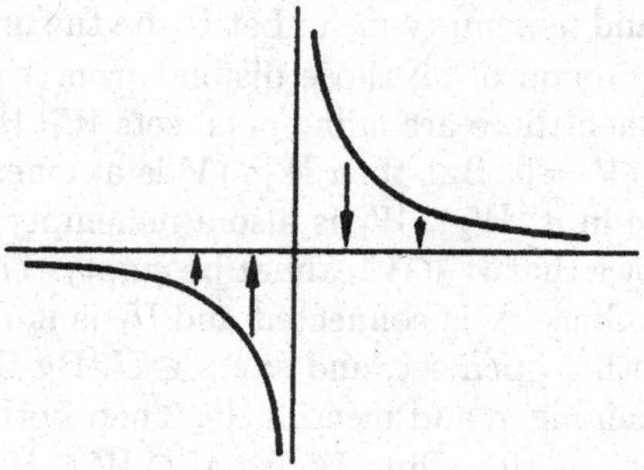

Projection of the hyperbola down to the axis is an isomorphism with X_f.

Not all open subsets of affine varieties are affine varieties. For instance, you cannot push the origin in $\mathbb{A}^2$ out to infinity, and $\mathbb{A}^2 - (0,0)$ is not an affine variety. In fact, no rational function is well-defined on $\mathbb{A}^2 - (0,0)$ but not at $(0,0)$; i.e., the intersection of the local rings $\underline{o}_x$, for all $x \in \mathbb{A}^2$, $x \neq (0,0)$, is contained in $\underline{o}_{(0,0)}$. Hence if we had any embedding $\mathbb{A}^2 - (0,0) \to k^n$, the coordinate functions giving this embedding would have to extend to all of $\mathbb{A}^2$, and so the image of $\mathbb{A}^2 - (0,0)$ would not be closed. There is an analogous statement about complex functions: a holomorphic function on $\mathbb{C} \times \mathbb{C} - (0,0)$ is necessarily holomorphic at $(0,0)$.

§5. Definition of prevarieties and morphisms

Definition 1. A topological space X plus a sheaf $\underline{o}_X$ of k-valued functions on X is a *prevariety* if

1) X is connected, and
2) there is a finite open covering $\{U_i\}$ of X such that for all i, $\left(U_i, \underline{o}_{X|U_i}\right)$ is an affine variety.

Definition 2. An open subset U of X is called an *open affine set* if $\left(U, \underline{o}_{X|U}\right)$ is an affine variety.

Note that the open affine sets are a basis of the topology. In fact, we know by Proposition 4, §4, that this is true within each of the open affines U_i, and they cover X.

Definition 3. A topological space is *irreducible* if it is not the union of two proper closed subsets (equivalently, the intersection of any two nonempty sets is nonempty).

Proposition 1. *Every prevariety X is an irreducible topological space.*

Proof. Let V be open and nonempty in X. Let U_1 be the union of all open affine sets meeting V, U_2 the union of all those disjoint from V; then $U_1 \cup U_2 = X$. Suppose $y \in U_1 \cap U_2$; then there are affine open sets W_1, W_2 containing y, such that $W_1 \cap V \neq \emptyset$, $W_2 \cap V = \emptyset$. But then $W_1 \cap V$ is a nonempty open set in the affine W_1, so it is dense in it; $W_2 \cap W_1$ is also a nonempty open set in W, so it meets $W_1 \cap V$. This shows that $W_2 \cap V$ cannot be empty. This is a contradiction, hence no such y exists. Since X is connected and U_1 is nonempty, $U_1 = X$.

Now let U be any other open set, and say $x \in U$. By the above, there is an affine open set W containing x and meeting V. Then both $V \cap W$ and $U \cap W$ are nonempty open sets in the affine W, so $V \cap W \cap U \neq \emptyset$, and *a fortiori* $U \cap V \neq \emptyset$. □

In particular, every open set is dense. Thus prevarieties are not like differentiable manifolds, which can have disjoint coordinate patches; to get a prevariety, we just put things around the edges of one affine piece.

Proposition 2. *If X is a prevariety, then the closed sets of X satisfy the descending chain condition, i.e., X is a noetherian space.*

Proof. Let $\{Z_i\}$ be a sequence of closed sets, such that $Z_1 \supset Z_2 \supset Z_3 \supset \dots$. Since X is covered by finitely many affines, it suffices to show that $\{U \cap Z_i\}$ stationary for each affine open U in X. The result in the affine case follows immediately from the fact that $\Gamma(U, \underline{o}_X)$ is noetherian as we noted in §3. □

In particular, every variety is quasi-compact.

Proposition 3. *Let X be a noetherian topological space. Then every closed set Z in X can be written uniquely as an irredundant union of finitely many irreducible closed sets (called the components of Z).*

Proof. Suppose Z is a minimal closed set for which the Proposition is false; this exists since X is noetherian. Then Z is not itself irreducible, so $Z = Z_1 \cup Z_2$, where Z_1, Z_2 are smaller and hence are unions of the required type. Then so is Z. □

Let X be a prevariety. Since X is an irreducible topological space, any 2 nonempty open subsets have a nonempty intersection. Therefore, all sheaves have "generic" stalks:

Definition 4. The *function field* $k(X)$ is the generic stalk of $\underline{o}_X$, i.e.

$$k(X) = \varinjlim_{\substack{\text{all non-empty} \\ \text{open } U}} \underline{o}_X(U) .$$

In fact, $k(X)$ equals the function field of each open affine set U in X, since the open subsets of U are cofinal. In particular, this shows that $k(X)$ is really a field. The elements of $k(X)$ are called *rational functions on* X, although they are, strictly speaking, only functions on open dense subsets of X.

Another type of $\varinjlim$ over $\underline{o}_X(U)$'s is sometimes very useful. This is intermediate between the $\varinjlim$ that leads to $\underline{o}_x$ and that which leads to $k(X)$. Let $Y \subset X$ be an irreducible closed subset of X. Then let:

$$\underline{o}_{Y,X} = \varinjlim_{\substack{\text{open sets } U \\ \text{such that} \\ U \cap Y \neq \emptyset}} \underline{o}_X(U) .$$

To express more simply the ring which you get in this way, fix one open affine $U \subset X$ which meets Y. Let R be the coordinate ring of U and $P = I(Y \cap U)$ the ideal in R determined by Y. Then:

$$\underline{o}_{Y,X} = \varinjlim_{\substack{\text{open sets } U_f \\ f \in R,\ f \notin P}} \underline{o}_X(U_f) = R_P .$$

In particular, $\underline{o}_{Y,X}$ is a local ring with quotient field $k(X)$ and residue field $k(Y)$.

Proposition 4. *An open subset of a prevariety is a prevariety.*

Proof. Let $U \subset X$ be open. Since X is irreducible, U is connected. U is of course a union of affine open subsets. But since X is noetherian, U is quasi-compact and hence it is covered by finitely many affines. □

Now let Y be a closed irreducible subset of a prevariety X. The sheaf $\underline{o}_X$ induces a sheaf $\underline{o}_Y$ on Y as follows:

If V is open in Y,

$$\underline{o}_Y(V) = \left\{ \begin{array}{l} k\text{-valued functions} \\ f \text{ on } V \end{array} \middle| \begin{array}{l} \forall x \in V,\ \exists \text{ a neighbourhood } U \text{ of } x \text{ in } X \\ \text{and a function } F \in \underline{o}_X(U) \text{ such that} \\ f = \text{ restriction to } U \cap V \text{ of } F \end{array} \right\} .$$

Proposition 5. *The pair $(Y, \underline{o}_Y)$ is a prevariety.*

Proof. This follows immediately from the definition and Proposition 3, §4. □

Combining Propositions 4 and 5, we can even give a prevariety structure to every *locally closed* subset of a prevariety X. The set of all prevarieties so obtained are called the *sub-prevarieties* of X.

Example Q. $\mathbb{P}_1$:

Take two copies U and V of $\mathbb{A}^1$. Let u, v be the coordinates on these 2 affine lines. Let $U_0 \subset U$ (resp. $V_0 \subset V$) be defined by $u \neq 0$ (resp. $v \neq 0$). Then $\Gamma(U, \underline{o}_U) = k[u]$, so $\Gamma(U_0, \underline{o}_U) = k[u, u^{-1}]$. Similarly $\Gamma(V_0, \underline{o}_V) = k[v, v^{-1}]$. Define a map $\varphi : U_0 \to V_0$ taking the point with coordinate $u = a$ to the point with coordinate $v = \frac{1}{a}$; this gives a map $\varphi^* : k[v, v^{-1}] \to k[u, u^{-1}]$ taking v to u^{-1}, v^{-1} to u. φ^* is an isomorphism of rings, so φ is an isomorphism of varieties

(φ has an inverse since φ^* does). Now we patch together U and V *via* φ, i.e., we form $U \cup V$ with U_0 and V_0 identified *via* φ. This has a sheaf on it, in the obvious way, and is a prevariety. The space is homeomorphic to $\mathbb{P}_1$, and we call it the variety $\mathbb{P}_1$. Our patching can be pictured as follows:

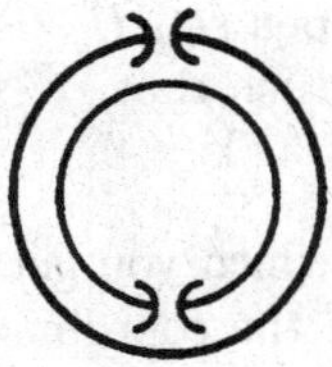

We could have patched U and V differently: $v \to u$, $v^{-1} \to u^{-1}$ also gives an isomorphism of U_0 onto V_0. But this is a silly way to patch; we are leaving out the same point each time:

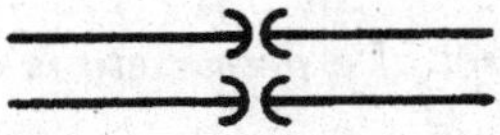

and the result is $\mathbb{A}^1$ with a point doubled:

This is a prevariety, of course, but, in fact, not a variety (cf. §6).

We could define all projective varieties by this kind of scissors and glue method, but there is a more intrinsic definition.

Definition of Projective Varieties. Let $P \subset k[X_0, \ldots, X_n]$ be a homogeneous prime ideal, $X = V(P) \subset \mathbb{P}_n(k)$. We want to make X (with the Zariski topology) into a prevariety. We do it by defining a function field, getting local rings, and intersecting them, just as for affine varieties.

The elements of $k[X_0, \ldots, X_n]$, even the homogeneous ones, do not give functions on X; but the ratio of any two having the same degree is a function. Since P is homogeneous, $R = k[X_0, \ldots, X_n]/P$ is in a natural way a graded ring $\bigoplus_{n=0}^{\infty} R_n$ and an integral domain. We let $k(X)$ be the zeroth graded piece of the localization of R with respect to homogeneous elements, i.e., $\{f/g \mid f, g \in R_n \text{ for the same } n\}$.

If $x \in X$, and $g \in R_n$ it makes sense to say $g(x) \neq 0$, even though g is not a function on X; for g changes by a nonzero factor as we change the homogeneous coordinates of x. Hence we can define a ring $\underline{o}_x$ in $k(X)$ as $\{f/g \in k(X) \mid g(x) \neq 0\}$. The set

$$m_x = \left\{ f/g \in k(X) \middle| \begin{matrix} f(x) = 0 \\ g(x) \neq 0 \end{matrix} \right\}$$

is clearly an ideal in the ring $\underline{o}_x$, and any element not in m_x is invertible in $\underline{o}_x$. Thus $\underline{o}_x$ is a local ring.

We now define a sheaf $\underline{o}_X$ on X by

$$\underline{o}_X(U) = \bigcap_{x \in U} \underline{o}_x, \qquad \text{for all } U \subset X \text{ open .}$$

We can identify $\underline{o}_X$ with a sheaf of k-valued functions. For suppose $x \in U$, and $a \in \Gamma(U, \underline{o}_X)$. Then a can be written f/g with $f, g \in R_n$, $g(x) \neq 0$, and we define $a(x) = f(x)/g(x)$. [That is, we lift f, g to F, G homogeneous polynomials of degree n in $k[X_0, \ldots, X_n]$, let $\tilde{x}$ be a set of homogeneous coordinates of x, and take $F(\tilde{x})(/G(\tilde{x})$. It is clear that this value is unchanged if we take a different set of homogeneous coordinates, or if we change F and G by members of P_n, so we have a well-defined function.] We still should check that if $a \in \Gamma(U, \underline{o}_x)$ and $a(x) = 0$ for all $x \in U$, then $a = 0$. But this also comes out of the next step, which consists in checking that $(X, \underline{o}_X)$ is locally isomorphic to an affine variety. In fact, we claim that for all i, $0 \leq i \leq n$,

$$(X \cap \mathbb{P}_n(k)_{X_i}, \quad \text{restriction of } \underline{o}_X)$$

is an affine variety. We will check this only for X_0, since the general case goes just the same.

For every homogeneous polynomial $F \in P$, $F/X_0^{\deg F}$ can be written as a polynomial F' in the variables $\frac{X_1}{X_0}, \ldots \frac{X_n}{X_0}$. Let $P' \subset k[Y_1, \ldots, Y_n]$ be the ideal generated by all these F'. We can map $k[Y_1, \ldots, Y_n]$ into a subring of $k(X)$ by taking Y_i to the function given by X_i/X_0; the kernel of this map is exactly P', so P' is prime. It is easy to see that we get an isomorphism $\varphi : X \cap \mathbb{P}_n(k)_{X_n} \to X' = V(P') \subset k^n$ by taking x to $\left(\frac{x_1}{x_0}, \ldots, \frac{x_n}{x_0}\right)$; φ is actually a homeomorphism (cf. Problem at end of §3).

Now for $x \in X \cap \mathbb{P}_n(k)_{X_0}$, the local ring $\underline{o}_x$ is the set of all elements of $k(X)$ having the form f/g for f, g in some R_m, $g(x) \neq 0$. X' has the affine coordinate ring $R' = k[Y_1, \ldots, Y_n]/P$. If $k(X')$ is its quotient field, $\underline{o}_{\varphi(x)}$ is the set of all elements in $k(X')$ having the form F/G for $F, G \in R'$, $G(\varphi(x)) \neq 0$. The map defined above taking R' into $k(X)$ extends to an isomorphism $k(X') \xrightarrow{\sim} k(X)$. I claim that this map takes $\underline{o}_{\varphi x}$ precisely onto $\underline{o}_x$. First of all it clearly maps $\underline{o}_{\varphi x}$ into $\underline{o}_x$; and if $f, g \in R_m$, $g(x) \neq 0$, then $f/g = f/X_0^m/g/X_0^m$ in $k(X)$ and f/X_0^m, g/X_0^m come from F, G in R' with $G(\varphi(x)) \neq 0$. Thus the local rings correspond; since the sheaves were defined by intersecting local rings, they also correspond, and $X \cap \mathbb{P}_n(k)_{X_0}$ is indeed an affine variety.

Definition 5. Let X and Y be prevarieties. A map $f : X \to Y$ is a *morphism* if f is continuous and, for all open sets V in Y,

$$g \in \Gamma(V, \underline{o}_Y) \Longrightarrow g \cdot f \in \Gamma(f^{-1}V, \underline{o}_X).$$

Proposition 6. *Let $f : X \to Y$ be any map. Let (V_i) be a collection of open affine subsets covering Y. Suppose that $\{U_i\}$ is an open covering of X such that 1) $f(U_i) \subset V_i$ and 2) f^* maps $\Gamma(V_i, \underline{o}_Y)$ into $\Gamma(U_i, \underline{o}_X)$. Then f is a morphism.*

Proof. We may assume the U_i are affine; for if $U \subset U_i$ is affine, f^* certainly maps $\Gamma(V_i, \underline{o}_Y)$ into $\Gamma(U, \underline{o}_X)$, and we can replace U_i by a set of affines that cover U_i. First of all, the restriction f_i of f to a map from U_i to V_i is a morphism. In fact, the homomorphism

$$f_i^* : \Gamma(V_i, \underline{o}_Y) \longrightarrow \Gamma(U_i, \underline{o}_X)$$

is also induced by some morphism $g_i : U_i \to V_i$ (Proposition 1, §3). And since the functions in $\Gamma(V_i, \underline{o}_Y)$ separate points, a map from U_i to V_i is determined by the contravariant map from $\Gamma(U_i, \underline{o}_X)$ to $\Gamma(V_i, \underline{o}_Y)$. Therefore $f_i = g_i$ and f_i is a morphism. In particular, f_i is continuous and this implies immediately that f itself is continuous. It remains to check that f^* always takes sections $\underline{o}_Y$ to sections $\underline{o}_X$. But if $V \subset Y$ is open, and $g \in \Gamma(V, \underline{o}_Y)$, then $g \cdot f$ is at least a section of $\underline{o}_X$ on the sets $f^{-1}(V \cap V_i)$, hence on the sets $f^{-1}(V) \cap U_i$. Since $\underline{o}_X$ is a sheaf, $g \cdot f$ is actually a section of $\underline{o}_X$ in $f^{-1}(V)$. □

To illustrate the meaning of our definitions, it seems worthwhile to work out in detail a non-trivial example. We shall reconsider Example F, §3. Let C be the plane cubic curve defined in homogeneous coordinates by:

$$zy^2 = x(x^2 - z^2).$$

Look first at $C \cap \mathbb{P}_2(k)_z$, with affine coordinates $X = x/z$, $Y = y/z$. The equation of C becomes:

$$Y^2 = X \cdot (X^2 - 1).$$

For all lines ℓ through the origin, we want to interchange the 2 points in $\ell \cap C$ (other than the origin). Start with a point $(a, b) \in C$. This is joined to the origin by the line

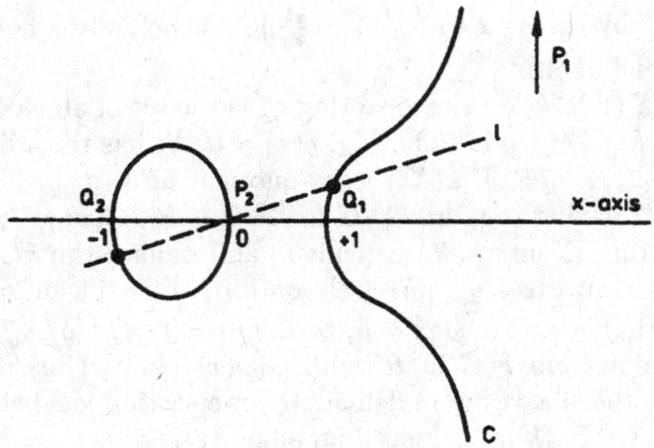

$$X = at$$
$$Y = bt.$$

Intersecting this with the cubic, we get the equation

$$b^2t^2 = at\,(a^2t^2 - 1)$$

or

$$0 = at(t-1)\,(a^2t+1)\ .$$

Thus the 2nd point of intersection is given by $t = -1/a^2$. In other words, the morphism on C is to be given by:

$$(a,b) \longmapsto (-1/a, -b/a^2)\ .$$

These are not polynomials, so at any rate they do not define a map from this affine piece of C into itself; this is as it should be, since as we can see from the drawing, we want the origin itself to go to the one point at infinity on the cubic.

To describe the subsets on which we *will* get a morphism, we must throw out the various "bad" points one at a time. We need names for them:

$P_1 = (0,1,0)$ (the only point at ∞ with respect to the affine piece X, Y)

$P_2 = (0,0,1)$ (the origin)

$\left.\begin{array}{l} Q_1 = (1,0,1) \\ Q_2 = (-1,0,1) \end{array}\right\}$ other points on the X-axis .

The morphism – call it f – should interchange P_1 and P_2, Q_1 and Q_2. Define

$$\begin{aligned} U_1 &= C - \{P_1, P_2\} \\ U_2 &= C - \{P_1, Q_1, Q_2\} \\ U_3 &= C - \{P_2, Q_1, Q_2\} \\ V_1 &= C - \{P_1\} = C \cap \mathbb{P}_2(k)_z \\ V_2 &= C - \{P_2, Q_1, Q_2\} = C \cap \mathbb{P}_2(k)_y\ . \end{aligned}$$

Then 1) U_1, U_2, U_3 is an open covering of C, 2) V_1, V_2 is an *affine* open covering of C, and 3) if f is defined set-theoretically as above, then $f(U_1) \subset V_1$, $f(U_2) \subset V_2$ and $f(U_3) \subset V_1$. Therefore, by Proposition 6, it suffices to check that

$$f^*\left[\Gamma(V_1, \underline{o}_C)\right] \subset \Gamma(U_1, \underline{o}_C) \cap \Gamma(U_3, \underline{o}_C)$$

and

$$f^*\left[\Gamma(V_2, \underline{o}_C)\right] \subset \Gamma(U_2, \underline{o}_C)\,,$$

and then it follows that f is a morphism. In more down to earth terms: note that X, Y are affine coordinates in V_1 and

$$\begin{aligned} S &= x/y \\ T &= z/y \end{aligned}$$

are affine coordinates in V_2. Then what we have to check is that f, described via one of these 2 sets of coordinates, is given by polynomials in U_1, U_2, and U_3.

In U_1: X, Y and $1/X$ are functions in $\Gamma(U_1, \underline{o}_C)$. Thus describing the image point also by coordinates X, Y, f is given by:

$$(a, b) \longmapsto (-1/a, -b/a^2)$$

and these are polynomials in $a, b, 1/a$.

In U_2: X, Y and $\frac{1}{x^2-1}$ are functions in $\Gamma(U_2, \underline{o}_C)$. Describe the image of the point (a, b) by coordinates S, T now:

$$\begin{aligned} S[f(a,b)] &= a/b \\ T[f(a,b)] &= -a^2/b\,. \end{aligned}$$

This does not look very promising until we use the relation $b^2 = a\cdot(a^2-1)$ to rewrite these as:

$$\begin{aligned} S &= b/a^2 - 1 \\ T &= -ab/a^2 - 1 \end{aligned}$$

which are polynomials in a, b, $1/a^2 - 1$.

In U_3: S and T are functions in $\Gamma(U_3, \underline{o}_C)$. Describe the image of the point $S = c$, $T = d$ by coordinates X, Y:

$$\begin{aligned} X[f(c,d)] &= -d/c \\ Y[f(c,d)] &= -d/c^2\,. \end{aligned}$$

However c and d are related by $d = c\cdot(c^2 - d^2)$, so we get:

$$\begin{aligned} X &= d^2 - c^2 \\ Y &= d\,(c^2 - d^2) - c\,. \end{aligned}$$

□

Problem. Generalize the result by which we covered $\mathbb{P}_n$ by open affine sets as follows: for all homogeneous polynomials $H \in k[X_0, \ldots, X_n]$ of positive degree, show that $\mathbb{P}_n(k)_H$ is an affine variety.

§6. Products and the Hausdorff Axiom

We want to define the product $X \times Y$ of any two prevarieties X, Y. Now we will certainly want to have $\mathbb{A}^n \times \mathbb{A}^m \cong \mathbb{A}^{n+m}$. But the product of the Zariski topologies in $\mathbb{A}^n$ and $\mathbb{A}^m$ does not give the Zariski topology in $\mathbb{A}^{n+m}$; in $\mathbb{A}^1 \times \mathbb{A}^1$, for instance, the only closed sets in the product topology are finite unions of horizontal and vertical lines. The only reliable way to find the correct definition is to use the general category-theoretic definition of product.

Definition 1. Let $\mathcal{C}$ be a category, X, Y objects in $\mathcal{C}$. An object Z plus two morphisms

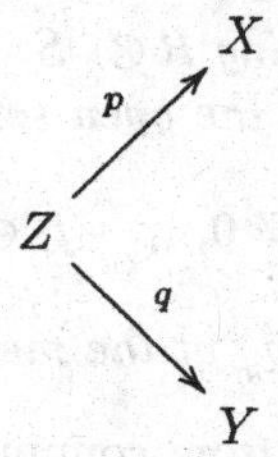

is a *product* if it has the following universal mapping property: for all objects W and morphisms

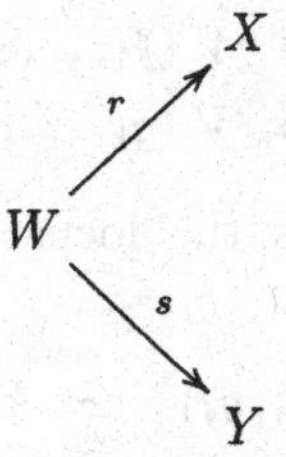

there is a unique morphism $t : W \to Z$ such that $r = p \cdot t$, $s = q \cdot t$, i.e., such that

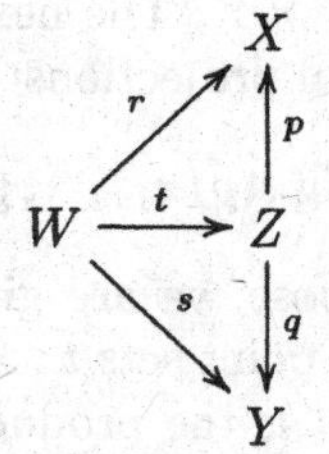

commutes. The induced morphism t in this situation will always be denoted (r, s). We call p and q the *projections* of the product onto its factors. Clearly a product, if it exists, is unique up to a unique isomorphism commuting with the projections.

The requirement of the definition can be rephrased to say $\hom(W,Z) \xrightarrow{\sim} \hom(W,X) \times \hom(W,Y)$ (under the obvious map induced by p and q).

We shall prove that products exist in the category of prevarieties over k. Note that we have no choice for the underlying set; for if $X \times Y$ is a product of the prevarieties X and Y, $X \times Y$ as a point set must be the usual product of the point sets X and Y. To see this, let W be a simple point; this is a prevariety ($\mathbb{A}^0$, in fact). The maps of W to any prevariety S clearly correspond to the points of S, and by definition $\hom(W, X \times Y) \simeq \hom(W,X) \times \hom(W,Y)$.

Proposition 1. *Let X and Y be affine varieties, with coordinate rings R and S. Then*

1) there is a product prevariety $X \times Y$.
2) $X \times Y$ is affine with coordinate ring $R \otimes_k S$.
3) a basis of the topology is given by the open sets

$$\sum f_i(x) g_i(y) \neq 0, \qquad f_i \in R, g_i \in S .$$

4) $\underline{o}_{(x,y)}$ is the localization of $\underline{o}_x \otimes_k \underline{o}_y$ at the maximal ideal $m_x \cdot \underline{o}_y + \underline{o}_x \cdot m_y$.

Proof. We recall the following result from commutative algebra: let R and S be integral domains over the algebraically closed field k. Then $R \otimes_k S$ is an integral domain. [Cf. Zariski-Samuel, vol. 1, Ch. 3, §15.]

Represent

$$\begin{aligned} X \subset k^{n_1} \quad &\text{as} \quad V(f_1, \ldots, f_{m_1}) \ . \\ Y \subset k^{n_2} \quad &\text{as} \quad V(g_1, \ldots, g_{m_2}) \ . \end{aligned}$$

Then the set $X \times Y \subset k^{n_1+n_2}$ is the locus of zeroes of $f_j(X_i), g_j(Y_i)$ in $k[X_1, \ldots, X_{n_1}, Y_1, \ldots, Y_{n_2}]$. Moreover

$$\begin{aligned} k[X_1, \ldots, X_{n_1}, Y_1, \ldots, Y_{n_2}] / (f_j, g_j) \ &\simeq \ k[X_i]/(f_j) \otimes_k k[Y_i]/(g_j) \\ &= \ R \otimes_k S. \end{aligned}$$

But $R \otimes_k S$ is an integral domain; hence (f_j, g_j) is prime, $X \times Y$ is irreducible, and $R \otimes_k S$ is its coordinate ring.

This gives us an affine variety $X \times Y$. The next step is to prove that it is a categorical product. We have natural projections

$$p, q : X \times Y \longrightarrow X, Y \quad [\text{e.g., } p(x_1, \ldots, y_{n_2}) = (x_1, \ldots, x_{n_1})]$$

which are clearly morphisms. Suppose we are given morphisms $f : Z \to X$, $s : Z \to Y$. There is just one map of point sets $t : Z \to X \times Y$ such that $r = p \cdot t$, $s = g \cdot t$ (since $X \times Y$ as a point set is the product of X and Y), and to verify the universal mapping property we need only check that t is always a morphism.

But this is simple. Since $X \times Y$ is affine, it suffices to check that $g \in \Gamma(X \times Y, \underline{o}_{X \times Y}) \Longrightarrow g \cdot t \in \Gamma(Z, \underline{o}_Z)$. Now $\Gamma(X \times Y, \underline{o}_{X \times Y})$ is generated by the images of $\Gamma(X, \underline{o}_X) = R$ and $\Gamma(Y, \underline{o}_Y) = S$. Both of these by composition with t

go into $\Gamma(Z, \underline{o}_Z)$ since r and s are morphisms; therefore all of $\Gamma(X \times Y, \underline{o}_{X \times Y})$ goes into $\Gamma(Z, \underline{o}_Z)$.

We have proved 1) and 2), and 3) follows from 2). Now $\underline{o}_{(x,y)}$ is the localization of $R \otimes_k S$ at the ideal of all functions vanishing at (x, y). Clearly $R \otimes_k S \subset \underline{o}_x \otimes_k \underline{o}_y \subset \underline{o}_{(x,y)}$, and therefore we can get $\underline{o}_{(x,y)}$ by localizing $\underline{o}_x \otimes_k \underline{o}_y$ at the ideal of all functions in it vanishing at (x, y). We claim that ideal is precisely $m_x \cdot \underline{o}_y + \underline{o}_x \cdot m_y$. Evidently all these functions do vanish there. Conversely, if we take any $h = \sum f_1 \otimes g_i \in \underline{o}_x \otimes_k \underline{o}_y$, with, say, $f_i(x) = \alpha_i$, $g_i(y) = \beta_i$, then we claim $h - \sum \alpha_i\beta_i \in m_x \cdot \underline{o}_y + \underline{o}_x \cdot m_y$. Indeed it equals

$$\sum f_i \otimes g_i - \sum \alpha_i \otimes \beta_i = \sum (f_i - \alpha_i) \otimes g_i + \sum \alpha_i \otimes (g_i - \beta_i) \in m_x \cdot \underline{o}_y + \underline{o}_x \cdot m_y\,.$$

□

We can now "glue together" these affine products to obtain:

Theorem 2. *Let X and Y be prevarieties over k. Then they have a product.*

Proof. We start, of course, with the product *set*. For all open affine $U \subset X$, $V \subset Y$ and all finite sets of elements $f_i \in \Gamma(U, \underline{o}_X)$, $g_i \in \Gamma(V, \underline{o}_Y)$ we form $(U \times V)_{\Sigma f_i g_i}$; these we take as a basis of the open sets. (They do form a basis, since $(U \times V)_{\Sigma f_i g_i} \cap (U' \times V')_{\Sigma f'_j g'_j}$ contains $(U'' \times V'')_{\Sigma f_i g_i \cdot \Sigma f'_j g'_j}$, where U'' [resp. V''] is an affine contained in $U \cap U'$ [resp. $V \cap V'$].) Note that on $U \times V$ this induces the topology of their true product.

Let K be the quotient field of $k(X) \otimes_k k(Y)$ [which, as before, is an integral domain]. For $x \in X$, $y \in Y$ we let $\underline{o}_{(x,y)} \subset K$ be the localization of $\underline{o}_x \otimes_k \underline{o}_y$ at the ideal $m_x \cdot \underline{o}_y + \underline{o}_x \cdot m_y$, and we set

$$\Gamma(U, \underline{o}_{X \times Y}) = \bigcap_{(x,y) \in U} \underline{o}_{(x,y)}.$$

This gives us a sheaf of functions. Furthermore, it coincides on each $U \times V$ (U, V affine) with the product of the affine varieties. Clearly $X \times Y$ is connected and covered by finitely many affines, so it is a prevariety.

Now suppose $Z \xrightarrow{r} X$, $Z \xrightarrow{s} Y$ are morphisms. Set-theoretically there is a unique map $(r, s) : Z \to X \times Y$ composing properly with the projections; we want to check that it is a morphism. For each U in X, V in Y affine, look at $Z_{U,V} = r^{-1}(U) \cap s^{-1}(V)$. These are open sets covering Z, and since being a morphism is a local property it is enough to prove $(r, s) \mid Z_{U.V}$ is a morphism. That is, we may assume $r(Z) \subset U$, $s(Z) \subset V$. But in the last proposition we saw that then the product map $Z \to U \times V$ is a morphism; and $U \times V$ is an open subprevariety of $X \times Y$, so the composite map $Z \to U \times V \to X \times Y$ is a morphism.

Remarks. I. If U is any open subprevariety of X, then $U \times Y$ is an open subprevariety of $X \times Y$.

II. If Z is a closed subprevariety of X, then $Z \times Y$ is a closed subprevariety of $X \times Y$. [It is enough to prove $(Z \cap U) \times V$ a closed subprevariety of $U \times V$ for U, V affine, and this is easily checked.]

Theorem 3. *The product of two projective varieties is a projective variety.*

Proof. Since a closed subvariety of a projective variety is a projective variety, it is enough to show $\mathbb{P}_n \times \mathbb{P}_m$ is a projective variety. In fact, we can embed it as a closed subvariety of $\mathbb{P}_{nm+n+m}$.

Take homogeneous coordinates $X_0, \ldots, X_n$ in $\mathbb{P}_n$, $Y_0, \ldots, Y_m$ in $\mathbb{P}_m$, and U_{ij} $(i = 0, \ldots, n,\ j = 0, \ldots, m)$ in $\mathbb{P}_{(n+1)(m+1)-1}$. Define a map:

$$I : \mathbb{P}_n \times \mathbb{P}_m \longrightarrow \mathbb{P}_{nm+n+m}$$

by

$$(x_0, \ldots, x_n) \times (y_0, \ldots, y_m) \longrightarrow \begin{array}{l}\text{a point with homogeneous coordinates}\\ U_{ij} = x_i y_j\end{array}$$

[This makes sense; multiplying all x_i for all y_j by λ multiplies all U_{ij} by λ; and some U_{ij} is nonzero.] Clearly

$$I^{-1}\left((\mathbb{P}_{nm+n+m})_{U_{ij}}\right) = (\mathbb{P}_n)_{X_i} \times (\mathbb{P}_m)_{Y_i} \quad .$$

We claim that first I is injective. Assume that for some $\lambda \neq 0$, $x_i y_j = \lambda x'_i y'_j$ for all i and j; then we want to prove that for some $\mu, \nu \neq 0$, $x_i = \mu x'_i$, $y_j = \nu y'_j$. We may by symmetry assume $x_0 \neq 0$, $y_0 \neq 0$. Then we have $0 \neq x_0 y_0 = x'_0 y'_0$, so $x'_0 \neq 0$, $y'_0 \neq 0$. We have $x_i/x_0 = x_i y_0 / x_0 y_0 = x'_i y'_0 / x'_0 y'_0 = x'_i / x'_0$ and similarly for y_j/y_0. This proves what we want.

Now we claim I is an isomorphism of $(\mathbb{P}_n)_{X_i} \times (\mathbb{P}_m)_{Y_j}$ onto a closed subvariety of $(\mathbb{P}_{n+m+nm})_{U_{ij}}$. We may assume $i = 0, j = 0$ for simplicity. On $(\mathbb{P}_n)_{X_0}$ we take affine coordinates $S_i = X_i/X_0$, $i = 1, \ldots, n$. On $(\mathbb{P}_m)_{Y_0}$ we take affine coordinates $T_j = Y_j/Y_0$, $j = 1, \ldots, m$. On $(\mathbb{P}_{n+m+nm})_{U_{00}}$ we take affine coordinates $R_{ij} = U_{ij}/U_{00}$, $i \geq 1$ or $j \geq 1$. In these coordinates, I takes (s, t) to the point

$$\begin{aligned} R_{ij} &= s_i t_j \qquad \text{if } i, j \geq 1 \\ R_{i0} &= s_i, \qquad R_{0j} = t_j \quad . \end{aligned}$$

Hence the image is the locus of points satisfying $R_{ij} = R_{i0} R_{0j}$ for all $i, j \geq 1$. This is certainly closed. Its affine coordinate ring is $k[R_{ij}] / (R_{ij} - R_{i0} R_{0j})$, which is clearly isomorphic to the polynomial ring $k[R_{i0}, R_{0j}]$. Under I^*, this is mapped isomorphically to $k[S_i, T_j]$ which is the affine coordinate ring of $(\mathbb{P}_n)_{X_0} \times (\mathbb{P}_m)_{Y_0}$. Hence we do have an isomorphism.

Let $Z = I(\mathbb{P}_n \times \mathbb{P}_m)$. Since $Z \cap (\mathbb{P}_{n+m+nm})_{U_{ij}}$ is closed for all i, j, Z is closed. I is a homeomorphism on each of these affines, so it is a homeomorphism globally, and in particular Z is irreducible. Thus Z is a projective variety. Since I is an isomorphism on each affine piece, it is an isomorphism globally and hence Z is isomorphic to $\mathbb{P}_n \times \mathbb{P}_m$. □

The classical example is the embedding $\mathbb{P}_1 \times \mathbb{P}_1 \to \mathbb{P}_3$. Take coordinates u, v on $\mathbb{P}_1$, s, t on $\mathbb{P}_1$, x, y, z, w on $\mathbb{P}_3$. I is defined by $x = us$, $y = vt$, $z = ut$, $w = vs$. It is easy to see that the image of I is the quadric $xy = zw$. [Thus, over an algebraically closed field all nondegenerate quadrics (those corresponding to nondegenerate quadric forms) are isomorphic to $\mathbb{P}_1 \times \mathbb{P}_1$.]

One can show in general that the homogeneous ideal of the image is generated by the elements $U_{ij}U_{i'j'} - U_{ij'}U_{i'j}$, so the image is an intersection of quadrics.

We now take up the main topic of this section – the Hausdorff axiom:

Definition 2. Let X be a prevariety. X is a *variety* if for all prevarieties Y and for all morphisms

$$Y \underset{g}{\overset{f}{\rightrightarrows}} X.$$

$\{y \in Y \mid f(y) = g(y)\}$ is a closed subset of Y.

One case of this criterion is particularly simple: let $Y = X \times X$, $f = p_1$, $g = p_2$. Also let

$$\Delta : X \longrightarrow X \times X$$

be a morphism $(\mathrm{id}_X, \mathrm{id}_X)$: Δ is called the diagonal morphism. Then

$$\Delta(X) = \{z \in X \times X \mid p_1(z) = p_2(z)\}.$$

Therefore the Hausdorff axiom implies $\Delta(X)$ is closed. But the converse is also true:

Proposition 4. *Let X be a prevariety. Then X is a variety if and only if $\Delta(X)$ is closed in $X \times X$.*

Proof. Suppose $f, g : Y \to X$ are given. Then induce a morphism $(f, g) : Y \to X \times X$. Since

$$\{y \in Y \mid f(y) = g(y)\} = (f, g)^{-1}[\Delta(X)],$$

$\Delta(X)$ being closed implies the Hausdorff axiom. □

Example R. Let X be:

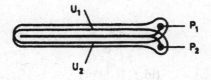

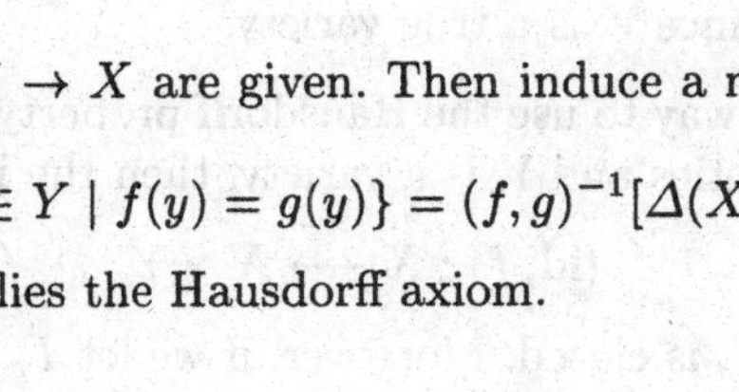

i.e., 2 copies U_1, U_2 of $\mathbb{A}^1$, say with coordinates x_1 and x_2, patched by the map $x_1 = x_2$ on the open sets $x_1 \neq 0$ and $x_2 \neq 0$. Consider the isomorphisms of $\mathbb{A}^1$ with each of the 2 copies:

$$\begin{array}{ccccc} & & U_1 & & \\ & \overset{\sim}{\nearrow} & & \cap & \\ \mathbb{A}^1 & & & & X \\ & \underset{\sim}{\searrow} & & \cup & \\ & & U_2 & & \end{array}$$

Call these $i_1, i_2 : \mathbb{A}^1 \to X$. Then

$$\{y \in \mathbb{A}^1 \mid i_1(y) = i_2(y)\} = \mathbb{A}^1 - \{0\}$$

is not closed in $\mathbb{A}^1$, hence X is not a variety.

Remarks. I. A subprevariety of a variety is a variety. A product of 2 varieties is a variety.

II. An affine variety is a variety: in fact, if X is affine, Y is an arbitrary prevariety, and $f, g : Y \to X$ are 2 morphisms, then

$$\{y \in Y \mid f(y) = g(y)\} = \left\{ \begin{array}{l} \text{locus of zeroes of the functions} \\ s.f. - s.g., \quad \text{all } s \in \Gamma(X, \underline{o}_X) \end{array} \right\}$$

and this set is closed.

III. From II., we can check that if $f, g : Y \to X$ are any morphisms of *any* prevarieties, then $\{y \in Y \mid f(y) = g(y)\}$ is always *locally closed.* In fact, call this set Z. If $z \in Z$, let V be an affine open neighbourhood of $f(z)$ $(= g(z))$. Then

$$Z \cap [f^{-1}(V) \cap g^{-1}(V)] = \left\{ z \in f^{-1}(V) \cap g^{-1}(V) \;\middle|\; \begin{array}{l} f(z) = g(z) \\ \text{in the affine} \\ \text{variety} \end{array} \right\}$$

and this set is closed since V is a true variety.

IV. Another useful way to use the Hausdorff property is this: if $f : X \to Y$ is a morphism of prevarieties and Y is a variety, then the image of the morphism

$$(\mathrm{id}, f) : X \longrightarrow X \times Y$$

which is the *graph* of f, is closed. Moreover, if we let Γ_g be this image, then Γ_f is even a closed subprevariety of $X \times Y$ isomorphic to X under the mutually inverse morphisms:

$$\begin{array}{c} \Gamma_f \\ (\mathrm{id}, f) \uparrow \downarrow p_{1|\Gamma_f} \\ X \end{array}$$

(here $p_1 : X \times Y \to X$ is the projection).

Proposition 5. *Let X be a prevariety. Assume that for all $x, y \in X$ there is an open affine U containing both x and y. Then X is a variety.*

Proof. Suppose $f, g : Y \to X$ are 2 morphisms such that $Z = \{y \in Y \mid f(y) = g(y)\}$ is not closed. Let $z \in \overline{Z}$, $x = f(z)$, $y = g(z)$. By assumption, there is an open affine V containing x and y. Let $U = f^{-1}(V) \cap g^{-1}(V)$: U is an open neighbourhood of Z. If f', g' are the restrictions of f, g to morphisms from U to V, then

$$Z \cap U = \{y \in U \mid f'(y) = g'(y)\}$$

is closed in U, since V, being affine, is a variety. Therefore $z \in Z \cap U$, and Z is closed. □

Corollary. *Every projective variety X is a variety.*

Proof. For all $x, y \in \mathbb{P}_n(k)$, there is a hyperplane not containing x or y, i.e., an element $H = \Sigma \alpha_i x_i \in k[X_0, \ldots, X_n]$ such that $x, y \in \mathbb{P}_n(k)_H$. But $\mathbb{P}_n(k)_H$ is affine, hence $X \cap \mathbb{P}_n(k)_H$ is open and affine in X. □

Proposition 6. *Let X be a variety, and let U, V be affine open subsets with coordinate rings R, S. Then $U \cap V$ is an affine open subset with coordinate ring $R \cdot S$ (the compositum being formed in $k(X)$).*

Proof. $U \times V$ is an affine open subset of $X \times X$ with coordinate ring $R \otimes_k S$. Let $Z = \Delta(X)$. Then $Z \cap (U \times V)$ is a closed subset of $U \times V$ isomorphic via Δ to $U \cap V$. Since Δ is an isomorphism, $Z \cap (U \times V)$ is irreducible, hence it is a closed subvariety of $U \times V$. Therefore it is affine, and its coordinate ring T is a quotient of the coordinate ring of $U \times V$. But it is also open in Z, hence $Z \subset k(Z)$, i.e., T is the image of $R \otimes_k S$ in $k(Z)$. Again since Δ is an isomorphism, this proves that

$$U \cap V = \Delta^{-1}\{Z \cap (U \times V)\}$$

is an open affine subvariety of X, and that its coordinate ring is $\Delta^*(R \otimes S)$ in $k(X)$. But $\Delta^*(f \otimes g) = f \cdot g$, hence $\Delta^*(R \otimes S) = R \cdot S$. □

Problem. Prove the following converse: let X be a prevariety, $\{U_i\}$ an affine open covering of X. Let R_i be the coordinate ring of U_i. Then if $U_i \cap U_j$ is an affine subset of X with coordinate ring $R_i \cdot R_j$, X is a variety.

Zariski and Chevalley have used a quite different definition of a variety which is useful for many questions, especially "birational" ones, i.e., questions dealing with various different prevarieties with a common function field.

Definition 3. Let $\mathcal{O} \subset \mathcal{O}'$ be local rings. We say $\mathcal{O}'$ *dominates* $\mathcal{O}$ or $\mathcal{O}' > \mathcal{O}$, if $m' \supset m$ (equivalently $m' \cap \mathcal{O} = m$) where m, m' are the maximal ideals of $\mathcal{O}$, $\mathcal{O}'$ respectively.

Local Criterion: Let X be a prevariety. Then X is a variety if and only if for all $x, y \in X$ such that $x \neq y$, there is *no* local ring $\mathcal{O} \subset k(X)$ such that $\mathcal{O} > \underline{o}_x$ and $\mathcal{O} > \underline{o}_y$.

The fact that this holds for varieties will be proven in Ch. II, §6. We omit the converse which we will not use. If one starts with this criterion, however, one can take the elegant Chevalley-Nagata definition of a variety: we identify a variety X with the set of local rings $\underline{o}_x$. For this definition, a variety is a finitely generated field extension K of k plus a collection of local subrings of K satisfying various conditions. [Notice: the topology can be recovered as follows – for each $f \in K$, let U_f be the set of those local rings in X containing f.]

Problem. Let $X \subset \mathbb{P}^n$ be defined by a homogeneous ideal $P \subset k[X_0, \ldots, X_n]$. Let P define the *affine* variety $X^* \subset \mathbb{A}^{n+1}$. Show that for all i,

$$\left\{ \begin{array}{l} \text{Subset of } X^* \\ \text{where } X_i \neq 0 \end{array} \right\} \cong \mathbb{A}^1 \times \left\{ \begin{array}{l} \text{Subset of } X \\ \text{where } X_i \neq 0 \end{array} \right\}.$$

§7. Dimension

Definition 1. Let X be a variety. $\dim X = \text{tr.d.}_k k(X)$.

If U is an open and nonempty set in X, $\dim U = \dim X$, since $k(U) = k(X)$. Since k is algebraically closed, the following are equivalent:

i) $\dim X = 0$
ii) $k(X) = k$
iii) X is a point.

Proposition 1. *Let* Y *be a proper closed subvariety of* X. *Then* $\dim Y < \dim X$.

Proof .

Lemma. *Let* R *be an integral domain over* k, $P \subset R$ *a prime. Then* $\text{tr.d.}_k R \geq \text{tr.d.}_k R/P$, *with equality only if* $P = \{0\}$ *or both sides are* ∞. *[By convention,* tr.d.R *is the* tr.d. *of the quotient field of* R.*]*

Proof. Say $P \neq 0$, $\text{tr.d.}_k R = n < \infty$. If the statement is false, there are n elements $x_1, \ldots, x_n$ in R such that their images $\bar{x}_i$ in R/P are algebraically independent. Let $0 \neq p \in P$. Then $p, x_1, \ldots, x_n$ cannot be algebraically independent over k, so there is a polynomial $P(Y, X_1, \ldots, X_n)$ over k such that $P(p, x_1, \ldots, x_n) = 0$. Since R is an integral domain, we may assume P is irreducible. The polynomial P cannot equal αY, $\alpha \in k$, since $p \neq 0$. Therefore P is not even a multiple of Y. But then $P(0, \bar{x}_1, \ldots, \bar{x}_n) = 0$ in R/P is a nontrivial relation on the $\bar{x}_1, \ldots, \bar{x}_n$.

Now choose $U \subset X$ an open affine set with $U \cap Y \neq \emptyset$. Let R be the coordinate ring of U, P the prime ideal corresponding to the closed set $U \cap Y$. Then $P \neq 0$, since $U \cap Y \neq U$. $k(X)$ is the quotient field of R, and $k(Y)$ is the quotient field of R/P. Therefore the Proposition follows immediately from the lemma. □

In the situation of this Proposition, $\dim X - \dim Y$ will be called the *codimension* of X in Y. This is half of what we want so that our definition gives a good dimension function. The other half is that it does not go down too much.

Theorem 2. *Let X be a variety, $U \subset X$ open, $g \in \Gamma(U, \underline{o}_X)$, Z an irreducible component of $\{x \in U \mid g(x) = 0\}$. Then if $g \not\equiv 0$, $\dim Z = \dim X - 1$.*

Proof. Take $U_0 \subset U$ an open affine set with $U_0 \cap Z \neq \emptyset$. Let $R = \Gamma(U_0, \underline{o}_X)$, $f = \mathrm{res}_{U,U_0} g \in R$. Then $Z \cap U_0$ corresponds to a prime $P \subset R$. Z is by hypothesis a maximal irreducible subset of the locus $g = 0$, so $Z \cap U_0$ is a maximal irreducible subset of the locus $f = 0$, i.e., P is a maximal prime containing f. Thus we have translated Theorem 2 into:

Algebraic Version (Krull's Principal Ideal Theorem). *Let R be a finitely generated integral domain over k, $f \in R$, P an isolated prime ideal of (f) (i.e., minimal among the prime ideals containing it). Then if $f \neq 0$, $\mathrm{tr.d.}_k R/P = \mathrm{tr.d.}_k R - 1$.*

This is a standard result in commutative algebra (cf. for example, Zariski-Samuel, vol. 2, Ch. 7, §7). However there is a fairly straightforward and geometric proof, using only the Noether normalization lemma and based on the Norm, so I think it is worthwhile proving the algebraic version too. This proof is due to J. Tate.

At this point, it is convenient to translate the results of §1 into the geometric framework that we have built up:

Definition 2. Let $f : X \to Y$ be a morphism of affine varieties. Let R and S be the coordinate rings of X and Y, and let $f^* : S \to R$ be the induced homeomorphism. Then f is *finite* if R is integrally dependent on the subring $f^*(S)$.

Note that the restriction of a finite morphism from X to Y to a closed subvariety Z of X is also finite. Examples N and O in §3 are finite morphisms, but example P is not. The main properties of finite morphisms are the following:

Proposition 2. *Let $f : X \to Z$ be a finite morphism of affine varieties.*

(1.) Then f is a closed map, i.e., maps closed sets onto closed sets.
(2.) For all $y \in Y$, $f^{-1}(y)$ is a finite set.
(3.) f is surjective if and only if the corresponding map f^* of coordinate rings is injective.

Proof. Let f be dual to the map $f^* : S \to R$ of coordinate rings. As usual, the maps V, I set up bijections between the points of X (resp. Y) and the maximal ideals m of R (resp. S). If $V(A)$ is a closed set in X, then $f(V(A))$ corresponds to the set of maximal ideals ${f^*}^{-1}(m)$, $m \subset R$ a maximal ideal containing A. But by the going-up theorem, whenever $R \supset S$ and R is integral over S every *prime* ideal $P \subset S$ is of the form $P' \cap S$ for some prime ideal $P' \subset R$ (cf. §1). If $B = {f^*}^{-1}(A)$, apply the going-up theorem to $S/B \subset R/A$; therefore $f(V(A))$

corresponds to the set of maximal ideals m such that $m \supset B$, i.e., to $V(B)$. Therefore, f is closed. Moreover, $f(X) = Y$ if and only if $\mathrm{Ker}(f^*) = (0)$. (2.) is equivalent to saying that for all maximal ideals $m \subset S$, there are only a *finite* number of maximal ideals $m' \subset R$ such that $m = {f^*}^{-1}(m')$. Since such m' all contain $f^*(m) \cdot R$, it is enough to check that $R/f^*(m) \cdot R$ contains only a finite number of maximal ideals. But $R/f^*(m) \cdot R$ is integral over S/m, i.e., it is a finite-dimensional algebra over k, so this is clear. □

Via this concept, we can state:

Geometric form of Noether's normalization lemma. *Let X be an affine variety of dimension n. Then there exists a finite surjective morphism π:*

$$X \xrightarrow{\pi} \mathbb{A}^n .$$

Proof of Theorem 2. We first reduce the proof to the case $P = \sqrt{(f)}$, i.e., geometrically, to the case $Z = V((f))$. To do this, look back at the way in which we related the geometric and algebraic versions of the Theorem. Note that the algebraic version is essentially identical to Theorem 2 in the case $X = U$ is affine with coordinate ring R. Suppose we have the decomposition

$$\sqrt{(f)} = P \cap P_1' \cap \ldots \cap P_t'$$

in R. Geometrically, if $Z_i' = V(P_i')$, this means that $Z, Z_1', \ldots, Z_t'$ are the components of $V((f))$. Now we pick an affine open $U_0 \subset X$ such that:

$$\begin{aligned} &U_0 \cap Z \neq \emptyset \\ &U_0 \cap Z_i' = \emptyset, \qquad i = 1, \ldots, t . \end{aligned}$$

For example, let $U_0 = X_g$, where:

$$g \in P_1' \cap \ldots \cap P_t', \qquad g \notin P .$$

Then replace X by U_0, R by $R_{(g)}$, and in the new set-up,

$$\begin{aligned} V_{U_0}((f)) &= V_X((f)) \cap U_0 \\ &= Z \cap U_0 \end{aligned}$$

is irreducible; hence in $R_{(g)}$, $\sqrt{(f)} = P \cdot R_{(g)}$ is prime.

Now use the normalization lemma to find a morphism as follows:

$$\begin{array}{ccc} X & \qquad & R \\ {\scriptstyle\pi}\big\downarrow & & \big\uparrow{\scriptstyle\pi^*} \\ \mathbb{A}^n & & k[X_1, \ldots, X_n] = S. \end{array}$$

Let K (resp. L) be the quotient field of R (resp. S). Then K/L is a finite algebraic extension. Let

$$f_0 = N_{K/L}(f) .$$

Then I claim $f_0 \in S$ and

$$(*) \qquad P \cap S = \sqrt{(f_0)} .$$

If we prove $(*)$, the theorem follows. For R/P is an integral extension of $S/S \cap P$, so

$$\text{tr.d.}_k R/P = \text{tr.d.}_k S/S \cap P .$$

But S is a UFD, so the primary decomposition of a principal ideal in S is just the product of decomposition of the generator into irreducible elements. Therefore $(*)$ implies that f_0 is a unit times f_{00}^{ℓ} for some integer ℓ and some irreducible f_{00}, and that $P \cap S = (f_{00})$. Hence

$$\text{tr.d.}_k S/S \cap P = \text{tr.d.}_k k[X_1, \ldots, X_n]/(f_{00}) = n^{-1}.$$

We check first that $f_0 \in P \cap S$. Let

$$Y^n + a_1 Y^{n-1} + \ldots + a_n = 0$$

be the irreducible equation satisfied by f over the field L. Then f_0 is a power $(a_n)^m$ of a_n. Moreover, all the a_i's are symmetric functions in the conjugates of f (in some normal extension of L): therefore the a_i are elements of L integrally dependent on S. Therefore $a_i \in S$. In particular, $f_0 = a_n^m \in S$, and since

$$\begin{aligned} 0 &= a_n^{m-1} \cdot (f^n + a_1 f^{n-1} + \ldots + a_n) \\ &= f \cdot (a_n^{m-1} f^{n-1} + a_n^{m-1} a_1 f^{n-2} + \ldots + a_n^{m-1} a_{n-1}) + f_0 , \end{aligned}$$

$f_0 \in P$ too.

Finally suppose $g \in P \cap S$. Then $g \in P = \sqrt{(f)}$, hence

$$g^n = f \cdot h, \qquad \begin{array}{l} \text{some integer } n , \\ \text{some } h \in R . \end{array}$$

Taking norms, we find that

$$\begin{aligned} g^{n \cdot [K:L]} &= N_{K/L}(g^n) \\ &= N_{K/L}(f) \cdot N_{K/L}(h) \in (f_0) \end{aligned}$$

since $N_{K/L} h$ is an element of S, by the reasoning used before. Therefore $g \in \sqrt{(f_0)}$, and $(*)$ is proven. □

Definition 3. Let X be a variety, and let $Z \subset X$ be a closed subset. Then Z has *pure dimension* r if each of its components has dimension r (similarly for *pure codimension* r).

The conclusion of Theorem 2 may be stated as: $V((g))$ has pure codimension 1, for any non-zero $g \in \Gamma(X, \underline{o}_X)$. The Theorem has an obvious converse: Suppose Z is an irreducible closed subset of a variety X of codimension 1. Then for all open sets U such that $Z \cap U \neq \emptyset$ and for all non-zero functions f on U vanishing on Z, $Z \cap U$ is a component of $f = 0$. In fact, if W were a component of $f = 0$ containing $Z \cap U$, then we have

$$\dim X > \dim W \geq \dim Z \cap U = \dim X - 1$$

hence $W = Z \cap U$ by Proposition 1.

Corollary 1. *Let X be a variety and Z a maximal closed irreducible subset, smaller than X itself. Then* $\dim Z = \dim X - 1$.

Corollary 2 (Topological characterization of dimension). *Suppose*

$$\emptyset \neq Z_1 \not\subseteq Z_2 \not\subseteq \ldots Z_1 \not\subseteq Z$$

is any maximal chain of closed irreducible subsets of X. Then $\dim X = r$.

Proof. Induction on $\dim X$.

Corollary 3. *Let X be a variety and let Z be a component of $V((f_1, \ldots, f_r))$, where $f_1, \ldots, f_r \in \Gamma(X, \underline{o}_X)$. Then* codim $Z \leq r$.

Proof. Induction on r. Z is an irreducible subset of $V((f_1, \ldots, f_{r-1}))$, so it is contained in some component Z' of $V((f_1, \ldots, f_{r-1}))$. Then Z is a component of $Z' \cap V((f_r))$, since $Z' \cap V((f_r)) \subset V((f_1, \ldots, f_r))$. By induction codim $Z' \leq r-1$. If $f_r \equiv 0$ on Z', $Z = Z'$. If f_r does not vanish identically on Z', then by the theorem, $\dim Z = \dim Z' - 1$, so codim $Z \leq r$. □

Of course, equality need not hold in the above result: e.g., take $f_1 = \ldots = f_1$, $r > 1$.

Corollary 4. *Let U be an affine variety, Z a closed irreducible subset. Let $r =$* codim Z. *Then there exist $f_1, \ldots, f_r$ in $R = \Gamma(U, \underline{o}_U)$ such that Z is a component of $V((f_1, \ldots, f_r))$.*

Proof. In fact, we prove the following. Let $Z_1 \supset Z_2 \supset \ldots \supset Z_r = Z$ be a chain of irreducibles with codim $Z_i = i$ (by Cor. 2). Then there are $f_1, \ldots, f_r$ in R such that Z_s is a component of $V((f_1, \ldots, f_s))$ and all components of $V((f_1, \ldots, f_s))$ have codim s.

We prove this by induction on s. For $s = 1$, take $f_1 \in I(Z_1)$, $f_1 \neq 0$, and we have just the converse of the theorem. Now say $f_1, \ldots, f_{s-1}$ have been chosen. Let $Z_{s-1} = Y_1, \ldots, Y_\ell$ be the components of $V((f_1, \ldots, f_{s-1}))$. For all i, $Z_s \not\supset Y_i$ (because of their dimensions), so $I(Y_i) \not\supset I(Z_s)$. Since the $I(Y_i)$ are prime, $\bigcup_{i=1}^{\ell} I(Y_i) \not\supset I(Z_s)$ [Zariski-Samuel, vol. 1, p. 215; Bourbaki, Ch. 2, p. 70]. Hence we can choose an element $f_s \in I(Z_s)$, $f_s \notin \bigcup_{i=1}^{\ell} I(Y_i)$.

If Y is any component of $V((f_1,\ldots,f_s))$, then (as in the proof of Cor. 3) Y is a component of $Y_i \cap V((f_s))$ for some i. Since f_s is not identically zero on Y_i, $\dim Y = \dim Y_i - 1$, so codim $Y = s$.

By the choice of f_s, $Z_s \subset V((f_1,\ldots,f_s))$. Being irreducible, Z_s is contained in some component of $V((f_1,\ldots,f_s))$, and it must equal this component since it has the same dimension. □

In the theory of local rings, it is shown that one can attach to every noetherian local ring $\mathcal{O}$ an integer called its Krull dimension. This number is defined as either

a) the length r of the longest chain of prime ideals:

$$P_0 \subsetneq P_1 \subsetneq \ldots \subsetneq P_n = M$$

$$(M \text{ the maximal ideal of } \mathcal{O})$$

or b) the least integer n such that there exist elements $f_1,\ldots,f_n \in M$ for which

$$M = \sqrt{(f_1,\ldots,f_n)} .$$

(Cf. Zariski-Samuel, vol. 1, pp. 240–242; vol. 2, p. 288. Also, cf. Serre, *Algèbre Locale*, Ch. 3B). Recall that in §5 we attached a local ring $\underline{o}_{Z,X}$ to every irreducible closed subset Z of every variety X. We now have:

Corollary 5. *The Krull dimension* $\underline{o}_{Z,X}$ *is* $\dim X - \dim Z$.

Proof. By Corollary 2, for all maximal chains of irreducible closed subvarieties:

$$Z = Z_n \subsetneq Z_{n-1} \subsetneq \ldots \subsetneq Z_0 = Z ,$$

$n = \dim X - \dim Z$. But it is not hard to check that there is an order reversing isomorphism between the set of irreducible closed subvarieties Y and X containing Z, and the set of prime ideals $P \subset \underline{o}_{Z,X}$. Or else, use the second definition of Krull dimension: first note that if $f_1,\ldots,f_n \in \underline{o}_{Z,X}$, then

$$M_{Z,X} = \sqrt{(f_1,\ldots,f_n)}$$

if and only if there is an open set $U \subset X$ such that

a) $U \cap Z \neq \emptyset$,
b) $f_1,\ldots,f_n \in \Gamma(U,\underline{o}_X)$,
c) $U \cap Z = V((f_1,\ldots,f_n))$.

Then using Corollaries 3 and 4, it follows that the smallest n for which such f_i's exist is $\dim X - \dim Z$. □

Suppose $Z \subset X$ is irreducible and of codimension 1. A natural question to ask is whether, for all $y \in Z$, there is some neighbourhood U of y in X and some function $f \in \Gamma(U,\underline{o}_X)$ such that $Z \cap U$ is not just a component of $f = 0$, but actually equal to the locus $f = 0$. More generally, if $Z \subset X$ is a closed subset of pure codimension r, one may ask whether, for all $y \in Z$, there is a neighbourhood U of y and functions $f_1,\ldots,f_r \in \Gamma(U,\underline{o}_X)$ such that

$$Z \cap U = V((f_1, \ldots, f_r)) .$$

This is unfortunately not always true even in the special case where Z is irreducible of codimension 1. A closed set Z with this property is often referred to as a *local set-theoretic complete intersection* and it has many other special properties. There is one case where we can say something however:

Proposition 4. *Let X be an affine variety with coordinate ring R. Assume R is a UFD. Then every closed subset $Z \subset X$ of pure codimension 1 equals $V((f))$ for some $f \in R$.*

Proof. Since R is a UFD, every minimal prime ideal of R is principal [i.e., say $P \subset R$ is minimal and prime. Let $f \in P$. Since P is prime, P contains one of the prime factors f' of f. By the UFD property, (f') is also a prime ideal and since $(f') \subset P$, we must have $(f') = P$]. Let $Z_1, \ldots, Z_t$ be the components of Z. Then $I((Z_1), \ldots, I(Z_t))$ are minimal prime ideals. If $I(Z_i) = (f_i)$, then

$$Z = V((f_1, \ldots, f_t)) .$$

□

Proposition 5. $\dim X \times Y = \dim X + \dim Y$.

The proof is easy.

The results and methods of this section all have projective formulations which give some global as well as some local information:

Let $X \subset \mathbb{P}_n(k)$ be a projective variety, and let $I(X) \subset k[X_0, \ldots, X_n]$ be its ideal.

Theorem 2*. *If $f \in k[X_0, \ldots, X_n]$ is homogeneous and not a constant and $f \notin I$, then $X \cap V((f))$ is non-empty and of pure codimension 1 in X, unless* $\dim X = 0$.

Proof. All this follows from Theorem 2, except for the fact that $X \cap V((f))$ is not empty if $\dim X \geq 1$. But let $X^* \subset \mathbb{A}^{n+1}$ be the cone over X, i.e., the affine variety defined by the ideal $I(X)$ in $k[X_0, \ldots, X_n]$. By the problem in §6, we know that $\dim X^* = \dim X + 1$, hence $\dim X \geq 2$. Let $V^*((f))$ be the locus $f = 0$ in $\mathbb{A}^{n+1}$. Since

$$(0, 0, \ldots, 0) \in X^* \cap V^*((f)) ,$$

therefore $X^* \cap V^*((f)) \neq \emptyset$ and by Theorem 2, $X^* \cap V^*((f))$ has a component of dimension at least 1. Therefore $X^* \cap V^*((f))$ contains points other than $(0, 0, \ldots, 0)$; but the affine coordinates of such points are homogeneous coordinates of points in $X \cap V((f))$. □

Corollary 3*. *If $f_1, \ldots, f_r \in k[X_0, \ldots, X_n]$ are homogeneous and not constants, then all components of $X \cap V((f_1, \ldots, f_r))$ have codimension at most r in X. And if* $\dim X \geq r$, *then $X \cap V((f_1, \ldots, f_r))$ is non-empty.*

Corollary 4*. *If $Y \subset X$ is a closed subvariety, (resp. Y is the empty subset), then there exist homogeneous non-constant elements $f_1, \ldots, f_r \in k[X_0, \ldots, X_n]$ where $r =$ codimension of Y (resp. $r = \dim X + 1$) such that Y is a component of $X \cap V((f_1, \ldots, f_r))$ (resp. $X \cap V((f_1, \ldots, f_r))) = \emptyset$.*

Proof. One can follow exactly the inductive proof of Corollary 4 above, using $k[X_0, \ldots, X_n]$ instead of the affine coordinate ring. In case $Y = \emptyset$, the last step is slightly different. By induction, we have $f_1, \ldots, f_r$ such that

$$X \cap V((f_1, \ldots, f_r))$$

is a finite set of points. Then let f_{r+1} be the equation of any hypersurface not containing any of these points. □

This implies, for instance, that every *space curve*, i.e., one-dimensional subvariety of $\mathbb{P}_3(k)$, is a component of an intersection $H_1 \cap H_2$ of 2 surfaces.

Proposition 4*. *Every closed subset of $\mathbb{P}_n(k)$ of pure codimension 1 is a hypersurface, i.e., equals $V((f))$ for some homogeneous element $f \in k[X_0, \ldots, X_n]$.*

Proof. Exactly the same as Proposition 4, using the fact that $k[X_0, \ldots, X_n]$ is a UFD. □

An interesting corollary of these results is the following global theorem:

Proposition 6. *Suppose*

$$\mathbb{P}_n \xrightarrow{f} \mathbb{P}_m$$

is a morphism. Assume $W = f(\mathbb{P}_m)$ is closed (actually this is always true as we will see in §9). Then either W is a single point, or

$$\dim W = n.$$

Proof. Let $r = \dim W$ and assume $1 \leq r \leq n-1$. By Cor. 4* applied to the empty subvariety Y of W, there are homogeneous non-constant elements

$$f_1, f_2, \ldots, f_{r+1} \in k[X_0, \ldots, X_m]$$

such that

$$W \cap V((f_1, \ldots, f_{r+1})) = \emptyset.$$

Also, by Cor. 2*, $W \cap V((f_i)) \neq \emptyset$ for all i. Let $Z_i = f^{-1}V((f_i))$. Then

$$Z_1 \cap \ldots \cap Z_{r+1} = \emptyset, \quad \text{and } Z_i \neq \emptyset, \qquad 1 \leq i \leq r+1.$$

Note that the hypersurface $V((f_i))$ in $\mathbb{P}_m$ is defined locally by the vanishing of a single function. Therefore the closed subset Z_i in $\mathbb{P}_n$ is also defined locally by the vanishing of a single function. Therefore $Z_i = \mathbb{P}_n$ or else Z_i is of pure codimension 1, hence by Cor. 4* a hypersurface. Since $r+1 \leq n$, the intersection of $r+1$ hypersurfaces in $\mathbb{P}_n$ cannot be empty because of Cor. 3*. This is a contradiction, so in fact $r = 0$ or $r = n$. □

§8. The fibres of a morphism

Let $f : X \to Y$ be a morphism of varieties. The purpose of this section is to study the family of closed subsets of X consisting of the sets $f^{-1}(y)$, $y \in Y$.

Definition 1. A morphism $f : X \to Y$ is *dominating* if its image is dense in Y, i.e., $Y = \overline{f(X)}$.

Proposition 1. *If $f : X \to Y$ is any morphism, let $Z = \overline{f(X)}$. Then Z is irreducible, the restricted morphism $f' : X \to Z$ is dominating and f'^* induces an injection*

$$k(Z) \overset{f'^*}{\hookrightarrow} k(X) .$$

Proof. Suppose $Z = W_1 \cup W_2$, where W_1 and W_2 are closed subsets. Then $X = f^{-1}(W_1) \cup f^{-1}(W_2)$. Since X is irreducible, $X = f^{-1}(W_1)$ or $f^{-1}(W_2)$, i.e., $f(X) \subset W_1$ or W_2. Therefore $Z = \overline{f(X)}$ is equal to W_1 or W_2: hence Z is irreducible. f^* is clearly dominating and since $f'(X)$ is dense in Z, for *all* nonempty open sets $U \subset Z$, $f^{-1}(U)$ is nonempty and open in X; therefore, we obtain a map:

$$k(Z) = \varinjlim_{\substack{U \subset Z \\ \text{open} \\ \text{nonempty}}} \Gamma(U, \underline{o}_Z) \xrightarrow{f'^*} \varinjlim_{\substack{V \subset X \\ \text{open} \\ \text{nonempty}}} \Gamma(V, \underline{o}_X) = k(X) .$$

□

This reduces the study of the fibres of an arbitrary morphism to the case of dominating morphisms. Note also that a finite morphism is dominating if and only if it is surjective.

Theorem 2. *Let $f : X \to Y$ be a dominating morphism of varieties and let $r = \dim X - \dim Y$. Let $W \subset Y$ be a closed irreducible subset and let Z be a component of $f^{-1}(W)$ that dominates W. Then*

$$\dim Z \geq \dim W + r$$

or

$$\text{codim } (Z \text{ in } X) \leq \text{codim } (W \text{ in } Y).$$

Proof. If U is an affine open subset of Y such that $U \cap W \neq \emptyset$, then to prove the theorem we may as well replace Y by U, X by $f^{-1}(U)$, W by $W \cap U$, and Z by $Z \cap f^{-1}(U)$. Therefore we assume that Y is affine. By Cor. 4 to Th. 2, §7, if $s = \text{codim } (W \text{ in } Y)$, there are functions $f_1, \ldots, f_s \in \Gamma(Y, \underline{o}_Y)$ such that W is a component of $V((f_1, \cdots, f_s))$. Let $g_i \in \Gamma(X, \underline{o}_X)$ denote the function $f^*(f_i)$. Then $Z \subset V((g_1, \cdots, g_s))$ and I claim Z is a component of $V((g_1, \cdots, g_s))$. Suppose

$$Z \subset Z' \subset V((g_1, \cdots, g_s)) ,$$

where Z' is a component of $V((g_1,\ldots,g_s))$. Then

$$W = \overline{f(Z)} \subset \overline{f(Z')} \subset V((f_1,\ldots,f_s)).$$

Since W is a component of $V((f_1,\ldots,f_s))$ and $\overline{f(Z')}$ is irreducible, it follows that $W = \overline{f(Z')}$. Therefore, $Z' \subset f^{-1}(W)$. But Z is a component of $f^{-1}(W)$. Therefore $Z = Z'$, and Z is also a component of $V((g_1,\ldots,g_s))$. By Cor. 3 to Th. 2, §7, this proves that codim $(Z \text{ in } X) \le s$. □

Corollary. *If Z is a component of $f^{-1}(y)$, for some $y \in Y$, then $\dim Z \ge r$.*

Theorem 3. *Let $f : X \to Y$ be a dominating morphism of varieties and let $r = \dim X - \dim Y$. Then there exists a nonempty open set $U \subset Y$ such that:*

i) $U \subset f(X)$
ii) for all irreducible closed subsets $W \subset Y$ such that $W \cap U \neq \emptyset$, and for all components Z of $f^{-1}(W)$ such that $Z \cap f^{-1}(U) \neq \emptyset$,

$$\dim Z = \dim W + r$$

or

$$\text{codim}\,(Z \text{ in } X) = \text{codim}\,(W \text{ in } Y)\,.$$

Proof. As in Theorem 2, we may as well replace Y by a nonempty open affine subset; therefore, assume that Y is affine. Moreover, we can also reduce the proof easily to the case where X is affine. In fact, cover X by affine open sets $\{X_i\}$ and let $f_i : X_i \to Y$ by the restriction of f. Let $U_i \subset Y$ satisfy i) and ii) of the theorem for f_i. Let $U = \cap U_i$. Then with this U, i) and ii) are correct for f itself.

Now assume X and Y are affine, and let R and S be their coordinate rings. f defines a homomorphism

$$f^* : S \longrightarrow R$$

which is an injection by Proposition 1. Let $K = k(Y)$, the quotient field of S. Apply the normalization lemma to the K-algebra $R \otimes_S K$. Note that $R \otimes_S K$ is just the localization of R with respect to the multiplicative system S^*, hence it is an integral domain with the same quotient field as R, i.e., $k(X)$. In particular,

$$\begin{aligned} \text{tr.d.}_K\,(R \otimes_S K) &= \text{tr.d.}_{k(Y)} k(X) \\ &= \text{tr.d.}_k k(X) - \text{tr.d.}_k k(Y) = r\,. \end{aligned}$$

Therefore, there exists a subring:

$$K[Y_1,\ldots,Y_r] \subset R \otimes_S K$$

such that $R \otimes_S K$ is integrally dependent on $K[Y_1,\ldots,Y_r]$. We can assume that the elements Y_i are actually in the subring R: for any element of $R \otimes_S K$ is the product of an element of R and a suitable "constant" in K. Now consider the 2 rings:

$$S[Y_1,\ldots,Y_r] \subset R\,.$$

R is not necessarily integral over $S[Y_1,\ldots,Y_r]$; however, if $\alpha \in R$, then α satisfies an equation:

$$X^n + P_1(Y_1,\ldots,Y_r)X^{n-1} + \ldots + P_n(Y_1,\ldots,Y_r) = 0\,,$$

where the P_i are polynomials with coefficients in K. If g is a common denominator of all the coefficients, α is integral over $S_{(g)}[Y_1,\ldots,Y_r]$. Applying this reasoning to a finite set of generators of R as an S-algebra, we can find some $g \in S$ such that $R_{(g)}$ is integral over $S_{(g)}[Y_1,\ldots,Y_r]$. Define $U \subset Y$ as Y_g, i.e., $\{y \in Y \mid g(y) \neq 0\}$. The subring $S_{(g)}[Y_1,\ldots,Y_r]$ in $R_{(g)}$ defines a factorization of f restricted to $f^{-1}(U)$:

$$\begin{array}{ccc} X & \supset & f^{-1}(u) \\ \Big| & & \downarrow \pi \\ f & & U \times \mathbb{A}^r \\ \Big\downarrow & & \downarrow p_1 \\ X & \supset & U \end{array}$$

where π is finite and surjective. This shows first of all that $U \subset f(X)$ which is (i). To show (ii), let $W \subset Y$ be an irreducible closed subset that meets U, and let $Z \subset X$ be a component of $f^{-1}(W)$ that meets $f^{-1}(U)$. It suffices to show that

$$\dim Z \leq \dim W + r$$

since the other inequality has been shown in Theorem 2. Let $Z_0 = Z \cap f^{-1}(U)$ and let $W_0 = W \cap U$. Then

$$\overline{\pi(Z_0)} \subset W_0 \times \mathbb{A}^r\,.$$

Therefore

$$\dim \overline{\pi(Z_0)} \leq \dim(W_0 \times \mathbb{A}^r) = \dim W + r\,.$$

The restriction π' of π to a map from Z_0 to $\overline{\pi(Z_0)}$ is still dominating and finite. Therefore it induces an inclusion of $k\left(\overline{\pi(Z_0)}\right)$ in $k(Z_0)$ such that $k(Z_0)$ is algebraic over $k\left(\overline{\pi(Z_0)}\right)$. Therefore

$$\dim Z \leq \dim \overline{\pi(Z_0)} \leq \dim W + r\,.$$

□

Corollary 1. *f as above. Then there is a nonempty open set $U \subset Y$ such that, for all $y \in Y$, $f^{-1}(y)$ is a nonempty "pure" r-dimensional set, i.e., all its components have dimension r.*

Theorems 2 and 3 together give a good qualitative picture of the structure of a morphism. We can work this out a bit by some simple inductions.

Definition 2. Let X be a variety. A subset A of X is *constructible* if it is a finite union of locally-closed subsets of X.

The constructible sets are easily seen to form a Boolean algebra of subsets of X; in fact, they are the smallest Boolean algebra containing all open sets. A typical constructible set which is not locally closed is

$$\{\mathbb{A}^2 - V((X))\} \cup \{(0,0)\}$$

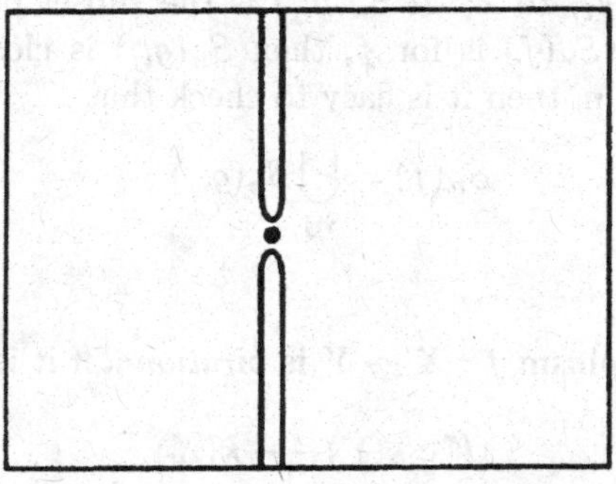

Corollary 2 (Chevalley). *Let $f : X \to Y$ be any morphism. Then the image of f is a constructible set in Y. More generally, f maps constructible sets in X to constructible sets in Y.*

Proof. The second statement follows immediately from the first. To prove the first, use induction on $\dim Y$.

a. If f is not dominating, let $Z = \overline{f(Y)}$. Then $f(Y) \subset Z$ and $\dim Z < \dim Y$, so the result follows by induction.
b. If f is dominating, let $U \subset Y$ be a nonempty open set as in Theorem 3. Let $Z_1, \ldots, Z_t$ be the components of $Y - U$ and let $W_{il}, \ldots, W_{is_i}$ be the components of $f^{-1}(Z_i)$. Let $g_{ij} : W_{ij} \to Z_i$ be the restriction of f. Since $\dim Z_i < \dim Y$, $g_{ij}(W_{ij})$ is constructible. Since
$$f(X) = U \cup \bigcup_{i,j} g_{ij}(W_{ij}) \,,$$
$f(X)$ is constructible too.

□

Corollary 3 (Upper semi-continuity of dimension). *Let $f : X \to Y$ be any morphism. For all $x \in X$, define*

$$e(x) = \left\{ \max(\dim Z) \;\middle|\; \begin{array}{l} Z \text{ a component of} \\ f^{-1}(f(x)) \text{ containing } x \end{array} \right\} .$$

Then e is upper semi-continuous, i.e., for all integers n

$$S_n(f) = \{x \in X \mid e(x) \geq n\}$$

is closed.

Proof. Again do an induction on $\dim Y$. Again, we may well assume that f is dominating. Let $U \subset Y$ be a set as in Theorem 3. Let $r = \dim X - \dim Y$. First of all, if $n \leq r$, then $S_n(f) = X$ by Theorem 2, so $S_n(f)$ is closed. Secondly, if $n > r$, then $S_n(f) \subset X - f^{-1}(U)$ by Theorem 3. Let $Z_1, \ldots, Z_t$ be the components of $Y - U$, $W_{i1}, \ldots, W_{is_i}$ the components of $f^{-1}(Z_i)$ and g_{ij} the restriction of f to a morphism from W_{ij} to Z_i. If $S_n(g_{ij})$ is the subset of W_{ij} defined for the morphism g_{ij}, just as $S_n(f)$ is for f, then $S_n(g_{ij})$ is closed by the induction hypothesis. But if $n > r$, then it is easy to check that

$$S_n(f) = \bigcup_{i,j} S_n(g_{ij}) \ ,$$

so $S_n(f)$ is closed too. □

Definition 3. A morphism $f : X \to Y$ is *birational* if it is dominating and the induced map

$$f^* : k(Y) \longrightarrow k(X)$$

is an isomorphism.

Theorem 4. *If $f : X \to Y$ is a birational morphism, then there is a nonempty open set $U \subset Y$ such that f restricts to an isomorphism from $f^{-1}(U)$ to U.*

Proof. We may as well assume that Y is affine with coordinate ring S. Let $U \subset X$ be any nonempty open affine set, with coordinate ring R. Let $W = \overline{f(X - U)}$. Since all components of $X - U$ are of lower dimension than X, also all components of W are of lower dimension than Y. Therefore W is a proper closed subset of Y. Pick $g \in S$ such that $g = 0$ on W, but $g \neq 0$. Then it follows that

$$f^{-1}(Y_g) \subset U \ .$$

If $g' = f * g$ is the induced element in R, then in fact

$$f^{-1}(Y_g) = U_{g'} \ ,$$

so by replacing Y by Y_g and X by $U_{g'}$, we have reduced the proof of the theorem to the case where Y and X are affine.

Now assume that R and S are the coordinate rings of X and Y. Then f defines the homomorphism

$$\begin{array}{ccc} R & \subset & k(X) \\ {\scriptstyle f^*}\uparrow & & \Big| \wr\!\!\! \cong \\ S & \subset & k(Y). \end{array}$$

Let $x_1, \ldots, x_n$ be a set of generators of R, and write $x_i = y_i/g$, $y_1, \ldots, y_n$, $g \in S$. Then f^* localizes to an isomorphism from $S_{(g)}$ to $R_{(g)}$. Therefore Y_g satisfies the requirements of the theorem. □

The theory developed in this section cries out for examples. Theorem 3 and its corollaries are illustrated in the following:

Example S. $\mathbb{A}^2 \xrightarrow{f} \mathbb{A}^2$ defined by:

$$f(x,y) = (xy, y) .$$

i) The image of f is the union of $(0,0)$ and

$$\left(\mathbb{A}^2\right)_y = \mathbb{A}^2 - \{\text{points where } y = 0\} .$$

This set is *not* locally closed.

ii) f is birational, and if $U = \mathbb{A}^2_y$, then $f^{-1}(U) = \mathbb{A}^2_y$ and the restriction of f to a map from $f^{-1}(U)$ to U is an isomorphism.

iii) On the other hand, $f^{-1}((0,0))$ is the whole line of points $(x, 0)$.

iv) $S_0(f) = \mathbb{A}^2$, $S_1(f) = \{(x,0)\}$, $S_2(f) = \emptyset$ (notation as in Corollary 3, Theorem 3).

To illustrate Theorem 4, look again at:

Examples O, P bis. In example O, §4, we defined a finite birational morphism

$$f : \mathbb{A}^1 \longrightarrow C ,$$

where C is the affine plane curve $X^3 = Y^2$. If $U = C - \{(0,0)\}$, then $f^{-1}(U) = \mathbb{A}^1 - \{(0)\}$, and $f^{-1}(U) \xrightarrow{\sim} U$. On the ring level:

$$k[T] \supsetneq k\left[T^2, T^3\right] ,$$

but

$$k\left[T, T^{-1}\right] = k\left[T^2, T^3, T^{-2}\right] .$$

In example P, we defined a finite birational morphism

$$f : \mathbb{A}^1 \longrightarrow D$$

and then considered its restriction

$$f' : \mathbb{A}^1 - \{1\} \longrightarrow D$$

to get a bijection. If $U = D - \{(0,0)\}$, then $f'^{-1}(U) = \mathbb{A}^1 - \{(1), (-1)\}$ and $f^{-1}(U) \xrightarrow{\sim} U$.

§9. Complete varieties

An affine variety can be embedded in a projective variety, by a birational inclusion. Can a projective variety be embedded birationally in anything even bigger? The answer is no; there is a type of variety, called complete, which in our algebraic theory plays the same role as compact spaces do in the theory of topological spaces. These are "maximal" and projective varieties turn out to be complete.

Recall the main result of classical elimination theory (which we will reprove later):

Given r polynomials, with coefficients in k:

$$f_1(x_0,\ldots,x_n;\ y_1,\ldots,y_m)$$
$$\cdots\cdots\cdots\cdots\cdots\cdots$$
$$f_r(x_0,\ldots,x_n;\ y_1,\ldots,y_m)\ ,$$

all of which are homogeneous in the variables $x_0,\ldots,x_n$, there is a second set of polynomials (with coefficients in k):

$$g_1(y_1,\ldots,y_m)$$
$$\cdots\cdots\cdots$$
$$g_\nu(y_1,\ldots,y_m)$$

such that for all m-tuples $(a_1,\ldots,a_m)$ in k, $g_i(a_1,\ldots,a_m) = 0$, all i if and only if there is a non-zero $(n+1)$-tuple $(b_0,\ldots,b_n)$ in k such that $f_i(b_0,\ldots,b_m;\ a_1,\ldots,a_m) = 0$, all i. (Cf. van der Waerden, §80). In our language, the equations $f_1 = \ldots = f_r = 0$ define a closed subset

$$X \subset \mathbb{P}_n \times \mathbb{A}^m\ .$$

Let p_2 be the projection of $\mathbb{P}_n \times \mathbb{A}^m$ onto $\mathbb{A}^m$. The conclusion asserted is that $p_2(X)$ is a closed subset of $\mathbb{A}^m$; in fact that

$$p_2(X) = V((g_1,\ldots,g_\nu))\ .$$

In other words, the theorem is:

$$p_2 : \mathbb{P}_n \times \mathbb{A}^m \longrightarrow \mathbb{A}^m \quad \text{is a } \textit{closed} \text{ map,}$$
i.e., it maps closed sets onto closed sets,

(modulo the fact that every closed subset of $\mathbb{P}_n \times \mathbb{A}^m$ is described by a set of equations $f_1,\ldots,f_r$ as above). This property easily implies the apparently stronger property – $p_2 : \mathbb{P}_n \times X \to X$ is closed, for all varieties X. This motivates:

Definition 1. A variety X is *complete* if for all varieties Y, the projection morphism

$$p_2 : X \times Y \longrightarrow Y$$

is a closed map.

The analogous property in the category of topological spaces characterizes compact spaces X, at least as long as X is a reasonable space – say completely regular or with a countable basis of open sets. This definition is very nice from a category-theoretic point of view. It gives the elementary properties of completeness very easily:

i) Let $f : X \to Y$ be a morphism, with X complete, then $f(X)$ is closed in Y and is complete.
ii) If X and Y are complete, then $X \times Y$ is complete.
iii) If X is complete and $Y \subset X$ is a closed subvariety, then Y is complete.
iv) An affine variety X is complete only if $\dim X = 0$, i.e., X consists in a single point.

[In fact, (iv) follows from (i) by embedding the affine variety X in its closure $\overline{X}$ in a suitable projective space; and noting that $\overline{X} - Y = \overline{X} \cap$ (hyperplane at ∞) is non-empty by Theorem 2*, §7].

It is harder to prove the main theorem of elimination theory:

Theorem 1. *$\mathbb{P}_n$ is complete.*

Proof (Grothendieck). We must show that for all varieties Y, $p_2 : \mathbb{P}_n \times Y \to Y$ is closed. The problem is clearly local on Y, so we can assume that Y is affine. Let $R = \Gamma(Y, \underline{o}_Y)$.

Note that $\mathbb{P}_n \times Y$ is covered by affine open sets $U_i = (\mathbb{P}_n)_{X_i} \times Y$, whose coordinate rings are $R\left[\frac{X_0}{X_i}, \ldots, \frac{X_n}{X_i}\right]$. Now suppose Z is a closed subset of $\mathbb{P}_n \times Y$. The first problem is to describe Z by a homogeneous ideal in the graded ring $S = R[X_0, \ldots, X_n]$ over R. Let S_m be the graded piece of degree m. Let $A_m \subset S_m$ be the vector space of homogeneous polynomials $f(X_0, \cdots, X_m)$, of degree m, coefficients in R, such that for all i,

$$f\left(\frac{X_0}{X_i}, \ldots, \frac{X_n}{X_i}\right) \in I(Z \cap U_i) \ .$$

Then $A = \Sigma A_m$ is a homogeneous ideal in S.

Lemma. *For all i and all $g \in I(Z \cap U_i)$, there is a polynomial $f \in A_m$ for some m such that*

$$g = f\left(\frac{X_0}{X_i}, \ldots, \frac{X_n}{X_i}\right) \ .$$

Proof. If m is large enough, $X_i^m \cdot g$ is a homogeneous polynomial $f' \in S_m$. To check whether $f' \in A_m$, look at the functions

$$g_j = \frac{f'}{X_j^m} \in R\left[\frac{X_0}{X_j}, \ldots, \frac{X_n}{X_j}\right] \ .$$

g_j is clearly zero on $Z \cap U_i \cap U_j$. And even if it is not zero on $Z \cap U_j$, $\frac{X_i}{X_j} \cdot g_j$ is zero there. Therefore $f = X_i \cdot f' \in A_{m+1}$ and this f does the trick. □

Now suppose $y \in Y - p_2(Z)$. Let $M = I(y)$ be the corresponding maximal ideal. Then $Z \cap U_i$ and $(\mathbb{P}_n)_{X_i} \times \{y\}$ are disjoint closed subsets of U_i. Therefore

$$I(Z \cap U_i) + M \cdot R\left[\frac{X_0}{X_i}, \dots, \frac{X_n}{X_i}\right] = R\left[\frac{X_0}{X_i}, \dots, \frac{X_n}{X_i}\right].$$

In particular

$$1 = a_i + \sum_j m_{ij} \cdot g_{ij} ,$$

where $a_i \in I(Z \cap U_i)$, $m_{ij} \in M$, and $g_{ij} \in R\left[\frac{X_0}{X_i}, \dots, \frac{X_n}{X_i}\right]$. If we multiply this equation through by a high power of X_i and use the lemma, it follows that

$$X_i^N = a_i' + \sum_j m_{ij} \cdot g_{ij}' ,$$

where $a_i' \in A_N$, $g_{ij} \in S_N$. (N may as well be chosen large enough to work for all i). In other words, $X_i^N \in A_N + M \cdot S_N$, for all i. Taking N even larger, it follows that all monomials in the X_i of degree N are in $A_N + M \cdot S_N$, i.e.,

$$(*) \qquad S_N = A_N + M \cdot S_N .$$

Now by the lemma of Nakayama applied to S_N/A_N there is an element $f \in R-M$ such that $f \cdot S_N \subset A_N$. In fact:

Nakayama's Lemma. *Let M be a finitely generated R-module, and let $A \subset R$ be an ideal such that*

$$M = A \cdot M .$$

Then there is an element $f \in 1 + A$ which annihilates M.

Proof. Let $m_1, \dots, m_n$ be generators of M as an R-module. By assumption,

$$m_i = \sum_{j=1}^n a_{ij} \cdot m_j$$

for suitable elements $a_{ij} \in A$. But then:

$$\sum_{j=1}^n (\delta_{ij} - a_{ij}) \cdot m_j = 0.$$

Solving these linear equations directly, it follows that

$$\det(\delta_{ij} - a_{ij}) \cdot m_k = 0, \qquad \text{all } k .$$

But if $f = \det(\delta_{ij} - a_{ij})$, then $1 - f \in A$. □

Now if $f \cdot S_n \subset A_N$, then $f \cdot X_i^N \in A_N$, hence $f \in I(Z \cap U_i)$. This shows that $f = 0$ at all points of $p_2(Z)$, i.e., $p_2(Z) \cap Y_f = \emptyset$.

This proves that the complement of $p_2(Z)$ contains a neighbourhood of every point in it, hence it is open. □

Putting the theorem and remark iii) together, it follows that every projective variety is complete. For some years people were not sure whether or not all complete varieties might not actually be projective varieties. In the next section, we will see that even if a complete variety is not projective, it can still be dominated by a projective variety with the same function field. Thus the problem is a "birational" one, i.e., concerned with the comparison of the collection of all varieties with a common function field. An example of a non-projective complete variety was first found by Nagata.

Theorem 1 can be proven also by valuation-theoretic methods, invented by Chevalley. This method is based on the:

Valuative Criterion. *A variety X is complete if and only if for all valuation rings $R \subset k(X)$ containing k and with quotient field $f(X)$, $R \supset \underline{o}_x$ for some $x \in X$.*

§10. Complex varieties

Suppose that our algebraically closed ground field k is given a topology making it into a topological field. The most interesting case of this is when $k = \mathbb{C}$, the complex numbers. However, we can make at least the first definition in complete generality. Namely, I claim that when k is a topological field, then there is a unique way to endow all varieties X over k with a new topology, which we will call the ***strong topology***, such that the following properties hold:

i) the strong topology is stronger than the Zariski-topology, i.e., a closed (resp. open) subset $Z \subset X$ is always strongly closed, (resp. strongly open).
ii) all morphisms are strongly continuous,
iii) if $Z \subset X$ is a locally closed subvariety, then the strong topology on Z is the one induced by the strong topology on X,
iv) the strong topology on $X \times Y$ is the product of the strong topologies on X and Y;
v) the strong topology on $\mathbb{A}^1$ is exactly the given topology on k.

(These are by no means independent requirements.)

We leave it to the reader to check that such a set of strong topologies exists; it is obvious that there is at most one such set. Note that all varieties X are Hausdorff spaces in their strong topology. In fact, if $\Delta : X \to X \times X$ is the diagonal map, then $\Delta(X)$ is strongly closed by (i). Since $X \times X$ has the product strong topology by (iv), this means exactly that X is a Hausdorff space.

From now on, suppose $k = \mathbb{C}$ with its usual topology. Then varieties not only have the strong topologies: they even have "strong structure sheaves", or in more conventional language, they are complex analytic spaces[3]. This means

[3] The standard definition of a complex analytic space is completely analogous to our definition of a variety: i.e., it is a Hausdorff topological space X, plus a sheaf of $\mathbb{C}$-valued continuous functions Ω_X on X, which is locally isomorphic to one of

that there is a unique collection of sheaves (in the strong topology) of strongly continuous $\mathbb{C}$-valued functions $\{\Omega_X\}$, one for each variety X, such that:

i) for each Zariski-open set $U \subset X$,

$$\varrho_X(U) \subset \Omega_X(U),$$

ii) all morphisms $f : X \to Y$ are "holomorphic", i.e., f^* takes sections of Ω_Y to sections of Ω_X,

iii) if $Z \subset X$ is a locally closed subvariety, then Ω_Z is the sheaf of $\mathbb{C}$-valued functions on Z induced by the sheaf Ω_X on X,

iv) if $X = \mathbb{A}^n \underset{\text{as a set}}{=} \mathbb{C}^n$, then Ω_X is the usual sheaf of holomorphic functions on $\mathbb{C}^n$.

(Again these properties are not independent.) We leave it to the reader to check that this set of sheaves exists: the uniqueness is obvious. Moreover, it follows immediately that (X, Ω_X) is a complex analytic space for all varieties X.

The first non-trivial comparison theorem relating the 2 topologies states that the strong topology is not "too strong":

Theorem 1. *Let X be a variety, and U a nonempty open subvariety. Then U is strongly dense in X.*

Proof (Based on suggestions of G. Stolzenberg). Since the theorem is a local statement (in the Zariski topology), we can suppose that X is affine. By Noether's normalization lemma (geometric form), there exists a finite surjective morphism

$$\pi : X \longrightarrow \mathbb{A}^n .$$

Let $Z = X - U$. Then $\pi(Z)$ is a Zariski closed subset of $\mathbb{A}^n$. Since all components of Z have dimension $< n$, so do all components of $\pi(Z)$, hence $\pi(Z)$ is even a proper closed subset of $\mathbb{A}^n$. In particular, there is a non-zero polynomial $g\,(X_1, \ldots, X_n)$ such that

$$\pi(Z) \subset \{(x_1, \ldots, x_n) \mid g\,(x_1, \ldots, x_n) = 0\} .$$

Now choose a point $x \in X - U$. Let's first represent $\pi(x)$ as a limit of points $y^{(i)} = \left(x_1^{(i)}, \ldots, x_n^{(i)}\right) \in \mathbb{A}^n$ such that $g\left(y^{(i)}\right) \neq 0$. To do this, choose any point $y^{(1)} \in \mathbb{A}^n$ such that $g\left(y^{(1)}\right) \neq 0$, and let

$$h(t) = g\left((1-t) \cdot \pi(x) + t \cdot y^{(1)}\right), \qquad t \in \mathbb{C}$$

(i.e., $\pi(x)$ and $y^{(1)}$ are regarded as *vectors*). Then $h \not\equiv 0$ since $h(1) \neq 0$. Therefore $h(t)$ has only a finite number of zeroes, and we can choose a sequence of numbers $t_i \in \mathbb{C}$ such that $t_i \to 0$, as $i \to \infty$, and $h(t_i) \neq 0$. Then let

---

the standard objects: i.e., the locus of zeroes in a polycylinder of a finite set of holomorphic functions, plus the sheaf of functions induced on it by the sheaf of holomorphic functions on the polycylinder. For details, see Gunning-Rossi, Ch. 5.

$$y^{(i)} = (1 - t_i) \cdot \pi(x) + t_i \cdot y^{(1)}\,.$$

Then $y^{(i)} \to \pi(x)$ strongly, as $i \to \infty$, and $g\left(y^{(i)}\right) \neq 0$.

The problem now is to lift each $y^{(i)}$ to a point $z^{(i)} \in X$ such that $z^{(i)} \to x$. Since $y^{(i)} \notin \pi(Z)$, all the points $z^{(i)}$ must be in U, hence it will follow that x is the strong closure of U. We will do this in 2 steps. First, let $\pi^{-1}(\pi(x)) = \{x, x_2, \ldots, x_n\}$. Choose a function $g \in \Gamma(X, \underline{o}_X)$ such that $g(x) = 0$, but $g(x_i) \neq 0$, $2 \leq i \leq n$. Let $F(X_1, \ldots, X_n, g) = 0$ be the irreducible equation satisfied by $X_1, \ldots, X_n$ and g in $\Gamma(X, \underline{o}_X)$. We shall work with the 3 rings and 3 affine varieties:

$$\begin{array}{ccccc}
 & & \Gamma(X,\underline{o}_X) & & X \\
 & & \cup & & \big\downarrow \pi_1 \\
k[X_1,\ldots,X_n,Y]/(F) & \cong & k[X_1,\ldots,X_n,g] & & V(F) \subset \mathbb{A}^{n+1} \\
 & & \cup & & \big\downarrow \pi_2 \\
 & & k[X_1,\ldots,X_n] & , & \mathbb{A}^n \qquad , \quad \pi = \pi_2 \cdot \pi_1 .
\end{array}$$

Since g is integrally dependent on $k[X_1, \ldots, X_n]$, F has the form:

$$F(X_1, \ldots, X_n, Y) = Y^d + A_1(X_1, \ldots, X_n) \cdot Y^{d-1} + \ldots + A_d(X_1, \ldots, X_n)\,.$$

Writing $(X_1, \ldots, X_n)$ as a vector, we abbreviate $F(X_1, \ldots, X_n, Y)$ to $F(X, Y)$. Now since $g(x) = 0$,

$$0 = F\big(\pi(x), g(x)\big) = A_d\big(\pi(x)\big).$$

Therefore, $A_d\left(y^{(i)}\right) \to 0$ as $i \to \infty$. On the other hand,

$$A_d\left(y^{(i)}\right) = \left\{ \begin{array}{l} \text{Product of the roots of the equation in } t \\ F\left(y^{(i)}, t\right) = 0 \end{array} \right\}.$$

Therefore we can find roots $t^{(i)}$ of $F\left(y^{(i)}, t\right) \equiv 0$ such that $t^{(i)} \to 0$. Then $\left(y^{(i)}, t^{(i)}\right)$ is a sequence of points of $V(F)$ converging strongly to $\pi_1(x)$. This is the 1st step.

Now choose generators $h_1, \ldots, h_N$ of the ring $\Gamma(X, \underline{o}_X)$. Via the h_i's, we can embed X in $\mathbb{A}^N$, so that its strong topology is induced by the strong topology in $\mathbb{A}^N$. Each h_i satisfies an equation of integral dependence:

$$h_i^m + a_{i1} \cdot h_i^{m-1} + \ldots + a_{im} = 0$$

with $a_{ij} \in k[X_1, \ldots, X_n, g]$. Therefore, if $\Sigma \in V(F)$ is a relatively compact subset, all the polynomials a_{ij} are bounded on Σ, so each of the functions h_i is bounded on $\pi_1^{-1}(\Sigma)$. Since $|h_i| \leq C$, all i, is a compact subset of $\mathbb{A}^N$, $\pi_1^{-1}(\Sigma)$ is a relatively compact subset of X. On the other hand, π_1 is a surjective map since π_1

is a finite morphism. So choose points $z^{(i)} \in X$ such that $\pi_1\left(z^{(i)}\right) = \left(y^{(i)}, t^{(i)}\right)$. Since the points $\left(y^{(i)}, t^{(i)}\right)$ converge in $V(F)$, they are a relatively compact set. Therefore $\left\{z^{(i)}\right\}$ is a relatively compact subset of X. Suppose they did not converge to x: then some subsequence $z^{(i_k)}$ would converge to some $x' \neq x$. Since

$$y^{(i_k)} = \pi\left(z^{(i_k)}\right) \longrightarrow \pi(x'), \qquad (\text{as } k \to \infty)$$

$\pi(x') = x$, so $x' = x_i$, for some $2 \le i \le n$. But then

$$t^{(i_k)} = g\left(z^{(i_k)}\right) \longrightarrow g(x') \neq 0 \qquad (\text{as } k \to \infty).$$

This is a contradiction, so $z^{(i)} \to x$. □

Corollary 1. *If $Z \subset X$ is a constructible subset of a variety, then the Zariski closure and the strong closure of Z are the same.*

The main result of this section is:

Theorem 2. *Let X be a variety over $\mathbb{C}$. Then X is complete if and only if X is compact in its strong topology.*

Proof. Suppose first that X is strongly compact. Let Y be another variety, let $p_2 : X \times Y \to Y$ be the projection, and let $Z \subset X \times Y$ be a closed subvariety. Since X is compact, p_2 is a proper map in the strong topology. Therefore p_2 takes strongly closed sets to strongly closed sets (cf. Bourbaki, *Topologie Générale*, Ch. I, §10). Therefore $p_2(Z)$ is strongly closed. Since it is also constructible (§8, Th. 3, Cor. 3), it is Zariski closed by the Cor. to Theorem 1.

Conversely, we must show that complete varieties are strongly compact. First of all, it is clear that $\mathbb{P}_n(\mathbb{C})$ is strongly compact. For example, it is a continuous image of the sphere in the space of homogeneous coordinates:

$$\begin{array}{c} \Sigma = \left\{(z_0, \ldots, z_n) \;\middle|\; \Sigma_i |z_i|^2 = 1\right\} \\ \Big\downarrow \text{surjective} \\ \mathbb{P}_n(\mathbb{C}) \end{array} .$$

Therefore all closed subvarieties of $\mathbb{P}_n(\mathbb{C})$ are strongly compact. The general case follows from:

Chow's Lemma. *Let X be a complete variety (over any algebraically closed field k). Then there exists a closed subvariety Y of $\mathbb{P}_n(k)$ for some n and a surjective birational morphism:*

$$\pi : Y \longrightarrow X .$$

Proof. Cover X by open affine subsets U_i with coordinate rings A_i for $1 \leq i \leq m$, and let $U^* = U_1 \cap \ldots \cap U_m$. Embed all the U_i's as closed subvarieties of $\mathbb{A}^n$ (for some n). With respect to the composite inclusion:

$$U_i \subset \mathbb{A}^n \subset \mathbb{P}_n(k)$$

U_i is a locally closed subvariety of $\mathbb{P}_n(k)$; let $\overline{U_i}$ be its closure in $\mathbb{P}_n(k)$. Note that $\overline{U}_1 \times \ldots \times \overline{U}_m$ is isomorphic to a closed subvariety of $\mathbb{P}_N(k)$ for some N by Theorem 3, §6.

Consider the composite morphism:

$$U^* \longrightarrow U^* \times \ldots \times U^* \subset \overline{U}_1 \times \ldots \times \overline{U}_m\,.$$

The first morphism is an isomorphism of U^* with a closed subvariety of $(U^*)^m$ – the "multidiagonal"; the second morphism is the product of all the inclusions $U^* \subset U_i \subset \overline{U}_i$, i.e., it is an isomorphism of $(U^*)^m$ with an open subvariety of $\overline{U}_1 \times \ldots \times \overline{U}_m$. Therefore the image is a locally closed subvariety of $\overline{U}_1 \times \ldots \times \overline{U}_m$ isomorphic to U^*. Let Y be the closure of the image. Y is certainly a projective variety and we will construct a morphism $\pi : Y \to X$.

To construct π, consider the morphism

$$U^* \xrightarrow{\Delta} U^* \times U^* \subset X \times Y$$

induced by a) the inclusion of U^* in Y, b) the inclusion of U^* in X. Let $\widetilde{Y}$ be the closure of the image. Since X and Y are complete, therefore $X \times Y$ and $\widetilde{Y}$ are complete. The projections of $X \times Y$ onto X and Y give the diagram:

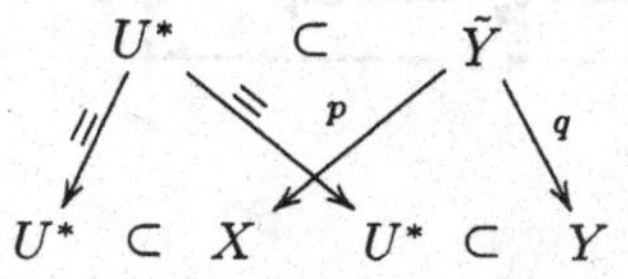

This shows that the projections p and q are both isomorphisms on U^*, hence they are birational morphisms. Moreover, since $\widetilde{Y}$ is complete, $p(\widetilde{Y})$ and $q(\widetilde{Y})$ are closed in X and Y respectively; since $p(\widetilde{Y}) \supset U^*$, $q(\widetilde{Y}) \supset U^*$, this implies that p and q are surjective. I claim

(∗) q is an isomorphism.

When (∗) is proven, we can set $\pi = p \cdot g^{-1}$ and everything is proven. $\widetilde{Y}$ is a closed subvariety of the product $X \times \overline{U}_1 \times \ldots \times \overline{U}_m$. We want to analyze its projection on the product $X \times \overline{U}_i$ of only 2 factors. Look at the diagram:

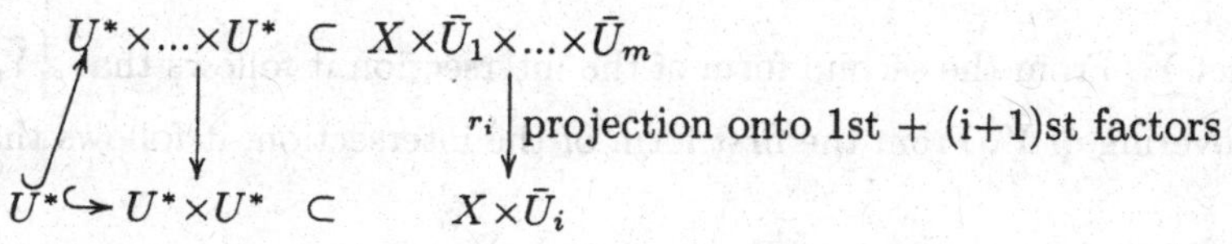

Since the projection $r_i(\widetilde{Y})$ is closed (since $\widetilde{Y}$ is complete) and contains the image of U^* as a dense subset, it follows that $r_i(\widetilde{Y})$ is just the closure of U^* in $X \times \overline{U}_i$ via the bottom arrows.

Sublemma. *Let S and T be varieties, with isomorphic open subsets $V_S \subset S$, $V_T \subset T$. For simplicity identify V_S with V_T and look at the morphism*

$$V \xrightarrow{\Delta} V \times V \subset S \times T .$$

If $\overline{V}$ is the closure of the image, then

$$\overline{V} \subset (S \times V) = \overline{V} \cap (V \times T) = \Delta(V) .$$

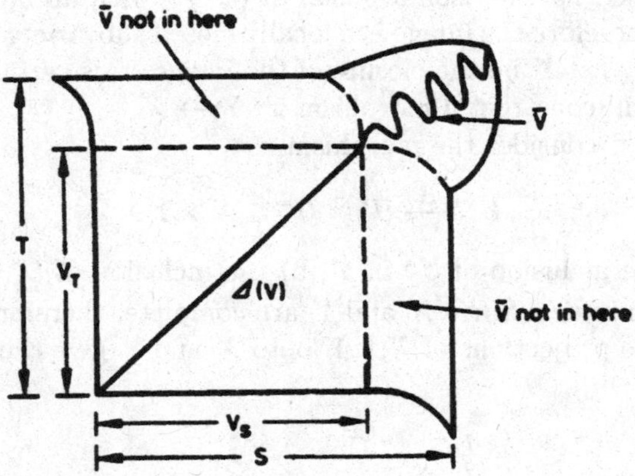

Proof. It suffices to show that $\Delta(V)$ is already closed in $V \times T$ and in $S \times V$. But $\Delta(V) \cap (V \times T)$, say, is just the graph of the inclusion morphism $V \to T$. Hence it is closed (cf. Remark II, following Def. 2, §6). □

Therefore

$$\begin{aligned} r_i(\widetilde{Y}) \cap (X \times U_i) &= r_i(\widetilde{Y}) \cap (U_i \times \overline{U}_i) \\ &= \{(x, x) \mid x \in U_i\} . \end{aligned}$$

Therefore

$$\widetilde{Y} \cap (X \times \overline{U}_1 \times \ldots \times U_i \times \ldots \times \overline{U}_m) = \widetilde{Y} \cap (U_i \times \overline{U}_1 \times \ldots \times \overline{U}_m) \quad .$$

Call this set $\widetilde{Y}_i$. From the second form of the intersection it follows that $\left\{\widetilde{Y}_i\right\}$ is an open covering of $\widetilde{Y}$. From the first form of the intersection, it follows that

$$\widetilde{Y}_i = q^{-1}(Y_i)$$

if

$$Y_i = Y \cap (\overline{U}_1 \times \ldots \times U_i \times \ldots \times \overline{U}_M) .$$

Since q is surjective, this implies that $\{Y_i\}$ must be an open covering of Y. But now define:

$$\sigma_i : Y_i \longrightarrow \widetilde{Y}_i$$
$$\sigma_i (u_1, \ldots, u_m) = (u_i, u_1, \ldots, u_m) ,$$

(which makes sense exactly because the ith component u_i is in U_i, hence is a point of X too). Then σ_i is an inverse of q restricted to $\widetilde{Y}_i$:

a) $q(\sigma_i (u_1, \ldots, u_m)) = q(u_i, u_1, \ldots, u_m) = (u_1, \ldots, u_m)$
b) by the sublemma, all points $(v, u_1, \ldots, u_m)$ of $\widetilde{Y}_i$ satisfy $v = u_i$, hence

$$\sigma_i (q (v, u_1, \ldots, u_m)) = (u_i, u_1, \ldots, u_m) = (v, u_1, \ldots, u_m) .$$

Therefore q is an isomorphism and π can be constructed.

QED for Chow's lemma and Th. 2.

II. Preschemes

The most satisfactory type of object yet derived in which to carry out all the operations natural to algebraic geometry is the prescheme. I fully agree that it is painful to go back again to the foundations and redefine our basic objects after we have built them up so carefully in the last chapter. The motivation for doing this comes from many directions. For one thing, we have so far neglected a very essential possibility inherent in our subject, which is, after all, a marriage of algebra and geometry: that is to examine and manipulate algebraically with the coefficients of the polynomials which define our varieties. It does not make much sense to say that a differentiable manifold is defined by integral equations; it makes good sense to say that an affine variety is the locus of zeroes of a set of integral polynomials. Another motivation for preschemes comes from the possibility of constructing via schemes an explicit and meaningful theory of infinitesimal objects. This is based on the idea of introducing nilpotent functions into the structure sheaf, whose values are everywhere zero, but which are still non-zero sections. Schemes with nilpotents are not only useful for many applications, but they come up inevitably when you examine the fibres of morphisms between quite nice varieties. Thirdly, it is only when you use schemes that the full analogy between arithmetic and geometric questions becomes explicit. For example, there is the connection given by the general theory of Dedekind domains which unites the theory of a) rings of integers in a number field and b) rings of algebraic functions in one complex variable. A much deeper connection is given by class field theory, between the tower of number fields and the tower of coverings of an algebraic curve defined over a finite field. The analogies suggested by this approach can be carried so far that they even give a definition of the higher homotopy groups of the integers, (i.e., of Spec(Z)): The vision of combined arithmetic-geometric objects goes back to Kronecker. It is interesting to read Felix Klein describing what to all intents is nothing but the theory of schemes:

> "Ich beschränke mich darauf, noch einmal das allgemeinste Problem, welches hier vorliegt, im Anschluß an Kroneckers Festschrift von 1881 zu charakterisieren. Es handelt sich nicht nur um die reinen Zahlkörper oder Körper, die von einem Parameter Z abhängen, oder um die Analogisierung dieser Körper, sondern es handelt sich schließlich darum, für Gebilde, die gleichzeitig arithmetisch und funktionentheoretisch sind, also von gegebenen algebraischen Zahlen und gegebenen algebraischen

Funktionen irgendwelcher Parameter algebraisch abhängen, dasselbe zu leisten, was mehr oder weniger vollständig in den einfachsten Fällen gelungen ist. Es bietet sich da ein ungeheurer Ausblick auf ein rein theoretisches Gebiet, welches durch seine allgemeinen Gesetzmäßigkeiten den größten ästhetischen Reiz ausübt, aber, wie wir nicht unterlassen dürfen hier zu bemerken, allen praktischen Anwendungen zunächst ganz fern liegt."[4]

§1. Spec (*R*)

It is possible to associate a "geometric" object to an arbitrary commutative ring R. This object will be called Spec (R). If R is a finitely generated integral domain over an algebraically closed field, Spec (R) will be very nearly the same as an affine variety associated to R in Chapter I. However in this section we will be completely indifferent to any special properties that R may or may not have – e.g., whether R has nilpotents or other zero-divisors in it or not; whether or not R has a large subfield over which it is finitely generated or even any subfield at all. We insist only that R be commutative and have a unit element 1.

First define a point set:

(I) Spec (R) = the set of prime ideals $P \not\subseteq R$ [R itself is not counted as a prime ideal but (0), if prime, is counted].

In order to have an unambiguous notation, we shall write $[P]$ for the element of Spec (R) given by the prime ideal P. This allows us to distinguish between the times when we think of P as an ideal in R, and the times when we think of $[P]$ as a point in Spec (R).

Secondly, define a topology on Spec (R), its *Zariski topology*:

(II) *Closed sets* = Sets of the form $\{[P] \mid P \supseteq A\}$ for some ideal $A \subseteq R$. This set will be denoted $V(A)$.

It is easy to verify that

$$i) \quad V\left(\sum_\alpha A_\alpha\right) = \bigcap_\alpha V(A_\alpha)$$

$$ii) \quad V(A \cap B) = V(A) \cup V(B) .$$

So the collection of closed subsets $\{V(A)\}$ does define a topology.

Thirdly, let

[4] *Die Entwicklung der Mathematik im* 19ten *Jahrhundert*, reprint by Chelsea Publ. Co., 1956, Ch. 7, p. 334.

(III) $$\text{Spec }(R)_f = \{[P] \mid f \notin P\}\,.$$

Since Spec $(R)_f$ = Spec $(R) - V((f))$, Spec $(R)_f$ is an open subset of Spec (R): we shall refer to these open subsets as the *distinguished open subsets.* They form a basis of the open sets of Spec (R) because any open subset Spec$(R) - V(A)$ is simply the union of the distinguished open sets Spec $(R)_f$, for all elements $f \in A$.

Spec (R) need not satisfy axiom T1. In fact, the closure of $[P]$ is exactly $V(P)$, i.e., $\{P' \mid P' \supseteq P\}$. Therefore $[P]$ is a closed point if and only if P is a maximal ideal. At the other extreme, when R is an integral domain, (O) is a prime ideal and $[(O)]$ is called the generic point of Spec (R) since its closure is the whole of Spec R. More generally, we define:

Definition 1. Suppose Z is an irreducible closed subset of Spec (R). Then a point $z \in Z$ is a *generic point* of Z if Z is the closure of z, i.e., every open subset Z_0 of Z contains z.

Proposition 1. *If $x \in$ Spec (R), then the closure of $\{x\}$ is irreducible and x is a generic point of this set. Conversely, every irreducible closed subset $Z \subset$ Spec (R) equals $V(P)$ for some prime ideal $P \subset R$, and $[P]$ is its unique generic point.*

Proof. Let Z be the closure of $\{x\}$. If $Z = W_1 \cup W_2$, where W_i is closed, then $x \in W_1$ or $x \in W_2$. In each case, W_1 or W_2 is a closed set containing x and contained in its closure, i.e., W_1 or W_2 equals Z.

Conversely, suppose $V(A)$ is irreducible. Since $V(A) = V(\sqrt{A})$, we may as well assume that $A = \sqrt{A}$. Then we claim A is prime. If not, there exist $f, g \in R$ such that $f \cdot g \in A$, $f \notin A$, $g \notin A$. Let $B = A + (f)$, $C = A + (g)$. Then $A = B \cap C$: in fact, if $h = \alpha f + a_1 = \beta g + a_2$ is an element of $B \cap C$ $(a_1, a_2 \in A;\ \alpha, \beta \in R)$, then
$$h^2 = \alpha\beta \cdot f \cdot g + a_1(\beta g + a_2) + a_2(\alpha f) \in A\,,$$
hence $h \in A$ also. Therefore
$$V(A) = V(B) \cup V(C)\,.$$
On the other hand, since $A = \sqrt{A}$, A is the intersection of the prime ideals P that contain it. In particular, there is a prime ideal P such that $P \supseteq A$, but $P \not\ni f$. Then $[P] \in V(A) - V(B)$, i.e., $V(A) \not\supseteq V(B)$. Similarly $V(A) \not\supseteq V(C)$. We conclude that $V(A)$ is not irreducible. This contradiction shows that A must have been a prime ideal. But then $V(A)$ is the closure of the point $[A]$ as mentioned above, so $[A]$ is a generic point of $V(A)$.

If $[P']$ were another generic point, then $[P'] \in V(A)$ implies $A \subseteq P'$. On the other hand, $[A]$ would also be in the closure of $[P']$, so $A \supseteq P'$. This proves that $A = P'$, hence there is only one generic point. □

Proposition 2. *Let $\{f_\alpha \mid \alpha \in S\}$ be a set of elements of R. Then*
$$\text{Spec }(R) = \bigcup_{\alpha \in S} \text{Spec }(R)_{f_\alpha}$$
if and only if $1 \in (\ldots, f_\alpha, \ldots)$, the ideal generated by the f_α's.

Proof. Spec (R) is the union of the Spec $(R)_{f_\alpha}$'s if and only if every point $[P]$ does *not* contain some f_α. This means that no prime ideal contains $(\ldots, f_\alpha, \ldots)$, and this happens if and only if

$$1 \in (\ldots, f_\alpha, \ldots) .$$

□

Notice that $1 \in (\ldots, f_\alpha, \ldots)$ if and only if there is a finite set $f_{\alpha_1}, \ldots, f_{\alpha_n}$ of the f_α's and elements $g_1, \ldots, g_n$ of R such that:

$$1 = \sum_{i=1}^{n} g_i \cdot f_{\alpha_i} .$$

This equation is the algebraic analog of the partitions of unity which are so useful in differential geometry. The fact that one always has such an equation whenever one is given a covering of Spec (R) by distinguished open subsets is the reason for the cohomological triviality of affine schemes, (the so-called Theorems A and B – cf. Chapter 6).

Corollary. Spec (R) *is quasi-compact.*

Proof. It suffices to check that every covering by distinguished open sets has a finite subcover. Because of the Proposition, this follows from the fact that $1 \in (\ldots, f_\alpha, \ldots) \Longrightarrow 1 \in (f_{\alpha_1}, \ldots, f_{\alpha_n})$ for some finite set $\alpha_1, \ldots, \alpha_n \in S$. □

An easy generalization of this argument shows that Spec $(R)_f$ is also quasi-compact. But unless R is noetherian, for example, there may be some open sets $U \subset$ Spec (R) which are not quasi-compact.

We shall now endow the topological space Spec (R) with a sheaf of rings, $\underline{o}_{\text{Spec }(R)}$, called its *structure sheaf.* For simplicity of notation, let $X = \text{Spec}(R)$. At first, we will work with the distinguished open sets of X. We need a few properties of these sets:

a) $X_f \cap X_g = X_{fg}$, all $f, g \in R$.
(This is easy to check.)

b) $X_f \supset X_g$ if and only if $g \in \sqrt{(f)}$.
(In fact, since $\sqrt{(f)} = \bigcap\{P \mid f \in P\}$, it follows that

$$\begin{aligned} g \notin \sqrt{(f)} &\iff \exists P, \quad f \in P, \quad g \notin P \\ &\iff \exists P, \quad [P] \notin X_f, \quad [P] \in X_g . \,) \end{aligned}$$

Now we want to associate the localization R_f of the ring R with respect to the multiplicative system $(f, f^2, f^3, \ldots)$ to the open set X_f. Note that if $X_g \subset X_f$, then by (b) there is a canonical map $R_f \to R_g$. (Explicitly, we know $g^n = h \cdot f$ for some h and n, and we map

$$\frac{a}{f^m} \longmapsto \frac{a \cdot h^m}{g^{nm}} \; . \;)$$

In particular, if $X_g = X_f$, we have canonical maps $R_f \to R_g$ and $R_g \to R_f$ which are inverse to each other, so we can identify R_f and R_g. Therefore we really can associate a ring R_f to each open set X_f. Furthermore, whenever $X_k \subset X_g \subset X_f$, we get a commutative diagram of canonical maps:

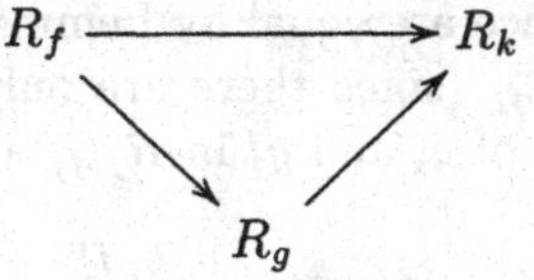

Furthermore, if $[P] \in X_f$, then $f \notin P$ and there is a natural map $R_f \to R_p$, since the multiplicative system $R - P$ contains the multiplicative system $\{f, f^2, \ldots\}$. It is easy to check in fact that

$$R_P = \varinjlim_{f \in R-P} R_f = \varinjlim_{\substack{f \text{ such that} \\ [P] \in X_f}} R_f \; .$$

Lemma 1. *Suppose* $X_f = \bigcup_{\alpha \in S} X_{f_\alpha}$. *If an element* $g \in R_f$ *has image* 0 *in all the rings* R_{f_α}, *then* $g = 0$.

Proof. Let $g = b/f^n$. Let $A = \{c \in R \mid cb = 0\}$. The following are clearly equivalent:

i) $g = 0$ in R_f
ii) $\exists m$ such that $f^m \cdot b = 0$ in R
iii) $f \in \sqrt{A}$
iv) $P \supset A \Rightarrow f \in P$.

Therefore if $g \neq 0$, we can choose a prime ideal $P \supset A$ with $f \notin P$, i.e., $[P] \in X_f$. Then $[P] \in X_{f_\alpha}$ for some α. Using the commutative diagram

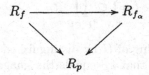

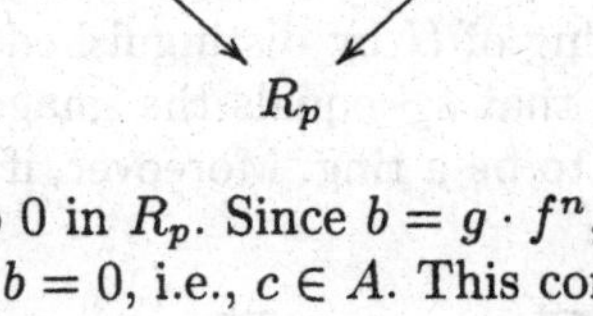

it folllows that g goes to 0 in R_p. Since $b = g \cdot f^n$, so does b. Therefore there is some $c \in R - P$ with $c \cdot b = 0$, i.e., $c \in A$. This contradicts the fact that $P \supset A$ and the lemma is proven. □

Lemma 2. *Suppose again that* $X_f = \bigcup_{\alpha \in S} X_{f_\alpha}$. *Suppose that we have elements* $g_\alpha \in R_{f_\alpha}$ *such that* g_α *and* g_β *have the same image in* $R_{f_\alpha f_\beta}$. *Then there is an element* $g \in R_f$ *which has the image* g_α *in* R_{f_α} *for all* α.

Proof. We prove this only for $f = 1$; the general case is left to the reader. It suffices to assume the covering is finite, say $X = X_{f_1} \cup \ldots \cup X_{f_k}$. For we may choose a finite subcovering of the X_{f_α}; and if $g \in R$ goes to $g_i \in R_{f_i}$, $1 \leq i \leq k$, then for all α both g_α and image (g) in R_{f_α} have the same images down in the rings $R_{f_\alpha f_i}$, $1 \leq i \leq k$, and hence are equal by Lemma 1.

Write out $g_i = b_i / f_i^n \in R_{f_i}$ (since there are only finite many i's, we can choose a single n). The images of g_i and g_j in $R_{f_i f_j}$ are

$$\frac{b_i f_j^n}{(f_i f_j)^n} \quad \text{and} \quad \frac{b_j f_i^n}{(f_i f_j)^n} .$$

These are equal by hypothesis, which means that there is an m_{ij} such that:

$$(f_i f_j)^{m_{ij}} \cdot \left[b_i f_j^n - b_j f_i^n\right] = 0 \quad \text{in } R.$$

Let M be bigger than all m_{ij}, set $N = m + M$, and $b_i' = b_i \cdot f_i^M$. Then we have $g_i = b_i' / f_i^N$ in R_{f_i} and $b_i' f_f^N - b_j' \cdot f_i^N = 0$ in R. On the other hand, we have

$$X = \bigcup_{i=1}^{k} X_{f_i} = \bigcup_{i=1}^{k} X_{f_i N} .$$

Therefore $1 \in (f_1^N, \ldots, f_K^N)$, so $1 = \sum h_i f_i^N$ for some $h_i \in R$. Let $g = \sum h_i b_i'$. Then

$$f_j^N \cdot g = \sum_{i=1}^{k} f_j^N \cdot b_i' \cdot h_i = \sum_{i=1}^{k} f_i^N \cdot b_j' \cdot h_i = b_j'$$

i.e., g goes to g_j in R_{f_j}. □

The lemmas show that assigning R_f to X_f gives us something as close to a sheaf as we can come when we are only considering a basis of open sets. Moreover there is one and only one way to extend this assignment to all open sets so as to get a sheaf. Explicitly, for each open $U \subset X$, let $\Gamma(U, \underline{o}_X)$ be the set of elements

$$\{s_P\} \in \prod_{[P] \in U} R_P$$

for which there exists a covering of U by distinguished open sets X_{f_α} together with elements $s_\alpha \in R_{f_\alpha}$ such that s_P equals the image of s_α in R_P whenever $[P] \in X_{f_\alpha}$. This is easily seen to be a ring. Moreover, if $V \subset U$, it is easy to see that the coordinate projection

$$\prod_{[P] \in U} R_P \longrightarrow \prod_{[P] \in V} R_P$$

takes $\Gamma(U, \underline{o}_X)$ into $\Gamma(V, \underline{o}_X)$. Taking this as the restriction map, we get a presheaf $\underline{o}_X$.

1.) $\underline{o}_X$ is a sheaf.

Proof. Suppose $U = \bigcup U_\beta$. If we have an element of $\Gamma(U, \underline{o}_X)$ going to 0 in all $\Gamma(U_\beta, \underline{o}_X)$, then its component at each R_P is 0, so it's 0. If we have elements $s_\beta \in \Gamma(U_\beta, \underline{o}_X)$ agreeing on overlaps, they clearly determine a unique element of $\prod_{[P]\in U} R_P$; this element is locally given by elements from R_f's since each of the s_β's was. □

2.) $\Gamma(X_f, \underline{o}_X) = R_f$.

Proof. We have a map $R_f \to \prod_{[P]\in X_f} R_P$ which is injective by Lemma 1 and which lands us in $\Gamma(X_f, \underline{o}_X)$ by definition. The map is surjective by Lemma 2. □

3.) The stalk of $\underline{o}_X$ at $[P]$ is R_P.

Proof. We can get the stalk by talking a direct limit over a basis of open sets containing $[P]$, so we get:

$$(\underline{o}_X)_{[P]} = \varinjlim_{[P]\in X_f} \Gamma(X_f, \underline{o}_X) = \varinjlim_{f\notin P} R_f = R_P \, .$$

□

Note that once we have Lemmas 1 and 2, the rest is purely sheaf-theoretic. One further point which is useful: since our topological space has non-closed points, we have some maps between the stalks $\underline{o}_x$ of $\underline{o}_X$. Suppose $P_1 \subset P_2$ are 2 prime ideals. Let $x_i = [P_i]$. Then $x_2 \in \overline{\{x_1\}}$, so every neighbourhood U of x_2 contains x_1; this gives us a map:

$$\underline{o}_{x_2} = \varinjlim_{x_2\in U} \Gamma(U, \underline{o}_X) \longrightarrow \varinjlim_{x_1\in V} \Gamma(V, \underline{o}_X) = \underline{o}_{x_1} \, .$$

Recalling that $\underline{o}_{x_i} \cong R_{P_i}$, we see that this is just the natural map $R_{P_2} \to R_{P_1}$.

Example A. Let k be a field. Then Spec (k) has just one point $[(o)]$, and the structure sheaf is just k sitting on that point.

More generally, let R be any commutative ring with descending chain condition. Then R is the direct sum of its primary subrings:

$$R = \bigoplus_{i=1}^{n} R_i$$

(cf. Zariski-Samuel, vol. 1, p. 205). Now, quite generally, the spectrum of a direct sum of rings is just the disjoint union of the spectra of its components (each being open in the whole):

$$\text{Spec}\,(R \oplus S) = \text{Spec}\,(R) \amalg \text{Spec}\,(S)\,.$$

In our case, Spec (R) is the union of the Spec (R_i); since R_i is primary, it has *one* prime ideal

$$M_i = \{x \in R_i \mid x \text{ nilpotent }\}\,,$$

and Spec (R_i) is one point. Therefore, Spec (R) itself consists in n points with the discrete topology. Moreover, the structure sheaf just consists in R_i sitting as the stalk on the ith point. Geometrically, the presence of nilpotents in M_i should be taken as meaning that the ith point is surrounded by some infinitesimal normal neighbourhood.

Example B. Spec $(k[X])$: the *affine line over k*. This is denoted $\mathbb{A}^1_k$. $k[X]$ has 2 types of prime ideals: (o) and $(f(X))$, f an an irreducible polynomial. Therefore Spec $(k[X])$ has one closed point for each monic irreducible polynomial, and one generic point $[(o)]$ whose closure is all of Spec $(k[X])$. Assume k is algebraically closed. Then the closed points are all of the form $[(X-a)]$: we call this "the point $X = a$", and we find that $\mathbb{A}^1_k$ is just the ordinary X-line together with a generic point. The most general proper closed set is just a finite union of closed points.

The stalk of $\underline{o}_{\mathbb{A}^1}$ at $[(X-a)]$ is:

$$k[X]_{(X-a)} = \left\{ \frac{f(X)}{g(X)} \;\middle|\; f, g \text{ polynomials}, \quad g(a) \neq 0 \right\}\,.$$

The stalk at $[(o)]$ is $k(X)$, which we called before the function field. Note that whereas previously (Ch. I, §3) the generic stalk was only analogous to the stalks at closed points, it is now just another case of the same construction.

Example C. Spec $(\mathbb{Z})$. $\mathbb{Z}$ is a P.I.D. like $k[X]$, and Spec $(\mathbb{Z})$ is usually visualized as a line:

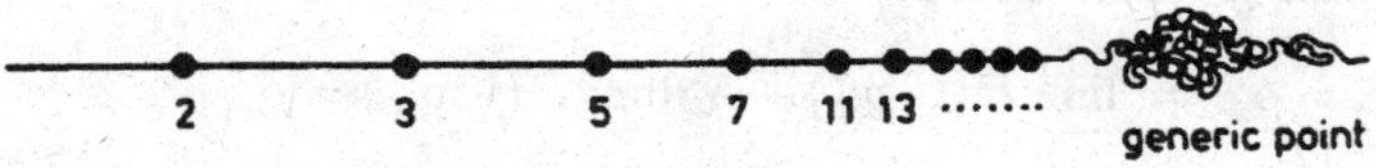

There is one closed point for each prime number, plus a generic point $[(o)]$. The stalk at $[(p)]$ is $\mathbb{Z}_{(p)}$ and at $[(o)]$ it is $\mathbb{Q}$, so $\mathbb{Q}$ is the "function field" of Spec $(\mathbb{Z})$. The non-empty open sets of Spec $(\mathbb{Z})$ are gotten by throwing away finitely many primes $p_1, \ldots, p_n$. If $m = \prod p_i$, then this is the distinguished open set Spec $(\mathbb{Z})_m$, and

$$\Gamma\left(\text{Spec}\,(\mathbb{Z})_m, \underline{o}_{\text{Spec}\,\mathbb{Z}}\right) = \left\{ \frac{a}{m^k} \middle| a \in \mathbb{Z}, \quad k \geq 0 \right\}\,.$$

The residue fields of the stalks $\bar{o}_x$ are $\mathbb{Z}/2\mathbb{Z}, \mathbb{Z}/3\mathbb{Z}, \ldots, \mathbb{Q}$: we get each prime field exactly once.

Example D. Almost identical comments apply to Spec (R) for any Dedekind domain R. In fact, all prime ideals are maximal or (0); hence again we have a "line" of closed points plus a generic point. (However, if R is not a P.I.D., we cannot conclude as before that all non-empty open sets are distinguished; we know only that they are gotten by throwing out a finite set of closed points.)

A very important case is when R is a principal valuation ring[5]. Such a ring has a unique maximal ideal M, hence Spec (R) has 2 points $[(o)]$ and $[M]$. Then $[(o)]$ is an open point and $[M]$ is closed:

Imagine it as the affine line after all but one of its closed points has been thrown away. Valuation rings should always be considered as generalized one-dimensional objects.

Example E. $\mathbb{A}_k^2 = $ Spec ($k[X, Y]$), k algebraically closed. We get the maximal ideals $(X - a, Y - b)$, the principal prime ideals $(f(X, Y))$, for f irreducible, and (0). By dimension theory there are no other prime ideals. The set of maximal ideals gives us a set of closed points isomorphic to the usual X, Y-plane. Then we must add one big generic point; and for every irreducible curve, a point generic in that curve but not sticking out of it:

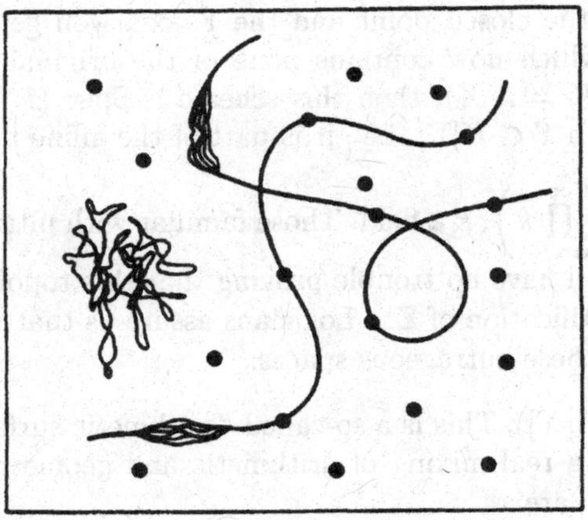

[5] i.e., a valuation ring with value group $\mathbb{Z}$. I call this principal because in the literature an ambiguity has arisen between those who call these rings "discrete valuation rings" and those who call them "discrete, rank 1 valuation rings".

To get a proper closed set, we take a finite number of irreducible curves, generic points and all, plus a finite number of closed points. Clearly, adding the non-closed points has not affected the topology much here: given a closed subset of the set of closed points in the old topology, there is a unique set of non-closed points to add to get a closed set in our new plane.

Example F. There is a somewhat more startling way to make a scheme out of the non-closed points in the plane. Let

$$\mathcal{O} = \left\{ \frac{f(X,Y)}{g(X,Y)} \;\middle|\; f,g \in k[X,Y],\;\; g(0,0) \neq 0 \right\} .$$

Then $\mathcal{O}$ is the stalk of $\underline{o}_X$ at $(0,0)$ if $X = \text{Spec}\,(k[X,Y])$. $\mathcal{O}$ has the maximal ideal (X,Y), the principal prime ideals $(f(X,Y))$ where f is irreducible and $f(0,0) = 0$, and (0). Therefore Spec $(\mathcal{O})$ has only one closed point:

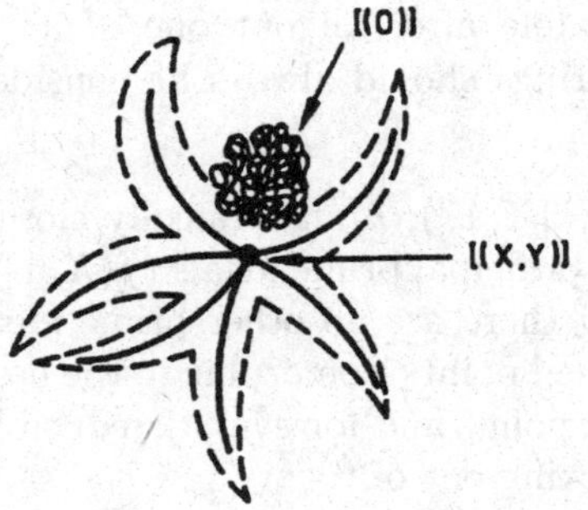

If you throw away the closed point and the Y-axis, you get the distinguished open set $X \neq 0$, which now contains none of the original closed points. On the other hand, if $K = k(X)$, then this scheme is Spec $(K[Y]_S)$ for a certain multiplicative system $S \subset K[Y]$, i.e., it is part of the affine line over K.

Example G. Spec $\left(\prod_{i=1}^{\infty} k\right)$, k a field. Those familiar with ultrafilters and similar far-out mysteries will have no trouble proving that this topological space is the Stone-Čech compactification of $\mathbb{Z}_+$. Logicians assure us that we can prove more theorems if we use these outrageous spaces.

Example H. Spec $(\mathbb{Z}[X])$. This is a so-called "arithmetic surface" and is the first example which has a real mixing of arithmetic and geometric properties. The prime ideals in $\mathbb{Z}[X]$ are:

i) (0).
ii) principal prime ideals (f), where f is either a prime p, or a $\mathbb{Q}$-irreducible polynomial written so that its coefficients have g.c.d. 1,

iii) maximal ideals $(p.f)$, p a prime and f a monic integral polynomial irreducible modulo p.

The whole should be pictured as follows:

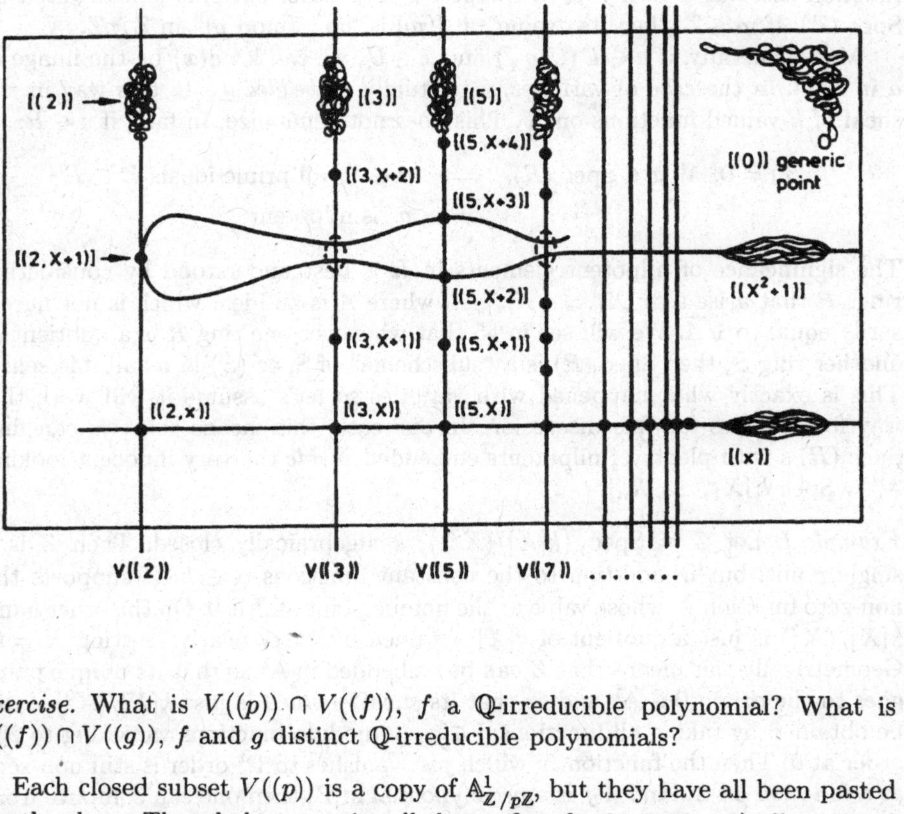

Exercise. What is $V((p)) \cap V((f))$, f a $\mathbb{Q}$-irreducible polynomial? What is $V((f)) \cap V((g))$, f and g distinct $\mathbb{Q}$-irreducible polynomials?

Each closed subset $V((p))$ is a copy of $\mathbb{A}^1_{\mathbb{Z}/p\mathbb{Z}}$, but they have all been pasted together here. The whole set-up is called a surface for 2 reasons: 1) all maximal chains of irreducible proper closed subsets have length 2, just as in $\mathbb{A}^2_k$. 2) If $\mathcal{O}$ is the local ring at a closed point x, then $\mathcal{O}$ has Krull dimension 2. In fact, if $x = [(p, f)]$, then its maximal ideal is generated by p and f, and there is *no single element* $g \in \mathcal{O}$ *such that* $m = \sqrt{(g)}$.

The elements of R can always, in a certain sense, be interpreted as functions on Spec (R). For all $x = [P] \in$ Spec (R), let

$$\begin{aligned} \Bbbk(x) &= R_P/P \cdot R_P \\ &= \text{quotient field } (R/P) \\ &= \text{residue field } (\underline{o}_x) \, . \end{aligned}$$

Then if $a \in R$, a induces an element of R_P, and then in $\Bbbk(x)$: we call this the *value* $a(x)$ of a at x. *These values lie in different fields at different points.* This generalizes what we did before for varieties. In that case, $\underline{o}_X$ was a sheaf of k-algebras when we associate to each $\alpha \in k$ the constant function on X with value α. And for all points x of the variety X, k is mapped isomorphically onto the residue field $\mathbb{k}(x)$ of $\underline{o}_x$. We could therefore pull all values back to a single field k, and the $\mathbb{k}(x)$-valued function associated to $a \in R$ above becomes the k-valued function associated to $a \in R$ in Chapter I. The other extreme is illustrated by Spec $(\mathbb{Z})$. If $m \in \mathbb{Z}$, then its "value" at $[(p)]$ is "$m \pmod p$" in $\mathbb{Z}/p\mathbb{Z}$.

More generally, if $a \in \Gamma(U, \underline{o}_X)$ and $x \in U$, we can let $a(x)$ be the image of a in $\mathbb{k}(x)$. In the case of varieties, we actually *embedded* $\underline{o}_X$ in this way in the sheaf of k-valued functions on X. This does not generalize. In fact, if $a \in R$:

$$\begin{aligned} a(x) = 0, \text{ all } x \in \text{Spec}(R) &\iff a \in P, \text{ all prime ideals } P \subset R \\ &\iff a \text{ is nilpotent .} \end{aligned}$$

The significance of nilpotent elements in R is best understood by considering rings R that arise as $k[X_1, \ldots, X_m]/A$, where A is an ideal which is not necessarily equal to $\sqrt{A}$. We will see in §5 that whenever one ring R is a quotient of another ring S, then Spec (R) is a "subscheme" of Spec (S) in a suitable sense. This is exactly what happened with varieties so let's assume it will work this way for schemes for this discussion. In our case, this means that we can find Spec (R)'s with plenty of nilpotents embedded *inside* the very innocent looking $\mathbb{A}^n_k = \text{Spec}\, k[X_1, \ldots, X_n]$.

Example I. Let $Z = \text{Spec}\ \left(k[X]/(X^2)\right)$, k algebraically closed. Then Z is a single point, but in addition to the constant functions $\alpha \in k$, Z supports the non-zero function X whose value at the unique point of Z is 0. On the other hand $k[X]/(X^2)$ is just a quotient of $k[X]$ obtained by "very nearly" setting $X = 0$. Geometrically this means that Z can be embedded in $\mathbb{A}^1$ so that its unique point goes to the origin $0 \in \mathbb{A}^1$ and so that its sheaf – which is just $k[X]/[X^2]$ – can be obtained by taking all functions $f \in \underline{o}_{0,\mathbb{A}^1}$ modulo functions vanishing to 2nd order at 0. Then the function X which just vanishes to 1st order is still non-zero as a function on Z. And if f is any polynomial in Z, then one can compute from the restriction of f to the subscheme Z not only its value at 0, but even its 1st derivative at 0. This example will be taken up more fully in §5.

In the most general case, Spec $(k[X_1, \ldots, X_n]/A)$ can be interpreted as the subset $V(A)$ in $\mathbb{A}^n_k$, but with a sheaf of rings on it suitably "fattened" so as to include the first few terms of the Taylor expansion of polynomials f in directions *normal* to $V(A)$. Another case would be Spec $\left(k[X,Y]/(g^2)\right)$, g being an irreducible polynomial. This Spec is the curve $g = 0$ with "multiplicity 2" and from the restriction of a polynomial f to it or, what is the same, from the image of f in $k[X,Y]/(g^2)$, one can reconstruct all 1st partials of f including the one normal to $g = 0$. Now by analogy if R is any ring with nilpotents, one can visualize Spec (R) as containing some extra *normal material*, in the direction of which one can for example take partial derivatives, but which is not actually tangent to a dimension present in the space Spec (R) itself. All this will, I hope, be much clearer in §5, where we will return to this discussion.

The following useful fact generalizes Prop. 4, Ch. I, §3:

Proposition 3. *Let R be a ring and let $f \in R$. Then the topological space* $\mathrm{Spec}\,(R)_f$ *together with the restriction of the sheaf of rings* $\underline{o}_{\mathrm{Spec}\,(R)}$ *to* $\mathrm{Spec}\,(R)_f$ *is isomorphic to* $\mathrm{Spec}\,(R_f)$ *together with the sheaf of rings* $\underline{o}_{\mathrm{Spec}\,(R_f)}$.

Proof. Let $i : R \to R_f$ be the canonical map. Then if P is a prime ideal of R, such that $f \notin P$, $i(P) \cdot R_f$ is a prime ideal of R_f; and if P is a prime ideal of R_f, $i^{-1}(P)$ is a prime ideal of R not containing f. These maps set up a bijection between $\mathrm{Spec}\,(R)_f$ and $\mathrm{Spec}\,(R_f)$ (cf. Zariski-Samuel, vol. 1, p. 223). The reader can check easily that this is a homeomorphism, and that in fact the open sets

$$\mathrm{Spec}\,(R)_{fg} \subset \mathrm{Spec}\,(R)_f$$

and

$$\mathrm{Spec}\,(R_f)_g \subset \mathrm{Spec}\,(R_f)$$

correspond to each other. But the sections of the structure sheaves $\underline{o}_{\mathrm{Spec}\,(R)}$ and $\underline{o}_{\mathrm{Spec}\,(R_f)}$ on these two open sets are both isomorphic to R_{fg}. Therefore, these rings of sections can be naturally identified with each other and this sets up an isomorphism of i) the restriction of $\underline{o}_{\mathrm{Spec}\,(R)}$ to $\mathrm{Spec}\,(R)_f$, and ii) $\underline{o}_{\mathrm{Spec}\,(R_f)}$ compatible with the homeomorphism of underlying spaces. □

§2. The category of preschemes

There is only one possible definition to make:

Definition 1. A *prescheme* is a topological space X, plus a sheaf of rings $\underline{o}_X$ on X, provided that there exists an open covering $\{U_\alpha\}$ of X such that each pair $\left(U_\alpha, \underline{o}_{X|U_\alpha}\right)$ is isomorphic to $\left(\mathrm{Spec}\,(R_\alpha), \underline{o}_{\mathrm{Spec}\,(R_\alpha)}\right)$ for some commutative ring R.

Definition 2. An *affine scheme* is a prescheme $(X, \underline{o}_X)$ isomorphic to $\left(\mathrm{Spec}\,(R), \underline{o}_{\mathrm{Spec}\,(R)}\right)$ for some ring R.

Notice that an open subset U of a prescheme X is also a prescheme if we take as its structure sheaf the restriction of $\underline{o}_X$ to U. To see this, first note that an affine scheme $(Y, \underline{o}_Y)$ has a basis of open sets U such that $\left(U, \underline{o}_{Y|U}\right)$ is again an affine scheme: i.e., take $U = Y_f = \{y \in Y \mid f(y) \neq 0\}$, where $f \in \Gamma(Y, \underline{o}_Y)$ and use the Prop. of the last section. Therefore if X is covered by open affines U_α, $U \cap U_\alpha$ can be further covered by open affines for any open set U. The definition of a morphism of preschemes is slightly less obvious.

Definition 3. If $(X, \underline{o}_X)$ and $(Y, \underline{o}_Y)$ are 2 preschemes, a *morphism* from X to Y is a continuous map $f : X \to Y$, plus a collection of homomorphisms:

$$\Gamma(V, \underline{o}_Y) \xrightarrow{f_V^*} \Gamma(f^{-1}(V), \underline{o}_X)$$

one for each open set $V \subset Y$, such that

a) whenever $V_1 \subset V_2$ are 2 open sets in Y, then the diagram:

$$\begin{array}{ccc}
\Gamma(V_2, \underline{o}_Y) & \xrightarrow{f^*_{V_2}} & \Gamma(f^{-1}(V_2), \underline{o}_X) \\
\Big\downarrow \text{res} & & \Big\downarrow \text{res} \\
\Gamma(V_1, \underline{o}_Y) & \xrightarrow[f^*_{V_1}]{} & \Gamma(f^{-1}(V_1), \underline{o}_X)
\end{array}$$

commutes, and

b) if $V \subset Y$ is open and $x \in f^{-1}(V)$, and $a \in \Gamma(V, \underline{o}_Y)$, then $a(f(x)) = 0$ implies $f^*_V(a)(x) = 0$.

The difficulty here is that, unlike the situation with varieties, it is now necessary to give explicitly the pull-back f^* on "functions" in the structure sheaf $\underline{o}_Y$ in addition to the map f itself. Of course for any set S there is always a natural pull-back from the sheaf of S-valued functions on Y to the sheaf of S-valued functions on X. But in the first place the function associated to a section of $\underline{o}_Y$ does not determine that section back again, and in the second place the values of these functions, and the corresponding functions on X lie in completely unrelated fields. Condition (b) expresses the only possible compatibility of f and f^*. This condition may be expressed in other ways, too. Suppose $x \in X$ and $y = f(x)$. For all open neighbourhoods U and V of x and y respectively such that $f(U) \subset V$, we have a homomorphism:

$$\Gamma(V, \underline{o}_Y) \xrightarrow{f^*_V} \Gamma(f^{-1}(V), \underline{o}_X) \xrightarrow{\text{res}} \Gamma(U, \underline{o}_X) \ .$$

Passing to the limit over such U and V, we get a homomorphism between stalks:

$$\begin{array}{ccc}
\varinjlim_{y \in V} \Gamma(V, \underline{o}_Y) & \longrightarrow & \varinjlim_{x \in U} \Gamma(U, \underline{o}_X) \\
\Big\| & & \Big\| \\
\underline{o}_{y,Y} & \overset{f^*_x}{\cdots\cdots\rightarrow} & \underline{o}_{x,X}
\end{array}$$

Notice that these stalks are *local* rings (since the stalks on affine schemes are always local rings). Then condition (b) asserts that f^*_x is a *local* homomorphism, i.e., equivalently

$$f^*_x(m_y) \subset m_x$$

or

$$m_y = (f^*_x)^{-1} m_x \ .$$

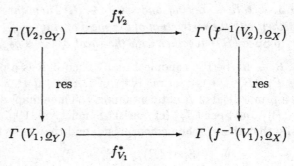

Whenever this is the case, f_x^* also induces a homomorphism:

$$\mathbb{k}(y) = \underline{o}_y/m_y \xrightarrow{\overline{f_X^*}} \underline{o}_x/m_x = \mathbb{k}(x) \ .$$

Call this k_x. By definition, k_x has the property:

(b') For all $a \in \Gamma(V, \underline{o}_Y)$, for some neighbourhood V of Y,

$$f_V^*(a)(x) = k_x[a(y)] \ .$$

In other words, if k_x is used to relate $\mathbb{k}(y)$ and $\mathbb{k}(x)$, then the pull-back f^* agrees with the pull-back of the functions associated to the sections of $\underline{o}_Y$.

Given 2 morphisms $X \xrightarrow{f} Y$ and $Y \xrightarrow{g} Z$, we can define their composition $g \cdot f : X \to Z$ in an obvious way. This gives us the category of preschemes. The first result that assures us that we are on the right track is:

Theorem 1. *Let X be a prescheme and let R be a ring. To every morphism $f : X \to \operatorname{Spec}(R)$, associate the homomorphism:*

$$R \cong \Gamma\left(\operatorname{Spec}(R), \underline{o}_{\operatorname{Spec}(R)}\right) \xrightarrow{f^*} \Gamma(X, \underline{o}_X) \ .$$

Then this induces a bijection between $\hom(X, \operatorname{Spec}(R))$ *and* $\hom(R, \Gamma(X, \underline{o}_X))$.

Proof. For all f's, let $A_f : R \to \Gamma(X, \underline{o}_X)$ denote the induced homomorphism. We first show that f is determined by $A_{f'}$. We must begin by showing how the map of point sets $X \to \operatorname{Spec}(R)$ is determined by $A_{f'}$. Suppose $x \in X$. The crucial fact we need is that a point of $\operatorname{Spec}(R)$ is determined by the ideal of elements of R vanishing at it (since $P = \{a \in R \mid a([P]) = 0\}$). Thus $f(x)$ is determined if we know $\{a \in R \mid a(f(x)) = 0\}$. But this equals $\{a \in R \mid f_x^*(a)(x) = 0\}$, and $f_x^*(a)$ is obtained by restricting $A_f(a)$ to $\underline{o}_x$. Therefore

$$f(x) = [\{a \in R \mid (A_f a)(x) = 0\}] \ .$$

Next we must show that the maps f_U^* are determined by A_f for all open sets $U \subset \operatorname{Spec}(R)$. Since f^* is a map of sheaves, it is enough to show this for a basis of open sets (in fact, if $U = \cup U_\alpha$ and $s \in \Gamma\left(U, \underline{o}_{\operatorname{Spec}(R)}\right)$, then $f_U^*(s)$ is determined by its restrictions to these sets $f^{-1}(U_\alpha)$, and these equal $f_{U_\alpha}^*(\mathrm{res}_{U,U_\alpha} s)$.) Now let $Y = \operatorname{Spec}(R)$ and consider f^* for the distinguished open set Y_b. It makes the diagram

$$\begin{array}{ccc}
\Gamma(f^{-1}(Y_b), \underline{o}_X) & \xleftarrow{\quad f_{Y_b}^* \quad} & \Gamma(Y_b, \underline{o}_Y) = R_b \\
\uparrow \text{res} & & \uparrow \text{res} \\
\Gamma(X, \underline{o}_X) & \xleftarrow{\quad A_f \quad} & \Gamma(Y, \underline{o}_Y) = R
\end{array}$$

commutative. Since these are ring homomorphisms, the map on the ring of fractions R_b is determined by that on R: thus A_f determined everything.

Finally any homomorphism $A : R \to \Gamma(X, \underline{o}_X)$ comes from some morphism f. To prove this, we first reduce to the case when X is affine. Cover X by open affine sets X_α. Then A induces homomorphisms

$$A_\alpha : R \longrightarrow \Gamma(X, \underline{o}_X) \xrightarrow{\text{res}} \Gamma(X_\alpha, \underline{o}_{X_\alpha}) .$$

Assuming the result in the affine case, there is a morphism $f_\alpha : X_\alpha \to \text{Spec}\,(R)$ such that $A_\alpha = A_{f_\alpha}$. On $X_\alpha \cap X_\beta$, f_α and f_β agree because the homomorphisms

$$\begin{array}{ccccc} & & \Gamma(X_\alpha, \underline{o}_X) & & \\ & \overset{A_\alpha}{\nearrow} & & \overset{\text{res}}{\searrow} & \\ R & & & & \Gamma(X_\alpha \cap X_\beta, \underline{o}_X) \\ & \underset{A_\beta}{\searrow} & & \underset{\text{res}}{\nearrow} & \\ & & \Gamma(X_\beta, \underline{o}_X) & & \end{array}$$

agree and we know that the morphism is determined by the homomorphism. Hence the f_α patch together to a morphism $f : X \to \text{Spec}\,(R)$, and one checks that A_f is exactly A.

Now let $A : R \to S$ be a homomorphism. We want a morphism $f : \text{Spec}\,(S) \to \text{Spec}\,(R)$. Following our earlier comments, we have no choice in defining f : for all points $[P] \in \text{Spec}\,(S)$,

$$f([P]) = \left[a^{-1}(P)\right] .$$

This is continuous since for all ideals $I \subseteq R$, $f^{-1}(V(I)) = V(A(I) \cdot S)$. Moreover if $U = \text{Spec}\,(R)_a$, then $f^{-1}(U) = \text{Spec}\,(S)_{A(a)}$, so for f_U^* we need a map $R_a \to S_{A(a)}$. We take the localization of A. These maps are then compatible with restriction, i.e.,

$$\begin{array}{ccc} R_a & \longrightarrow & S_{A(a)} \\ \downarrow & & \downarrow \\ R_{ab} & \longrightarrow & S_{A(a)\cdot A(b)} \end{array}$$

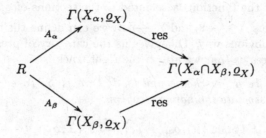

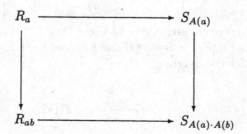

commutes. Hence they determine a sheaf map (in fact, if $U = \cup U_\alpha$, U_α distinguished, and $s \in \Gamma\left(U, \underline{o}_{\text{Spec}\,(R)}\right)$ then the elements $f_{U_\alpha}^*\,(\text{res}_{U,U_\alpha} s)$ patch together to give an element $f_U^*(s)$ in $\Gamma\left(f^{-1}(U), \underline{o}_{\text{Spec}\,(S)}\right)$). From our definition of f, it follows easily that f^* on $\underline{o}_{[A^{-1}P]}$ takes the maximal ideal $m_{[A^{-1}P]}$ into $m_{[P]}$. □

Corollary 1. *The category of affine schemes is isomorphic to the category of commutative rings with unit, with arrows reversed.*

Corollary 2. Spec ($\mathbb{Z}$) *is the final object in the category of preschemes, i.e., for every prescheme* X, *there is a unique morphism* $f : X \to$ Spec ($\mathbb{Z}$).

This is very important because it shows us that every prescheme X is a kind of fibred object, with one fibre for each prime p, and one over the generic point of Spec ($\mathbb{Z}$). More concretely, this fibering is given by the function

$$x \longmapsto \text{char.}\ (\Bbbk(x))$$

associating to each x the characteristic of its residue field. If $X =$ Spec (R), it is given by

$$[P] \longmapsto \text{the prime defined by the ideal } P \cap \mathbb{Z}\ .$$

Each particular fibre of X over Spec ($\mathbb{Z}$) will turn out to be a prescheme in its own right whose structure sheaf is a sheaf of k-algebras, where $k = \mathbb{Q}$ or $\mathbb{Z}/p\mathbb{Z}$: in other words, a more geometric object.

Prop. 1 of the last section generalizes to arbitrary preschemes:

Proposition 2. *Let* X *be a prescheme, and* $Z \subset X$ *an irreducible closed subset. Then there is one and only one point* $z \in Z$ *such that* $Z = \overline{\{z\}}$.

Proof. Let $U \subset X$ be an open affine set such that $Z \cap U \neq \emptyset$. Then any point $z \in Z$ dense in Z must be in $Z \cap U$; and a point $z \in Z \cap U$ whose closure contains $Z \cap U$ is also dense in Z. Therefore it suffices to prove the theorem for the closed subset $Z \cap U$. But by Prop. 1 of §4, there is a unique $z \in Z \cap U$ dense in $Z \cap U$. □

This point z will be called the *generic point* of Z. As a result of this Proposition, there is a $1-1$ correspondence between the points of X and the irreducible closed subsets of X.

A local version of this is often useful. If $x \in X$, then there is a $1-1$ correspondence between the following sets:

i) irreducible closed subsets $Z \subset U$, such that $x \in Z$,
ii) points $z \in X$ such that $x \in \overline{\{z\}}$,
iii) prime ideals $P \subset \underline{o}_{x,X}$.

The proof is left to the reader. We want to give next 2 examples of preschemes which involve non-trivial patching:

Example F bis. Let $\mathcal{O}$ be an arbitrary noetherian local ring. Since $\mathcal{O}$ has a unique maximal ideal M, Spec $(\mathcal{O})$ has a unique closed point $x = [M]$. Let X be the open subscheme Spec $(\mathcal{O}) - \{x\}$. The closed points of X correspond to prime ideals $P \not\subseteq M$ such that there are no prime ideals between P and M. X is only very rarely affine itself: In fact, to cover X by affines, choose elements $f_1, \ldots, f_n \in M$ such that

$$M = \sqrt{(f_1, \ldots, f_n)} \ .$$

Then

$$X = \bigcup_{i=1}^{n} \text{Spec } (\mathcal{O}_{f_i}) \ ,$$

where Spec $(\mathcal{O}_{f_i})$ and Spec $(\mathcal{O}_{f_j})$ are patched along Spec $(\mathcal{O}_{f_i f_j})$. If $\mathcal{O} = \mathbb{C}[[X_1, \ldots, X_n]]$ for example, the resulting X turns out to have topological properties identical to the ordinary $(2n-1)$-sphere.

Example J. $\mathbb{P}^n_{\mathbb{Z}}$. This prescheme is to be the union of $(n+1)$-copies $U_0, U_1, \ldots, U_n$ of integral affine space:

$$A^n_{\mathbb{Z}} = \text{Spec } \mathbb{Z}\,[X_1, \ldots, X_n] \ .$$

To simplify notation, introduce variables X_{ij}, $0 \le i, j \le n$ and $i \ne j$ and set

$$U_i = \text{Spec } \mathbb{Z}\,[X_{0i}, X_{1i}, \ldots, X_{i-1,i}, X_{i+1,i}, \ldots, X_{ni}] \ .$$

Then $U_i \cap U_j$, as a subset of U_i, is to be the open set $(U_i)_{X_{ij}}$; and as a subset of U_j it is to be the open set $(U_j)_{X_{ij}}$. Since

$$\begin{aligned} (U_i)_{X_{ji}} &= \text{Spec } \mathbb{Z}\,[X_{0i}, \ldots, X_{ni}, X_{ji}^{-1}] \\ (U_j)_{X_{ij}} &= \text{Spec } \mathbb{Z}\,[X_{0j}, \ldots, X_{nj}, X_{ij}^{-1}] \end{aligned}$$

these 2 preschemes can be identified by the map of rings:

$$\begin{aligned} X_{kj} &= X_{ki}/X_{ji}, \qquad 0 \le k \le n, \quad k \ne i, k \\ X_{ij} &= 1/X_{ji} \end{aligned}$$

with inverse

$$\begin{aligned} X_{ki} &= X_{kj}/X_{ij}, \qquad 0 \le k \le n, \quad k \ne i, j \\ X_{ji} &= 1/X_{ij} \ . \end{aligned}$$

This identification of variables is usually abbreviated by introducing $n+1$ variables X_i, $0 \le i \le n$, and replacing X_{ij} by X_i/X_j. Then as usual

$$U_i = \text{Spec } \mathbb{Z}\,[X_0/X_i, \ldots, X_n/X_i]$$

and the identification of U_i and U_j is given by the identity map between rings $\mathbb{Z}\,[X_0/X_i, \ldots, X_n/X_i, X_i/X_j]$ and $\mathbb{Z}\,[X_0/X_j, \ldots, X_n/X_j, X_j/X_i]$. Then if $F(X_0, \ldots, X_n)$ is a homogeneous integral polynomial, one can talk of the set $F = 0$, where, by definition,

$$\{F = 0\} \cap U_i = V\left((F/X_i^d)\right)$$

if $d =$ degree (F).

There is one exceedingly important and very elementary existence theorem in the category of preschemes. This asserts that arbitrary fibre products exist. It is much easier to prove this statement than to understand all the applications that it has. It is a much more far-reaching fact than the mere existence of plain products was in the category of prevarieties, and we will see some of the key uses of this theorem in the rest of this chapter.

Recall that if morphisms:

$$\begin{array}{ccc} X & & Y \\ & {\searrow}^{r} \quad {}^{s}{\swarrow} & \\ & S & \end{array}$$

are given, a fibre product is a commutative diagram

$$\begin{array}{ccc} & X\times_S Y & \\ {}^{p_1}\swarrow & & \searrow^{p_2} \\ X & & Y \\ \searrow & & \swarrow^{s} \\ & S & \end{array}$$

with the obvious universal property: i.e., given any commutative diagram

$$\begin{array}{ccc} & Z & \\ {}^{q_1}\swarrow & & \searrow^{q_2} \\ X & & Y \\ {}_{r}\searrow & & \swarrow_{s} \\ & S & \end{array}$$

there is a unique morphism $t : Z \to X \times_S Y$ such that $q_1 = p_1 \cdot t$, $q_2 = p_2 \cdot t$. The fibre product is unique up to canonical isomorphism.

Theorem 3a. *If A and B are C-algebras, let the diagram of affine schemes:*

$$\begin{array}{ccc} & \mathit{Spec}\,(A\otimes_C B) & \\ {}^{p_1}\swarrow & & \searrow^{p_2} \\ \mathit{Spec}\,(A) & & \mathit{Spec}\,(B) \\ {}_{r}\searrow & & \swarrow_{s} \\ & \mathit{Spec}\,(C) & \end{array}$$

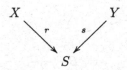

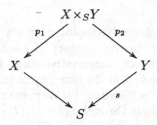

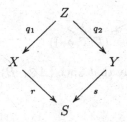

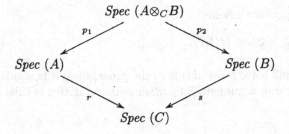

be defined by the canonical homomorphisms $C \to A$, $C \to B$, $A \to A \otimes_C B$, $B \to A \otimes_C B$. This makes Spec $(A \otimes_C B)$ *into a fibre product.*

Theorem 3b. *Given any morphisms $r : X \to S$, $s : Y \to S$, a fibre product exists.*

Proof of 3a. It is well known that in the diagram (of solid arrows):

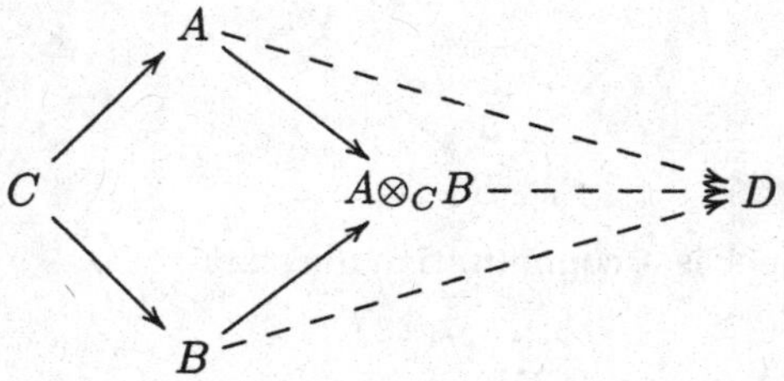

the tensor product has the universal mapping property indicated by dotted arrows, i.e., is the "direct sum" in the category of commutative C-algebras, or the "fibre sum" in the category of commutative rings. Dually, this means that Spec $(A \otimes_C B)$ is the fibre product in the category of affine schemes. But if T is an arbitrary prescheme, then by Theorem 1, every morphism of T into an affine scheme Spec (E) factors uniquely through Spec $(\Gamma(T, \underline{o}_T))$:

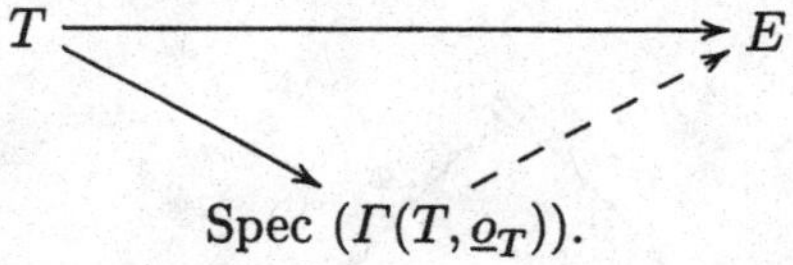

Using this, it follows immediately that Spec $(A \otimes_C B)$ is the fibre product in the category of all preschemes. □

Note for example, that if

$$\mathbb{A}^n_{\mathbb{Z}} = \text{Spec } (\mathbb{Z}[X_1, \ldots, X_n]),$$

then for all rings R:

$$\mathbb{A}^n_{\mathbb{Z}} \times \text{Spec } (R) = \text{Spec } (R(X_1, \ldots, X_n]).$$

We name this important scheme:

Definition 4. $\mathbb{A}^n_R = \text{Spec } (R[X_1, \ldots, X_n])$.

Proof of 3b. We shall leave most of this to the rader, since it is almost completely a mechanical patching argument. The main point to notice is this: suppose

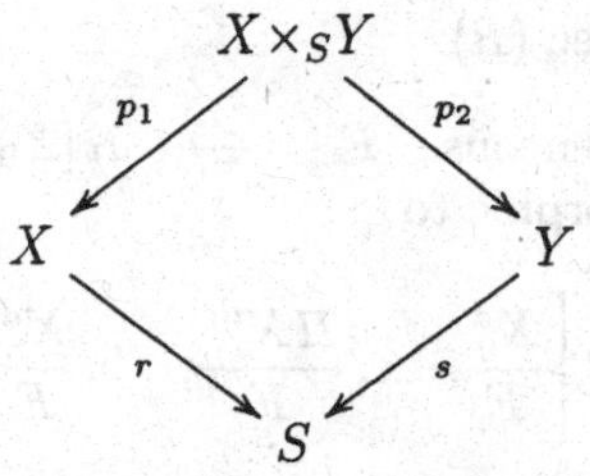

is some fibre product and suppose that $X_0 \subset X$, $Y_0 \subset Y$ and $S_0 \subset S$ are open subsets. Assume that $r(X_0) \subset S_0$ and $s(Y_0) \subset S_0$. Then the open subset

$$p_1^{-1}(X_0) \cap p_2^{-1}(Y_0) \subset X \times_s Y$$

is always the fibre product of X_0 and Y_0 over S_0. This being so, it is clear how we must set about constructing a fibre product: 1st cover S by open affines:

$$\text{Spec }(C_k) = W_k \subset S.$$

Next, cover $r^{-1}(W_k)$ and $s^{-1}(W_k)$ by open affines:

$$\begin{aligned} \text{Spec }(A_{k,i}) &= U_{k,i} \subset X, \\ \text{Spec }(B_{k,j}) &= V_{k,j} \subset Y. \end{aligned}$$

Then the affine schemes:

$$\text{Spec }(A_{k,i} \otimes_{C_k} B_{k,j}) = \Phi_{k,i,j}$$

must make an open affine covering of $X \times_S Y$ if it exists at all. To patch together $\Phi_{k,i,j}$ and $\Phi_{k',i',j'}$, let p_1, p_2 and p_1', p_2' stand for the canonical projections of $\Phi_{k,i,j}$ and $\phi_{k',i',j'}$ onto its factors. Then one must next check that the open subsets:

$$p_1^{-1}(u_{k,i} \cap U_{k',i'}) \cap p_2^{-1}(V_{k,j} \cap V_{k',j'}) \subset \Phi_{k,i,j}$$

and

$$p_1'^{-1}(U_{k',i'} \cap U_{k,i}) \cap p_2'^{-1}(V_{k',j'} \cap V_{k,j}) \subset \Phi_{k',i',j'}$$

are both fibre products of $U_{k,i} \cap U_{k',i'}$ and $V_{k,j} \cap V_{k',j'}$ over S. Hence they are canonically isomorphic and can be patched. Then you have to check that everything is consistent at triple overlaps. Finally you have to check the universal mapping property. All this is at worst confusing and at best obvious: in the former case, cf. EGA, Ch. 1, pp. 106–107 for assistance. □

Example J bis. To illustrate this patching, what is the scheme:

$$\mathbb{P}^n_{\mathbb{Z}} \times \text{Spec }(R)$$

R any ring? Since $\mathbb{P}^n_{\mathbb{Z}}$ is the union of $(n+1)$-affines

$$U_i = \text{Spec }\mathbb{Z}\,[X_0/X_i, \ldots, X_n/X_i]\ ,$$

$\mathbb{P}^n_{\mathbb{Z}} \times \text{Spec }(R)$ is just the union of the affines

$$U_i' = \text{Spec }R\,[X_0/X_i, \ldots, X_n/X_i]$$

with the standard patching. All homogeneous polynomials $F\,(X_0, \ldots, X_n)$ over R define closed subsets $F = 0$ of $\mathbb{P}^n_{\mathbb{Z}} \times \text{Spec }(R)$.

Definition 5. $\mathbb{P}^n_R = \mathbb{P}^n_{\mathbb{Z}} \times \text{Spec}\,(R)$.

Problem. For all homogeneous $F \in R[X_0, \ldots, X_n]$, show that $\{x \in \mathbb{P}^n_R \mid F(X) \neq 0\}$ is isomorphic to

$$\text{Spec}\, R\left[\frac{X_0^d}{F}, \ldots, \frac{\Pi X_i^{d_i}}{F}, \ldots, \frac{X_n^d}{F}\right]$$

where $\sum d_i = d = \text{degree}\,(F)$.

Here is an example of how one can use the Universal Mapping Property of fibre products to prove something tangible:

Proposition 4. *If $r : X \to S$ is surjective, then $p_2 : X \times_S Y \to Y$ is surjective.*

Proof. Let $y \in Y$ be any point; since r is surjective, there is a point $x \in X$ such that $r(x) = s(y)$. The maps r^* and s^* define inclusions of fields

$$\mathbf{k}(x) \hookleftarrow \mathbf{k}(s(y)) \hookrightarrow \mathbf{k}(y).$$

Let Ω be a composition of these fields. Define morphisms $\alpha : \text{Spec}\,\Omega \to X$, $\beta : \text{Spec}\,\Omega \to Y$ by a) mapping Spec (Ω) – which is one point – to x and y respectively, b) defining α^* and β^* to be the compositions

$$\underline{o}_{x,X} \longrightarrow \mathbf{k}(x) \longrightarrow \Omega$$

and

$$\underline{o}_{y,Y} \longrightarrow \mathbf{k}(y) \longrightarrow \Omega$$

respectively. Then $r \cdot \alpha = s \cdot \beta$. Therefore there is a morphism $\gamma : \text{Spec}\,\Omega \to X \times_S Y$ such that $p_2 \cdot \gamma = \beta$. Let $z = \text{Image}\,(\gamma)$. Then $p_2(z) = y$. □

§3. Varieties and preschemes

I want to make it crystal clear how our new category of preschemes contains our old category of varieties. When this is done, we shall *redefine the term variety*, and use it to mean instead the corresponding prescheme (or more precisely, prescheme/k).

Definition 1. Let R be a ring. Then a *prescheme X over R* is a morphism $\pi : X \to \text{Spec}\,(R)$.

Note that by Theorem 1, §2, this is the same as giving $\Gamma(X, \underline{o}_X)$ an R-algebra structure. This then gives all the rings $\Gamma(U, \underline{o}_X)$ R-algebra structures in such a way that the restriction maps are R-algebra homomorphisms. For instance, if $R = k$ is a field, we will have an injection of k into each $\Gamma(U, \underline{o}_X)$. Another example: take $R = \mathbb{Z}$. Then every prescheme is a prescheme over $\mathbb{Z}$ in exactly one way.

Definition 2. Let X and Y be preschemes over R. An R-morphism from X to Y is a morphism $f : X \to Y$ such that

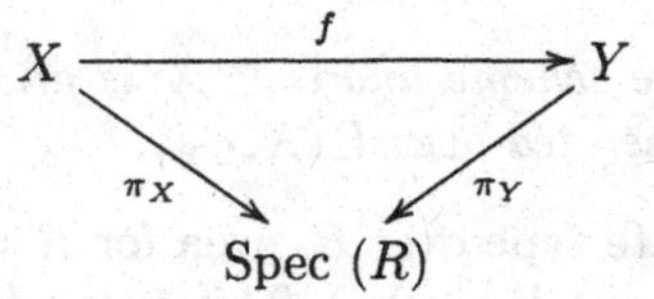

commutes.

Alternatively, this just means that the maps f_V^* ($V \subset Y$ open) are all R-algebra homomorphisms. Definitions 1 and 2 give us the *category of preschemes over R*.

Definition 3. A prescheme X over R is of *finite type over* R, if X is quasi-compact and for all open affine subsets $U \subset X$, $\Gamma(U, \underline{o}_X)$ is a finitely generated R-algebra.

Proposition 1. *Let X be a prescheme over R. If there exists a finite affine open covering $\{U_i\}$, $1 \leq u \leq n$, of X such that $R_i = \Gamma(U_i, \underline{o}_X)$ is a finitely generated R-algebra, then X is of finite type over R.*

This proposition is the archetype of a large number of similar propositions all of which assert: to check a property that refers to some behavior for all open affines somewhere, it suffices to check it for an open affine covering there.

Proof. Since $U_i = \text{Spec}(R_i)$, U_i is quasi-compact, hence X is quasi-compact (since there are only a finite number of U_i's). Now suppose $U = \text{Spec}(S) \subset X$ is any affine open subset. If $f_i \in R_i$, then $(U_i)_{f_i} = \text{Spec}\ R_i[1/f_i]$, and $R_i[1/f_i]$ is also a finitely generated R-algebra. But the $(U_i)_{f_i}$ are a basis of the topology on X and U is quasi-compact (since it is affine), so U is a union of finitely many sets $U_i' = \text{Spec}\ R_i'$, with R_i' a finitely generated R-algebra.

Since $U_i' \subset U$, we have a map $\text{res} : S \to R_i'$. If $f \in S$, then $U_f \cap U_i' = (U_i')_{\text{res}\ f}$ (deleting the prime ideals of S containing f clearly gives the same result on U_i' as throwing out the primes of R_i' containing res f). We can cover U_i' with (finitely many) smaller open sets of the form U_f, since the latter are a basis of the topology on U. Moreover if $U_f \subset U_i'$, $\Gamma(U_f, \underline{o}_X) = \Gamma\left((U_i')_{\text{res}\ f}, \underline{o}_X\right) = R_i'[1/\text{res}\ f]$ which is again finitely generated.

Thus we may assume we have U covered by finitely many sets U_{f_i}, $f_i \in S$, such that $\Gamma(U_{f_i}, \underline{o}_X) = S[1/f_i]$ is finitely generated. We now construct a subring

$\widetilde{S}$ of S. Put in $\widetilde{S}$ all the f_i, and enough other elements so that the image of $\widetilde{S}$, together with $1/f_i$, generates $S[1/f_i]$; only finitely many elements are needed. Finally, since the U_{f_i} cover U, we can write $1 = \Sigma f_i g_i$ with $g_i \in S$; put the g_i in $\widetilde{S}$ also. Now we claim that $\widetilde{S} = S$ (hence S is finitely generated). For take $\alpha \in S$. Then in $S[1/f_i]$, $\alpha = \frac{\text{elt. of } \widetilde{S}}{(f_i^{n_i})}$, i.e., $f_i^{n_i}\alpha = \beta_i$ in $S[1/f_i]$ for some $\beta_i \in \widetilde{S}$. That means $f_i^{k_i}(f_i^{n_i}\alpha - \beta_i) = 0$ in S, so $f_i^{n_i+k_i}\alpha \in \widetilde{S}$. Taking N large enough, we have $f_i^N \alpha \in \widetilde{S}$, for all i. But $\alpha = 1 \cdot \alpha = (\Sigma f_i g_i)^k \alpha$. Taking k large enough, we can make every term in $(\Sigma f_i g_i)^k$ contain some f_i^N, so α times it will be in $\widetilde{S}$. □

Corollary. *If X is an affine scheme and $U \subset X$ is an open affine subscheme, then $\Gamma(U, \underline{o}_X)$ is finitely generated over $\Gamma(X, \underline{o}_X)$.*

Warning. X may be of finite type over R, even for $R = \mathbb{C}$, and yet $\Gamma(X, \underline{o}_X)$ may not be a finitely generated R-algebra. This is the fact that Hilbert's 14th Problem was completely false; Zariski gave a systematic way of constructing counterexamples.

Definition 4. A prescheme X is *reduced* if $\underline{o}_X$ contains no nilpotent sections, i.e., $\Gamma(U, \underline{o}_X)$ has no nilpotent elements for all open sets $U \subset X$.

One can check that this holds if and only if all the stalks $\underline{o}_{x,X}$ have no nilpotents, and that it also holds if there is a covering of X by open affine sets U_i such that $\Gamma(U_i, \underline{o}_X)$ has no nilpotents.

Theorem 2. *Let k be an algebraically closed field. Then there is an equivalence of categories between:*

1) the category of reduced, irreducible preschemes of finite type over k, and k-morphisms.
2) the category of prevarieties over k and morphisms of these (as in Ch. I).

Proof. We shall construct functors in both directions which are, up to canonical identifications, inverse to each other.

A) Suppose X is a prevariety. For all irreducible closed sets $W \subset X$ with $\dim W > 0$, let $[W]$ be a symbol. Let $\mathcal{X}$ be the union of X and this collection of symbols $[W]$. For all open sets $U \subset X$, let U^* be the union of U and the set of symbols $[W]$ for which $W \cap U \neq \emptyset$. (The idea is that if U meets W at all, it meets it in an open dense subset, so U should contain the generic point $[W]$ of W). It is easy to see that

$$\begin{aligned} \left(\bigcup U_\alpha\right)^* &= \bigcup U_\alpha^* \\ (U_1 \cap U_2)^* &= U_1^* \cap U_2^* \\ U^* \cap X &= U\,. \end{aligned}$$

Therefore we have a topology on $\mathcal{X}$ which induces on the subset X its Zariski topology. Moreover, $U \to U^*$ and $U^* \to U^* \cap X$ set up a *bijection* between the set of all open subsets of X and the set of all open subsets of $\mathcal{X}$. Therefore finally we can just push the sheaf $\underline{o}_X$ across via

$$\Gamma(U^*, \underline{o}_{\mathcal{X}}) \underset{\text{def}}{=} \Gamma(U, \underline{o}_X).$$

Then $(\mathcal{X}, \underline{o}_{\mathcal{X}})$ is a ringed space with k-algebra structure. We leave it to the reader to check that $(\mathcal{X}, \underline{o}_{\mathcal{X}})$ is a prescheme of finite type over k: in fact it has the same affine coordinate rings as X had. It is reduced since $\underline{o}_X$ has no nilpotent elements; it is irreducible since $[X]$ is in every non-empty open set U^* and is therefore a generic point.

To get a functor, suppose $F : X_1 \to X_2$ is a morphism in the category of prevarieties. Extend F to map:

$$\mathcal{F} : \mathcal{X}_1 \longrightarrow \mathcal{X}_2$$

($\mathcal{X}_i$ the prescheme associated to X_i) as follows:

1) $\mathcal{F} = F$ on X_1
2) For all irreducible closed subsets $W \subset X_1$ ($\dim W > 0$), let

$$\mathcal{F}([W]) = \begin{cases} F(W) \text{ if this is a single point} \\ [\text{closure of } F(W)] \text{ if } \dim \overline{F(W)} > 0 \end{cases} .$$

If U_2^* is any open subset of $\mathcal{X}_2$, then one checks that

$$(*) \qquad \mathcal{F}^{-1}(U_2^*) = F^{-1}(U_2)^* \quad .$$

Therefore $\mathcal{F}$ is continuous. Finally, to define the required homomorphism of k-algebras:

$$\begin{array}{ccc} \Gamma(\mathcal{F}^{-1}(U_2^*), \underline{o}_{\mathcal{X}_1}) & \xleftarrow{\ \mathcal{F}^*_{U_2^*}\ } & \Gamma(U_2^*, \underline{o}_{\mathcal{X}_2}) \\ \Big\| & & \Big\| \\ \Gamma(F^{-1}(U_2), \underline{o}_{X_1}) & & \Gamma(U_2, \underline{o}_{X_2}) \end{array}$$

we just use the "pull-back" of functions, i.e., an element $s \in \Gamma(U_2, \underline{o}_{X_2})$ is a k-valued function on U_2, and we define

$$\mathcal{F}^*_{U_2^*}(s) = s \cdot F \quad .$$

B) To go backwards, we need the following:

Lemma 1. *Let X be a prescheme of finite type over k, and let $x \in X$. Then x is closed if and only if the composition*

$$k \longrightarrow \underline{o}_{x,X} \longrightarrow \mathbb{k}(x)$$

is surjective.

Proof. If we prove this for affine X's, it follows easily in the general case. On the other hand, if $X = \text{Spec}\,(R)$, the lemma asserts that a prime ideal $P \subset R$ is maximal if and only if k maps onto R/P. The non-obvious implication is: P maximal $\Rightarrow k \cong R/P$, and this is just the Nullstellensatz. □

This lemma implies the following for preschemes X of finite type over k: a point $x \in X$ is closed if it is closed in some open neighbourhood of itself. This is definitely false for general preschemes: cf. Example D.

Lemma 2. *Let X be a prescheme of finite type over k. Then the closed points of X are dense in every closed subset of X.*

Proof. It suffices to prove that every locally closed subset Y of X contains a closed point. Let U be an open subset of X such that $Y \cap U$ is non-empty and closed in U. Let $y \in Y \cap U$ be a closed point of $Y \cap U$. Then (y) is still closed in U, and by Lemma 1, $\{y\}$ is still closed in X. □

Now let $\mathcal{X}$ be a reduced irreducible prescheme of finite type over k. Let X be the set of closed points of $\mathcal{X}$, with the induced topology. Then the map

$$U \longrightarrow U \cap V$$

sets up a bijection between the set of open sets of $\mathcal{X}$, and the set of open sets of X, in view of Lemma 2. Via this bijection, we can carry the sheaf $\underline{o}_{\mathcal{X}}$ over to a sheaf $\underline{o}_X$ on X, i.e., via

$$\begin{aligned} \Gamma\,(U \cap X, \underline{o}_X) &= \Gamma\,(U, \underline{o}_{\mathcal{X}}) \\ &\text{for all } U \subset \mathcal{X} \text{ open}\,. \end{aligned}$$

By Lemma 1, we can give sections of $\underline{o}_X$ values in k at each point of X, i.e., if $U \subset X$ is open, $f \in \Gamma\,(U, \underline{o}_X)$ and $x \in U$, then let

$$f(x) = \left\{ \begin{array}{l} \text{the element } \alpha \in k \text{ such that} \\ f - \alpha \in m_x, \text{ the maximal ideal of } \underline{o}_x \end{array} \right\} .$$

Note that if $f(x) = 0$, all $x \in U$, then $f = 0$. (In fact, it suffices to check this for affine U's; and then if $U^* \subset \mathcal{X}$ is the open affine such that $U^* \cap X = U$, f is also identically zero on U^* since U is dense in U^*; therefore as we saw in §1, f is nilpotent, hence 0 since $\mathcal{X}$ is reduced.) Now it's easy to verify that $(X, \underline{o}_X)$ is a variety.

To get a functor, suppose $\mathcal{F} : \mathcal{X}_1 \to \mathcal{X}_2$ is a k-morphism of preschemes of the type being considered. If $X \in \mathcal{X}_1$, recall that $\mathcal{F}^*$ defines an injection of k-extension fields:

$$\begin{array}{ccc} \mathbb{k}(x) & \xleftarrow{\ \mathcal{F}_x^*\ } & \mathbb{k}(f(x)) \\ & \nwarrow \quad \nearrow & \\ & k. & \end{array}$$

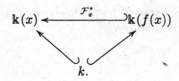

Therefore $f = \mathbb{k}(x)$ implies $k = \mathbb{k}(f(x))$, hence by Lemma 1 $\mathcal{F}$ takes closed points to closed points. Therefore $\mathcal{F}$ restricts to a continuous map $F : X_1 \to X_2$ of the corresponding prevarieties. One checks easily that F is actually a morphism.

It remains to put (A) and (B) together by showing that our 2 functors are mutually inverse. We leave this to the reader. □

Note carefully:

Re-definition 5. If k is an algebraically closed field, a *prevariety over* k is a reduced and irreducible prescheme of finite type over k. In the few occasions when we have to refer back to our old notation of prevariety, we will call them *old prevarieties.*

A detail that must be checked is:

Proposition 3. *The product of 2 prevarieties X and Y over k, in the category of prevarieties over k, is the same as their fibre product* $X \underset{\text{Spec }(k)}{\times} Y$, *in the category of all preschemes.*

Proof. This follows easily from Statement (2), Prop. 1, Ch. 1, §5, and we will omit the details. □

Here is an application which illustrates the use of the generic point of a variety. What we do here should be compared with the proof of Chow's lemma, §9, Ch. I, where we were using these techniques but in disguise. Suppose X and Y are 2 prevarieties over k, and suppose we are given k-isomorphisms

$$k(X) \underset{\alpha}{\xrightarrow{\sim}} K$$

$$k(Y) \underset{\beta}{\xrightarrow{\sim}} K$$

of their function fields with a third field K. Then α and β define morphisms:

$$\text{Spec } K \xrightarrow{A} X$$

$$\text{Spec } K \xrightarrow{B} Y$$

by requiring that the images of A and B be the generic points x, y of X and Y (Spec (K) consists in one point) and that $A^* : \underline{o}_x = k(X) \to K$ be α, and that $B^* : \underline{o}_y = k(Y) \to K$ be β. By the proposition, then, we get a morphism:

$$\text{Spec } (K) \xrightarrow{(A,B)} X \times_k Y.$$

Let t be the image point and let $T = \overline{\{t\}}$, a closed subset of $X \times Y$. T is irreducible, as it is the closure of a point, hence T is a closed subprevariety of $X \times Y$. Moreover, the function field $k(T)$ of T is isomorphic via $(A, B)^*$ to K also. We have constructed a diagram:

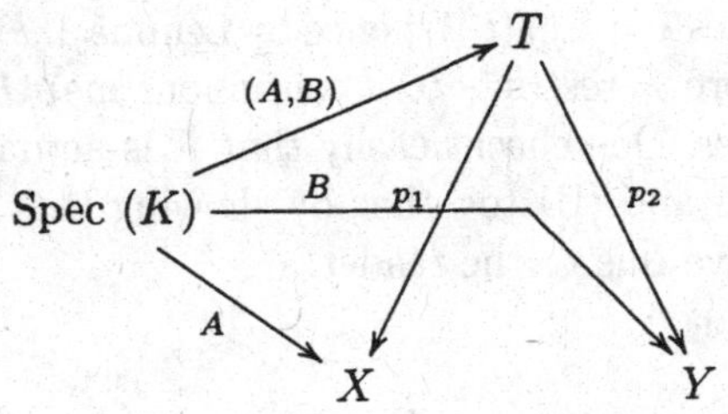

in which p_1 and p_2 are birational morphisms of prevarieties.

We describe this situation by saying that $\beta^{-1} \cdot \alpha$ is a *birational correspondence* between X and Y and that T is its *graph*. If we use Th. 4 of §7, Ch. I, we can see what is happening more clearly:

a) There is a non-empty open set $U \subset X$ such that $p_1^{-1}(U)$ is isomorphic to U via p_1.
b) There is a non-empty open set $V \subset Y$ such that $p_2^{-1}(V)$ is isomorphic to V via p_2.

Set

$$\begin{aligned} W_0 &= p_1^{-1}(U) \cap p_2^{-1}(V), \\ U_0 &= p_1(W_0), \\ V_0 &= p_2(W_0) \ . \end{aligned}$$

Then p_1 and p_2 define isomorphisms of the 3 open sets U_0, V_0, W_0 in X, Y and T. In other words, the isomorphism $\beta^{-1} \cdot \alpha$ of the function fields of X and Y extends to an isomorphism between the open dense subsets U_0, V_0 of X and Y; and T is obtained by taking the graph of its isomorphism in $X \times Y$ and closing it up. T itself is, in general, a many-many correspondence between X and Y.

To see what's happening explicitly, we must as usual look at affine pieces. Cover X by $U_i = \text{Spec } A_i$, Y by $V_j = \text{Spec } B_j$. Then $X \times Y$ is covered by $U_i \times V_j = \text{Spec } (A_i \otimes_j B_j)$. For all i, j, $t \in U_i \times V_j$, as its image in X [resp. Y] is the generic point and so lies in all U_i [resp. V_j]. Then $T \cap (U_i \times V_j)$ is the closure of $\{t\}$ in $U_i \times Y_j$; these are an open affine covering of T, and we want to see what they are.

Spec $(K) \longrightarrow U_i \times V_j$ corresponds to $K \overset{\varphi_{ij}}{\longleftarrow} A_i \otimes B_j$, and φ_{ij} is just the composition $A_i \otimes B_j \longrightarrow k(X) \otimes k(Y) \overset{(\alpha,\beta)}{\longrightarrow} K$. Then t is $[\varphi_{ij}^{-1}(0)]$, so its closure is $V\left(\varphi_{ij}^{-1}(0)\right)$, and $T \cap (U_i \times V_j)$ is the affine variety Spec $\left(A_i \otimes B_j / \varphi_{ij}^{-1}(0)\right)$. But this is just Spec of the composite ring $\alpha(A_i) \cdot \beta(B_j) \subset K$ that we get by pushing the two rings together.

Example K. Here is one of the oldest and prettiest examples of this definition. Take $X = \mathbb{P}_2$, and let x_0, x_1, x_2 and y_0, y_1, y_2 be homogeneous coordinates in X and Y. Let $U_0 \subset X$ and $V_0 \subset Y$ be defined as the open sets $x_0 \cdot x_1 \cdot x_2 \neq 0$ and $y_0 \cdot y_1 \cdot y_2 \neq 0$. Define an isomorphism between U_0 and V_0 by the map:

$$y_0 = 1/x_0, \qquad y_1 = 1/x_2, \qquad y_2 = 1/x_2 \ .$$

In fact, this is just an extension of the isomorphism of function fields:

$$\left(\frac{x_1}{x_0}, \frac{x_2}{x_0}\right) \xrightarrow{\sim} k\left(\frac{y_1}{y_0}, \frac{y_2}{y_0}\right)$$

$$\begin{aligned} x_1/x_0 &\longmapsto y_1/y_1 \\ x_2/x_0 &\longmapsto y_0/y_2 \ . \end{aligned}$$

But the closure $T \subset \mathbb{P}_2 \times \mathbb{P}_2$ of this isomorphism is a very remarkable affair. We leave it to the reader to work out the full set-theoretic many-many correspondence that we get: let's just check one piece of T to get an idea.

Let

$$\begin{aligned} U &= \operatorname{Spec} k\left[\frac{x_1}{x_0}, \frac{x_2}{x_0}\right] = X_{x_0} \\ V &= \operatorname{Spec} k\left[\frac{y_0}{y_1}, \frac{y_2}{y_1}\right] = Y_{y_1} \ . \end{aligned}$$

What is $T \cap (U \times V)$? In any case, it is

$$\begin{array}{ccccccccc} & & A & & B & & C & & D \\ & & \| & & \| & & \| & & \| \\ \operatorname{Spec} & k & \Big[\dfrac{x_1}{x_0} & , & \dfrac{x_2}{x_0} & , & \dfrac{y_0}{y_1} & , & \dfrac{y_2}{y_1} \Big] \Big/ P \end{array}$$

where P is the ideal of relations we get by putting all functions in the same function field. In particular, $A - C \in P$ and $BD - C \in P$. But dividing out by $A - C$ and $BD - C$, the quotient ring is $k[B, D]$, already an integral domain of the right dimension. So $P = (A - C, BD - C)$ and $T \cap (U \times V) = \operatorname{Spec}\,(k[B, D])$. Finally, the projection gives:

$$\begin{array}{ccc} & T\cap(U\times V) & \\ {}^{p_1}\swarrow & & \searrow^{p_2} \\ U & & V \end{array} \qquad\qquad \begin{array}{ccc} & k[B,D] & \\ {}^{p_1^*}\nearrow & & \nwarrow^{p_2^*} \\ k[A,B] & & k[C,D] \end{array}$$

where $p_1^*(A) = B \cdot D$, $p_1^*(C) = B \cdot D$. Thus the line $B = 0$ in $T \cap (U \times V)$ collapses to the point $A = B = 0$ in U; and the line $D = 0$ in $T \cap (U \times V)$ collapses to the point $C = D = 0$ in V:

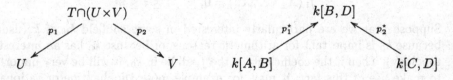

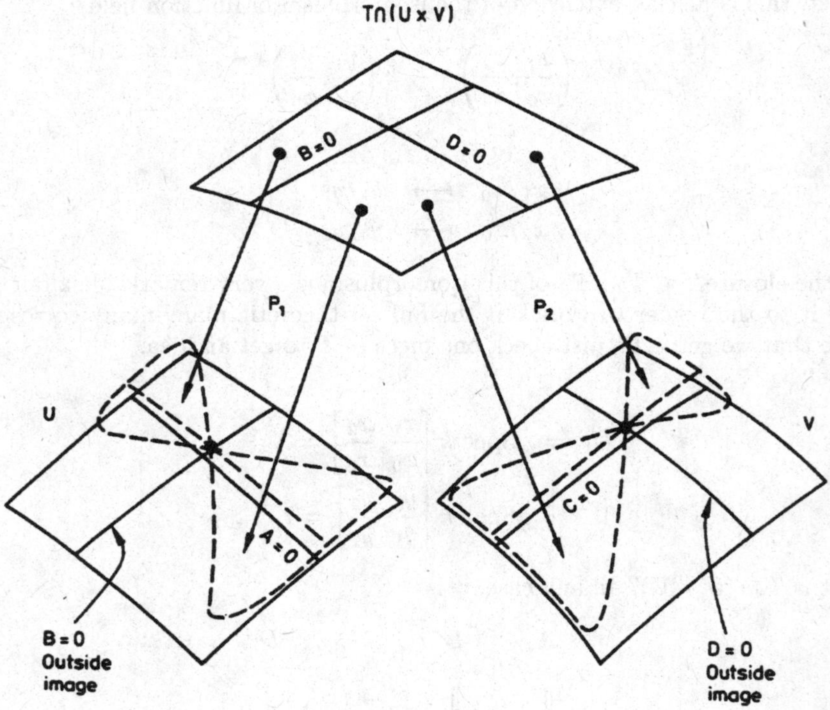

§4. Fields of definition

Let k be an algebraically closed field and let X be a closed subvariety of $\mathbb{A}^n$, defined by equations:

$$f_i(X_1, \ldots, X_n) = 0, \qquad 1 \leq i \leq m .$$

Suppose that we are particularly interested in some subfield k_0 of k (usually because k_0 is important for arithmetic reasons, or because k_0 has an interesting topology). Then if the coefficients of the f_i all lie in k_0, it will be very important to make use of this fact: it may, for example, pose Diophantine or rationality questions. When this happens one says that X is "defined over k_0". Let

$$A = (f_1, \ldots, f_m) = I(X) .$$

Then if the coefficients of the f_i lie in k_0, it follows firstly that $A = k \cdot A_0$ where $A_0 = A \cap k_0[X_1, \ldots, X_n]$; and secondly, if $R_0 = k_0[X_1, \ldots, X_n]/A_0$, then the affine ring $R = k[X_1, \ldots, X_n]/A$ of X is of the form $R_0 \otimes_{k_0} k$. This shows us the scheme-theoretic significance of X being defined over k_0: there exists an affine scheme $X_0 = \text{Spec}(R_0)$ of finite type over k_0 such that

$$(*) \qquad X = X_0 \times_{\mathrm{Spec}\,(k_0)} \mathrm{Spec}\,(k) \ .$$

In fact, conversely, suppose X is an affine variety over k, and that X_0 is a prescheme of finite type over k_0 such that $(*)$ holds. Then let $R_0 = \Gamma\left(X_0, \underline{o}_{X_0}\right)$ and write R_0 as a quotient of a polynomial ring: $k_0[X_1, \ldots, X_n](f_1, \ldots, f_m)$. It follows that if $R = \Gamma(X, \underline{o}_X)$, then $R \cong k[X_1, \ldots, X_n]/(f_1, \ldots, f_m)$, hence X is isomorphic to the closed subvariety of $\mathbb{A}^n$ defined by $f_i = 0$, $1 \le i \le m$. And since the coefficients of the f_i are in k_0, X is "defined over k_0" in our original sense.

On the other hand, notice that the relationship given by $(*)$ between X and X_0 does not involve any specific affine embedding and, in fact, can be considered for any X and X_0, not necessarily affine. It suggests that by considering preschemes of finite type over k_0 we can set up a whole k_0-geometry even when k_0 is not algebraically closed. In this k_0-geometry we can make definitions of a Diophantine type that have no analog over k. But whenever we want to visualize what is going on, we can form fibre products with Spec (k) and obtain a "classical" geometric set-up. In fact, the easiest way to think of a prescheme X_0 over k_0 is as a prescheme X over k, plus an extra "k_0-structure" given by expressing X as a fibre product $X_0 \times_{\mathrm{Spec}\,(k_0)} \mathrm{Spec}\,(k)$. Conversely, we will be able to express a given prevariety X as such a fibre product whenever it is defined for us by equations all of whose coefficients lie in k_0.

Assume that we are given any prescheme X_0 over k_0 and that we define X as $X_0 \times_{\mathrm{Spec}\,(k_0)} \mathrm{Spec}\,(k)$: we shall often write $X = X_0 \times_{k_0} k$ for simplicity. Explicitly, if X_0 is the union of open affine sets $(U_0)_i = \mathrm{Spec}\ (R_i)$, then X is just the union of the affine schemes

$$U_i = \mathrm{Spec}\ (R_i \otimes_{k_0} k) \ .$$

Definition 1. Let σ be an automorphism of k over k_0. The conjugation map $\sigma_X : X \to X$ is the underlying map of the morphism:

$$\begin{array}{ccc} X_0 \times_{\mathrm{Spec}\,(k_0)} \mathrm{Spec}\,(k) & \xrightarrow{\ 1_{X_0} \times \phi\ } & X_0 \times_{\mathrm{Spec}\,(k_0)} \mathrm{Spec}\,(k) \\ \Vert & & \Vert \\ X & & X \end{array}$$

where $\phi : \mathrm{Spec}\ (k) \to \mathrm{Spec}\ (k)$ is defined by taking σ^{-1} as the homomorphism $\phi^* : k \to k$.

One checks immediately that $(\sigma \cdot \tau)_X = \sigma_X \cdot \tau_X$, i.e., the Galois group of k/k_0 is acting on the topological space X. In particular, each σ_X is a homeomorphism of X. In simple cases, this conjugation is exactly what you expect it to be:

Assume

$$\left[\begin{array}{l} X_0 = \mathrm{Spec}\ k_0[X_1, \ldots, X_n]/(f_1, \ldots, f_m) \\ X = \mathrm{Spec}\ k[X_1, \ldots, X_n]/(f_1, \ldots, f_m) \\ x \in X \text{ is the closed point } X_1 = \alpha_1, \ldots, X_n = \alpha_n \\ \text{where } \alpha_1, \ldots, \alpha_n \in k;\ f_i(\alpha_1, \ldots, \alpha_n) = 0, \text{ all } i \ . \end{array}\right.$$

Then

$$\sigma_X(x) = \{\text{the closed point } X_1 = \sigma(\alpha_1), \ldots, X_n = \sigma(\alpha_n)\} .$$

Proof. By definition $\sigma_X(x) = (1_{X_0} \times \phi)(x)$, and $(1_{X_0} \times \phi)^*$ maps $g(X_1, \ldots, X_n)$ to $g^{\sigma^{-1}}(X_1, \ldots, X_n)$, where $g^{\sigma^{-1}}$ is the polynomial g with σ^{-1} applied to its coefficients. But $x = [(X_1 - \alpha_1, \ldots, X_n - \alpha_n)]$, so if $\sigma_X(x) = [P]$,

$$P = (1_{X_0} \times \phi)^{*^{-1}} (X_1 - \alpha_1, \ldots, X_n - \alpha_n) .$$

Since $(1_{X_0} \times \phi)(X_i - \sigma\alpha_i) = X_i - \alpha_i$, it follows that $P = (X_1 - \sigma\alpha_1, \ldots, X_n - \sigma\alpha_n)$ □

Theorem 1. *Let X_0 be a prescheme over k_0, let $X = X_0 \times_{k_0} k$, and let $p : X \to X_0$ be the projection. Assume that k is an algebraic closure of k_0. Then*

1) *p is surjective and both open and closed (i.e., maps open/closed sets to open/closed sets).*
2) *For all $x, y \in X$, $p(x) = p(y)$ if and only if $x = \sigma_X(y)$, some $\sigma \in \mathrm{Gal}(k/k_0)$. In other words, for all $x \in X_0$, $p^{-1}(x)$ is an orbit of $\mathrm{Gal}(k/k_0)$. Moreover, $p^{-1}(x)$ is a finite set.*

Proof. Since all these results are local on X_0, we may as well replace X_0 by an open affine subset U_0 and replace X by $p^{-1}(U_0)$. Therefore assume $X_0 = \mathrm{Spec}\,(R)$, $X = \mathrm{Spec}\,(R \otimes_{k_0} k)$. First of all p is surjective by Prop. 4, §2. Secondly, I claim p is closed: let $V(A)$ be any closed subset of X, where A is an ideal in $R \otimes_{k_0} k$. Let $A_0 = A \cap R$, and consider the pair of rings:

$$\begin{array}{c} R \otimes_{k_0} k/A \\ \cup \\ R/A_0 . \end{array}$$

Since k is algebraically dependent on k_0, $R \otimes_{k_0} k/A$ is integrally dependent on R/A_0. By the going-up theorem, every prime ideal $P_0 \subset R/A_0$ is of the form $P \cap (R/A_0)$ for some prime ideal $P \subset R \otimes_{k_0} k/A$. Therefore every prime ideal $P_0 \subset R$ such that $P_0 \supset A_0$ is of the form $P \cap R$ for some prime ideal $P \subset R \otimes_{k_0} k$ such that $P \supset A$: this means that $p(V(A)) = V(A_0)$, hence p is closed.

Thirdly, let's prove (2). We must show that if $P_1, P_2 \subset R \otimes_{k_0} k$ are 2 prime ideals, then

$$P_1 \cap R = P_2 \cap R \iff P_2 = (1_R \otimes \sigma)(P_1), \quad \text{some } \sigma \in \mathrm{Gal}(k/k_0).$$

$\Leftarrow$ is obvious. Now assume $P_0 = P_1 \cap R = P_2 \cap R$. Let K be the quotient field of R/P_0. Consider the diagram:

$$\begin{array}{ccc} R \otimes_{k_0} k & \xrightarrow{\ j\ } & K \otimes_{k_0} k \\ \cup & & \cup \\ R & \longrightarrow & K . \end{array}$$

Since $K \otimes_{k_0} k$ is the localization of $(R \otimes_{k_0} k)/P_0 \cdot (R \otimes_{k_0} k)$ with respect to the multiplicative system $R - P_0$, and since a) $P_i \supset P_0 \cdot (R \otimes_{k_0} k)$ and (b) P_i is disjoint from $R - P_0$, it follows that $P'_i = j(P_i) \cdot K \otimes_{k_0} k$ is a prime ideal and that $P_i = j^{-1}(P'_i)$. Therefore it suffices to prove that $P'_2 = (1_K \otimes \sigma)(P'_1)$ and it will follow that $P_2 = (1_R \otimes \sigma)(P_1)$. Now consider the integral domains $K \otimes_{k_0} k/P'_i$ over K: let Ω be a big extension field of K containing both of them. We obtain a picture like this:

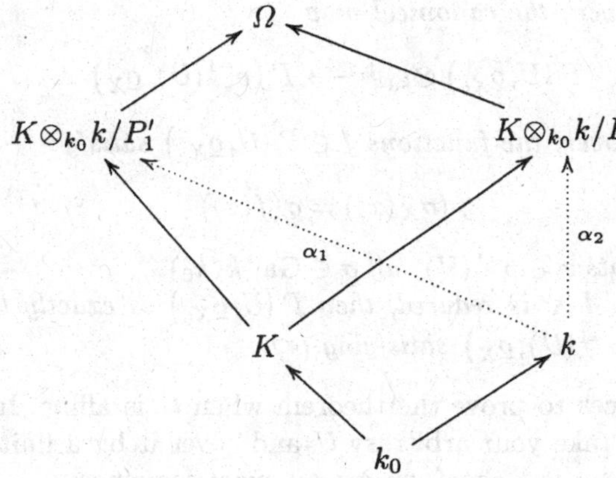

(Consider the solid arrows as inclusion maps to simplify notation.) Then $\alpha_1(k)$ and $\alpha_2(k)$ must both be equal to the algebraic closure of k_0 in Ω. Therefore there is an automorphism $\sigma \in \mathrm{Gal}(k/k_0)$ such that $\alpha_2 = \alpha_1 \cdot \sigma$. But then if $x_i \in K$, $y_i \in k$,

$$\begin{aligned} \sum x_i \otimes y_i \in P'_2 \quad &\Longleftrightarrow \quad \sum x_i \cdot \alpha_2(y_i) = 0 \text{ in } \Omega \\ &\Longleftrightarrow \quad \sum x_i \cdot \alpha_1(\sigma(y_i)) = 0 \text{ in } \Omega \\ &\Longleftrightarrow \quad \sum x_i \otimes \sigma(y_i) \in P'_i \end{aligned}$$

so $(1_K \otimes \sigma)(P'_2) = P'_1$.

Moreover, if $P \subset R \otimes_{k_0} k$ is a prime ideal, then P has only a finite number of distinct conjugates: in fact, let P be generated by $f_1, \ldots, f_m$ and let $f_i = \sum_j f_{ij} \otimes \alpha_{ij}$. Then P is left fixed by all σ's which leave the α_{ij}'s fixed, and this is a subgroup in $\mathrm{Gal}(k/k_0)$ of finite index.

Finally, p is also an open map. In fact, let $U \subset X$ be any open set. Then

$$U' = \bigcup_{\sigma \in \mathrm{Gal}(k/k_0)} \sigma_X(U)$$

is also open. But by (2), $p(U) = p(U')$ and $U' = p^{-1}(p(U'))$. Therefore $X_0 - p(U) = p(X - U')$ which is closed since p is a closed map. Therefore $p(U)$ is open. □

Corollary. *X_0, as a topological space, is the quotient of X by the action of* $\mathrm{Gal}(k/k_0)$.

Definition 2. The k_0-*topology* on X is the set of $\mathrm{Gal}(k/k_0)$-invariant open sets, i.e., the set of open sets $\{p^{-1}(U) \mid U \text{ open in } X_0\}$.

Theorem 2. *Let X_0 be a prescheme over k_0, let $X = X_0 \otimes_{k_0} k$ and let $p : X \to X_0$ be the projection. Assume that X_0 is of finite type over k_0 (hence X is of finite type over k).*

1) For all $U \subset X_0$ open, the canonical map

$$\Gamma\left(U, \underline{o}_{X_0}\right) \otimes_{k_0} k \longrightarrow \Gamma\left(p^{-1}(U), \underline{o}_X\right)$$

is bijective. Moreover, the functions $f \in \Gamma\left(U, \underline{o}_{X_0}\right)$ satisfy

$$(*) \qquad f\left(\sigma_X(x)\right) = \sigma(f(x))$$

for all closed points $x \in p^{-1}(U)$, all $\sigma \in \mathrm{Gal}(k/k_0)$.

2) If k_0 is perfect and X is reduced, then $\Gamma\left(U, \underline{o}_{X_0}\right)$ is exactly the subring of elements $f \in \Gamma\left(p^{-1}(U), \underline{o}_X\right)$ satisfying $(*)$.

Proof. Again it suffices to prove the theorem when U is affine. In fact, if it is proven in that case, take your arbitrary U and cover it by a finite set of open affine sets U_i. Since $\underline{o}_{X_0}$ is a sheaf, we get an exact sequence:

$$0 \longrightarrow \Gamma\left(U, \underline{o}_{X_0}\right) \xrightarrow{\text{res}} \prod_j \Gamma\left(U_i, \underline{o}_{X_0}\right) \xrightarrow{\text{res}} \prod_{i,j} \Gamma\left(U_i \cap U_j, \underline{o}_{X_0}\right)$$

where the 2nd arrow maps $\{s_i\}$ to $\left\{\mathrm{res}_{U_i, U_i \cap U_j}(s_i) - \mathrm{res}_{U_j, U_i \cap U_j}(s_j)\right\}$. Tensoring with k over k_0, we get a diagram:

$$\begin{array}{ccccccc}
0 \to & \Gamma\left(U, \underline{o}_{X_0}\right) \otimes_{k_0} k & \to & \prod_i \left(\Gamma\left(\left(U_i, \underline{o}_X\right) \otimes_{k_0} k\right)\right. & \to & \prod_{i,j} \left(\Gamma\left(U_i \cap U_j, \underline{o}_{X_0}\right) \otimes_{k_0} k\right) \\
& \downarrow \alpha & & \downarrow \beta & & \downarrow \gamma \\
0 \to & \Gamma\left(p^{-1}(U) \underline{o}_X\right) & \to & \prod_i \Gamma\left(p^{-1}(U_i), \underline{o}_X\right) & \to & \prod_{i,j} \Gamma\left(p^{-1}\left(U_i \cap U_j\right), \otimes_X\right) & .
\end{array}$$

Now we know β is bijective by the affine case. This implies that α is injective. Since this is now proven for every open U, it follows also that γ is injective. Therefore α is even bijective. The remaining results follow immediately for an arbitrary U once they are proven in the affine case.

Now assume $U = \mathrm{Spec}\ (R)$, hence $p^{-1}(U) = \mathrm{Spec}\ (R \otimes_{k_0} k)$. Then

$$\begin{array}{ccc} \Gamma(U, \underline{o}_{X_0}) \otimes_{k_0} k & \longrightarrow & \Gamma(p^{-1}(U), \underline{o}_X) \\ \| & & \| \\ R \otimes_{k_0} k & & R \otimes_{k_0} k \end{array}$$

is bijective by the very definition of fibre product. Moreover, suppose $M \subset R \otimes_{k_0} k$ is a maximal ideal. We know $\mathbb{k}([M]) = R \otimes_{k_0} k/M = k$ since R is of finite type over k_0. If $\sigma \in \mathrm{Gal}(k/k_0)$, then $\sigma_X([M]) = \left[\left(1_R \otimes \sigma^{-1}\right)^{-1} M\right] = [1_R \otimes \sigma(M)]$. Therefore $(*)$ asserts:

$$f \in R \Longrightarrow f \bmod (1_R \otimes \sigma(M)) = \sigma\{f \bmod M\} \ .$$

But say $f = \alpha + g$, $\alpha \in k$, $g \in M$ so that $f \bmod M = \alpha$. Then

$$\begin{aligned} f &= (1_R \otimes \sigma) f \\ &= \sigma(\alpha) + (1_R \otimes \sigma) g \end{aligned}$$

hence $f \bmod (1_R \otimes \sigma(M)) = \sigma(\alpha)$.

Now assume that $R \otimes_{k_0} k$ has no nilpotent elements and that k_0 is perfect. Since $R \otimes_{k_0} k$ has no nilpotent elements and is finitely generated over k,

$$(0) = \bigcap \{M \mid M \text{ maximal ideal in } R \otimes_{k_0} k\} \ .$$

Suppose $f \in R \otimes_{k_0} k$ satisfies $(*)$. Following backwards the argument we just gave shows that this means exactly that $f - (1_R \otimes \sigma) f$ is in every maximal ideal M. But therefore $f = (1_R \otimes \sigma) f$, all $\sigma \in \mathrm{Gal}(k/k_0)$. Since k_0 is perfect, k_0 is the set of $\alpha \in k$ invariant under $\mathrm{Gal}(k/k_0)$. Therefore R is the set of $\alpha \in R \otimes_{k_0} k$ invariant under $\mathrm{Gal}(k/k_0)$. Therefore $f \in R$. □

It follows that X_0, as a prescheme over k_0, can be reconstructed from 3 things:

i) the prescheme X over k
ii) the action of $\mathrm{Gal}(k/k_0)$ on X via conjugation
iii) the subsheaf of $\underline{o}_X$, defined only on the k_0-open sets, which is $\underline{o}_{X_0}$.

We call $\underline{o}_{X_0}$, regarded as a sheaf in the k_0-topology on X, the sheaf of k_0-rational functions. It is the analog of the ring of polynomials with coefficients in k_0. Moreover in good cases (X reduced, k_0 perfect), (i) and (ii) alone suffice to give us back X_0. Therefore, presenting X as a fibre product $X_0 \times_{k_0} k$ is the same thing as endowing X with a k_0-*structure* consisting of (1) the conjugation action $\{\sigma_X\}$, and (if necessary) (2) the subsheaf $\underline{o}_{X_0}$ of k_0-rational functions.

Note that if $x \in X$ is closed, then $p(x)$ is closed in X_0; and if $y \in X_0$ is closed, then $p^{-1}(y)$ is a finite set of conjugate closed points of X. It is worthwhile making this situation more precise:

Proposition 3. *Let $y \in X_0$ be a closed point. Then $\mathbb{k}(y)$ is a finite algebraic extension of k_0, and there is a natural bijection between the set of points $x \in p^{-1}(y)$, and the set of k_0-isomorphisms of $\mathbb{k}(y)$ into k.*

Proof. For all $x \in p^{-1}(u)$, we get maps:

$$\begin{array}{ccc} k & \xrightarrow{\ \sim\ } & \mathbb{k}(x) \\ \cup & & \uparrow p_x^* \\ k_0 & \longrightarrow & \mathbb{k}(y) \end{array}$$

hence we get a k_0-isomorphism of $\mathbb{k}(y)$ into k. Since $p^{-1}(y) \neq \emptyset$, one such exists and $\mathbb{k}(y)$ is algebraic over k_0. Also for any $y \in X_0$, $\mathbb{k}(y)$ is a finitely generated field extension of k_0, since X_0/k_0 is of finite type. Conversely, if we are given

$$\begin{array}{ccc} \mathbb{k}(y) & \xrightarrow{\ \Phi\ } & k \\ & \supset \quad \subset & \\ & k_o & \end{array}$$

define $\psi : \text{Spec }(k) \to X_0$ by Image$(\psi) = y$, and $\psi^* : \underline{o}_{y,Y} \to k$ given by ϕ. By the functorial meaning of fibre product, we get a morphism

$$(\phi, 1_{\text{Spec }(k)}) : \text{Spec }(k) \longrightarrow X_0 \times_{\text{Spec }(k_0)} \text{Spec }(k) = X \quad .$$

If x is the image point, this gives an inverse to the first procedure. Hence the set of X's, ϕ's are isomorphic. □

Definition 3. A closed point $y \in X_0$ is *rational over* k_0 if $\mathbb{k}(y) \cong k_0$.

Corollary. *If k_0 is perfect and $x \in X$, then $p(x)$ is rational over k_0 if and only if x is left fixed by all conjugations.*

Example L. The circle. Take $k = \mathbb{C}$, $k_0 = \mathbb{R}$,

$$X_0 = \text{Spec } \left(\mathbb{R}[X,Y]/\left(X^2 + Y^2 - 1\right)\right) \ .$$

Then X is a complex affine conic, with 2 points at infinity. Aside from its generic point, it looks like this:

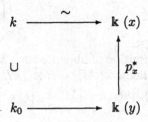

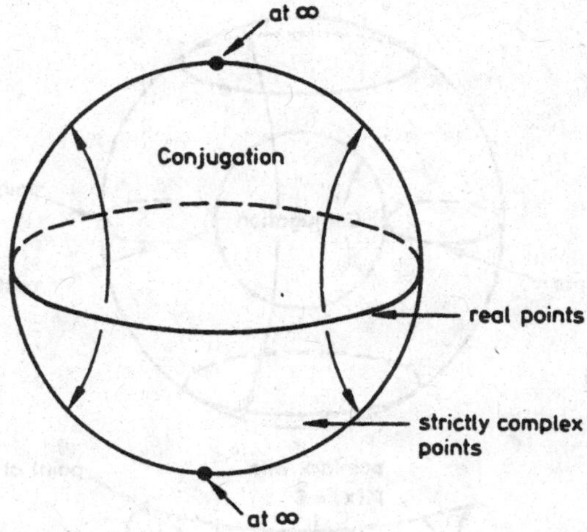

X_0, aside from its generic point, is the quotient space by conjugation:

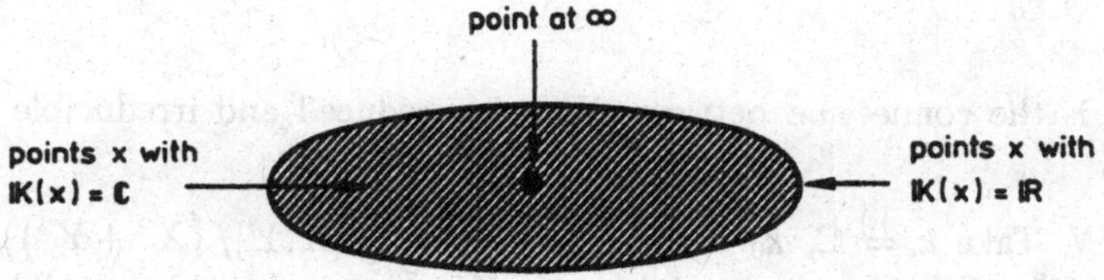

Here the boundary of the disc is the circle of points rational over $\mathbb{R}$ corresponding to the maximal ideals

$$(X - \alpha, Y - \beta)$$

where $\alpha, \beta \in \mathbb{R}$ and $\alpha^2 + \beta^2 = 1$, i.e., the real locus defined by $X^2 + Y^2 = 1$. The interior points (α, β), where $\alpha, \beta \in \mathbb{R}$ and $\alpha^2 + \beta^2 < 1$, correspond to the maximal ideals:

$$(X^2 + Y^2 - 1, \alpha X + \beta Y - 1)$$

and are not rational over $\mathbb{R}$.

(To prove this, note that over $\mathbb{C}$, every pair of conjugate complex roots of $X^2 + Y^2 = 1$ is the intersection of $X^2 + Y^2 = 1$ with a unique real line whose real points lie *outside* the circle.)

Example M. One further example of this type the details of which we leave to the reader: $k_0 = \mathbb{R}$, $k = \mathbb{C}$, X_0 the real plane curve $Y^2 = X(X^2 - 1)$.

X:

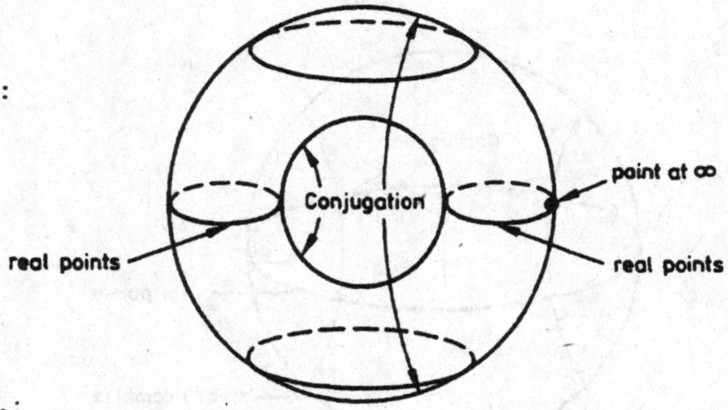

X_0:

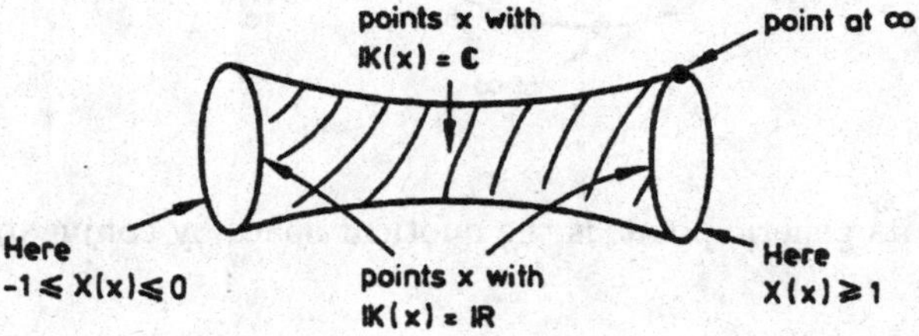

What is the connection between X_0 being reduced and irreducible and X being so?

Example N. Take $k = \mathbb{C}$, $k_0 = \mathbb{R}$, $X_0 = \text{Spec } \left(\mathbb{R}[X,Y]/\left(X^2 + Y^2\right)\right)$. Since $\mathbb{R}[X,Y]/\left(X^2 - Y^2\right)$ is an integral domain, X_0 is reduced and irreducible. However,

$$X = \text{Spec}\big(\mathbb{C}[X,Y]/(X+iY)\cdot(X-iY)\big)$$

is the union of 2 lines. Therefore X is reduced, but not irreducible. Note incidentally that X_0 has exactly one point rational/$\mathbb{R}$: namely the intersection of two lines.

Example O. Take k_0 imperfect of characteristic p, $a \in k_0 - k_0^p$. Let $\ell(X,Y)$ be a linear polynomial and let

$$\begin{aligned} X_0 &= \text{Spec}\,(k_0[X,Y]/\,(\ell(X,Y)^p - a)) \\ X &= \text{Spec}\left(k[X,Y]/\left(\ell(X,Y) - a^{1/p}\right)^p\right) . \end{aligned}$$

X_0 is reduced and irreducible, but X is just a "p-fold line". Set-theoretically, X is the line $\ell = a^{1/p}$, but $\ell - a^{1/p}$ is a non-zero nilpotent function on X.

Proposition 4. *Let X_0 be a reduced and irreducible prescheme over k_0. Let k be an algebraically closed extension of k_0. Let $x \in X_0$ be its generic point and let $k_0(X_0)$ be the stalk of $\underline{o}_{X_0}$ at x. Then*

i) $X_0 \times_{k_0} k$ *is reduced* $\Longleftrightarrow$ $k_0(X_0)$ *is separable over* k_0

ii) $X_0 \times_{k_0} k$ *is irreducible* $\iff k_0$ *is separably algebraically closed in* $k_0(X_0)$.

Proof. We may as well assume $X_0 = \text{Spec}\,(R)$, R an integral domain over k_0 with quotient field $k_0(X_0)$. For any extension $k \supset k_0$, notice:

$$\left.\begin{array}{l} k_0(X_0) \otimes_{k_0} k \text{ satisfies } (0) = \sqrt{(0)} \\ [\text{resp. } \sqrt{(0)} \text{ prime}] \end{array}\right\} \iff \left\{\begin{array}{l} R \otimes_{k_0} k \text{ satisfies } (0) = \sqrt{(0)} \\ [\text{ resp. } \sqrt{(0)} \text{ prime}] \end{array}\right.$$

$$\iff \left\{\begin{array}{l} X_0 \times_{k_0} k \text{ reduced} \\ [\text{ resp. irreducible}] \end{array}\right. .$$

In fact, $R \otimes_{k_0} k \subset k_0(X_0) \otimes_{k_0} k$, and the latter is the localization of $R \otimes_{k_0} k$ with respect to the multiplicative system of non 0-divisors $a \otimes 1$, $a \in R$, $a \neq 0$. This gives the 1st "$\iff$" easily.

Then (i) follows from the assertion:

Given $L \supset k_0$, $k \supset k_0$, k algebraically closed, then
$L \otimes_{k_0} k$ has no nilpotents $\iff L$ separable$/k_0$.

(Cf. Zariski-Samuel, vol. I, Ch. 3, Th. 39; Bourbaki, *Modules et Anneaux Semi-simples*, §7.3.) Moreover (ii) follows from:

Given $L \supset k_0$, $k \supset k_0 \cdot k$ algebraically closed, then
$L \otimes_{k_0} k/\sqrt{(0)}$ is a domain $\iff k_0$ separably algebraically closed in L

(Cf. Zariski-Samuel, vol. I, Ch. 3, Th. 38 + 40; Bourbaki, *Modules et Anneaux Semi-simples*, §7.3.) □

Definition 4. Let X_0 be a prescheme of finite type over k_0. Then X_0 is a *prevariety over* k_0 if it is reduced and irreducible and if a) its function field $k_0(X_0)$ is separable over k_0, and b) k_0 is algebraically closed in $k_0(X_0)$. (Compare Lang, Ch. 3, §§1–2.)

§5. Closed subpreschemes

We begin with a digression on some general facts about sheaves. Let $\mathcal{F}, \mathcal{G}$ be sheaves of abelian groups on the topological space X, $\varphi : \mathcal{F} \to \mathcal{G}$ a morphism. For U open, we let $\mathcal{K}(U) = \ker(\mathcal{F}(U) \to \mathcal{G}(U))$; it is easy to check that $\mathcal{K}$ is a sheaf, which we call the *kernel* of φ. We say φ is *injective* of $\mathcal{K} = 0$, i.e., if $\mathcal{F}(U) \to \mathcal{G}(U)$ is injective for all U. It is easily seen that this holds if and only if all the maps $\varphi_x : \mathcal{F}_x \to \mathcal{G}_x$ on stalks are injective.

The cokernel is a little tricky to define. The presheaf $\mathcal{L}^*$ given by $\mathcal{L}^*(U) = \mathcal{G}(U)/\varphi(\mathcal{F}(U))$ need not be a sheaf; we let $\mathcal{L}$ be the associated sheaf, and call it the *cokernel* of φ. Recall that by definition of $\mathcal{L}$, there is a presheaf map $\mathcal{L}^* \to \mathcal{L}$ inducing isomorphisms on all stalks. Hence we have $\mathcal{L}_x = \mathcal{G}_x/\varphi_*(\mathcal{F}_x)$ for all x. We say then that φ is *surjective* if $\mathcal{L} = 0$, or equivalently if $\varphi_x : \mathcal{F}_x \to \mathcal{G}_x$ is surjective for all x. This *does not imply* that φ_U is surjective for all U.

One can check that these agree with the usual categorical definitions, and that the category of sheaves of abelian groups on X is an abelian category. In particular, this includes the assertion:

Proposition 1. *If* $\varphi : \mathcal{F} \to \mathcal{G}$ *is injective and surjective, then it is an isomorphism.*

Proof. As φ is injective, we have $\mathcal{F}(U) \to \mathcal{G}(U)$ injective for all U, and we simply must show these maps are onto. Let $s \in \mathcal{G}(U)$. For all $x \in U$, let s_x be the image of s in $\mathcal{G}_x$. φ_x is by hypothesis an isomorphism, so there is a unique $t_x \in \mathcal{F}_x$ such that $\varphi_x(t_x) = s_x$. There is an open neighbourhood $U_x \subset U$ of x such that t_x is the image in $\mathcal{F}_x$ of an element $t^x \in \mathcal{F}(U_x)$. Then s and $\varphi(t^x)$ induce the same element in $\mathcal{G}_x$, so they agree locally; replacing U_x by a smaller neighbourhood, we may assume $\text{res}_{U,U_x} s = \varphi(t^x)$ in $\mathcal{G}(U_x)$. Now the U_x are a covering of U. On $U_{x_1} \cap U_{x_2}$, $\varphi(t^{x_1} - t^{x_2}) = s - s = 0$ in $\mathcal{G}(U_{x_1} \cap U_{x_2})$; since φ is injective, $t^{x_1} = t^{x_2}$ on $U_{x_1} \cap U_{x_2}$. Hence the t^x patch together to a $t \in \mathcal{F}(U)$. Then $\varphi(t)$ and s agree on each U_x, so $\varphi(t) = s$. □

Suppose $Y \subset X$ is a closed subset. Then the sheaves of abelian groups on Y correspond bijectively to the sheaves of abelian groups on X such that $\mathcal{F}_x = \{0\}$ if $x \notin Y$. We actually have canonical functors in both directions: If $\mathcal{F}_0$ is a sheaf on Y, we *extend it by zero* by defining, for $U \subset X$ open,

$$\mathcal{F}(U) = \begin{cases} \{0\}, & U \cap Y = \emptyset \\ \mathcal{F}_0(U \cap Y), & U \cap Y \neq \emptyset \end{cases} .$$

The inverse to this is given by restricting a sheaf on X to one on Y[6]. Since we are dealing with sheaves $\mathcal{F}$ such that $\mathcal{F}_x = (0)$, $x \notin Y$, it is easy to check that the restriction $\mathcal{F}'$ of $\mathcal{F}$ to Y satisfies $\Gamma(U, \mathcal{F}) = \Gamma(U \cap Y, \mathcal{F}')$, all open U in X, hence it is an inverse to the process of extending by 0.

Now let X be a prescheme; we want to define the concept of a closed subprescheme. Let Y be a closed subset of the underlying space of X. Since we may have nilpotent sections in $\underline{o}_X$ that are not determined by their values on points, we cannot "restrict" sections of $\underline{o}_X$ to Y; we must specify the sheaf $\underline{o}_Y$ explicitly. If $\underline{o}_Y$ is a sheaf on Y, we can extend it by zero; then we want there to be given a map $\underline{o}_X \to \underline{o}_Y$; also this map should be surjective since an allowable function on the subobject should extend *locally* to the ambient space. [Again beware – $\underline{o}_X(U) \to \underline{o}_Y(U)$ need not be surjective for all U; cf. Ch. I, §4]. Thus we want:

Definition 1. A *closed prescheme* of X is

[6] Recall that if Y is any subspace of X and $\mathcal{F}$ is a sheaf on X, then the *restriction* $\mathcal{F}'$ of $\mathcal{F}$ to Y is defined by:
For all U open in Y,

$$\Gamma(U, \mathcal{F}') = \left\{ s \in \prod_{x \in U} \mathcal{F}_x \;\middle|\; \begin{array}{l} \text{For all } x \in U, \exists \text{ a neighbourhood } V_x \text{ of } x \text{ in } X \\ \text{and } t \in \Gamma(V_x, \mathcal{F}) \text{ such that } t \text{ and } s \text{ give the} \\ \text{same element of } \mathcal{F}_y, \text{ all } y \in V_x \cap U. \end{array} \right\}$$

1) a closed set $Y \subset X$,
2) a sheaf of rings $\underline{o}_Y$ on Y such that $(Y, \underline{o}_Y)$ is a prescheme,
3) a surjective homomorphism π from $\underline{o}_X$ to $\underline{o}_Y$ (extended by zero).

Let $\mathcal{Q} = \ker \pi$, so $\mathcal{Q}$ is a sheaf of ideals in $\underline{o}_X$, i.e., $\mathcal{Q}$ is a subsheaf of $\underline{o}_X$ such that for all U, $\mathcal{Q}(U)$ is an ideal in $\underline{o}_X(U)$: we shall call such an object an $\underline{o}_X$-ideal, for short. We claim that $\mathcal{Q}$ determines the closed subpreschemes up to canonical isomorphism. For first of all, $Y = \{x \in X \mid \mathcal{Q}_x \neq \underline{o}_x\}$, as the sequence $0 \to \mathcal{Q}_x \to \underline{o}_{x,X} \to \underline{o}_{x,Y} \to 0$ is exact; and secondly $\underline{o}_Y$ extended by zero is canonically isomorphic to the cokernel of $\mathcal{Q} \to \underline{o}_X$. Thus a closed subprescheme is really just an $\underline{o}_X$-ideal. [But not all $\underline{o}_X$-ideals can occur; we will return to that later.]

Definition 2. A morphism of preschemes $f : Y \to X$ is a closed immersion if

(1) f is injective,
(2) f is closed, i.e., $Z \subset Y$ closed $\Longrightarrow f(Z)$ closed,
(3) $f_y^* : \underline{o}_{f(y)} \to \underline{o}_y$ is surjective for all $y \in Y$.

Note that this is equivalent to saying that f factors via (1) an isomorphism of Y with a closed subprescheme of X, followed by (2) the canonical injection of the closed subprescheme into X. The canonical example of a closed subprescheme is this:

Proposition 2. *Let R be a ring, $A \subset R$ an ideal. The canonical map $\pi : R \to R/A$ defines a morphism f : Spec $(R/A) \to$ Spec (R). Then f is a closed immersion and the corresponding sheaf of ideals $\mathcal{Q}$ satisfies:*

i) $\Gamma(\text{Spec }(R)_f, \mathcal{Q}) = A \cdot R_f$
ii) $\mathcal{Q}_x = A \cdot \underline{o}_{x,\text{Spec }(R)}$.

Proof. By definition, if $[P] \in$ Spec (R/A), $f([P]) = [\pi^{-1}P]$. But $P \to \pi^{-1}(P)$ is a bijection between the primes of R/A and the primes of R containing A, so f is an injection with image $V(A)$. More generally, if $B \supset A$ is any ideal and $\overline{B} = B/A$, then $f\left(V(\overline{B})\right) = V(B)$. Therefore f is closed.

Suppose $x = [\overline{P}]$, $f(x) = [P]$. We get the diagram:

$$\begin{array}{c} \underline{o}_{f(x),\text{Spec }(R)} = R_P \\ \Big\downarrow f_x^* \\ \underline{o}_{x,\text{Spec }(R/A)} = (R/A)_{\overline{P}} = R_P/A \cdot R_P \quad . \end{array}$$

Thus f_x^* is the localization of $\pi : R \to R/A$, hence it is surjective. Therefore f is a closed immersion. Also, it follows that the stalks of the kernel $\mathcal{Q}$ are $A \cdot R_P$. Finally we have

$$
\begin{aligned}
&\Gamma\left((\mathrm{Spec}\ R)_f, \mathcal{Q}\right)\\
&\quad = \ker\left\{\Gamma\left((\mathrm{Spec}\ R)_f, \underline{o}_{\mathrm{Spec}\ (R)}\right) \longrightarrow \left((\mathrm{Spec}\ R)_f, \underline{o}_{\mathrm{Spec}\ (R/A)}\right)\right\}\\
&\quad = \ker\left\{R_f \longrightarrow (R/A)_{\pi(f)}\right\}\\
&\quad = A \cdot R_f\ .
\end{aligned}
$$

□

We will now prove that the converse of this is true too! Thus, while the Nullstellensatz gave us a correspondence between those ideals A such that $A = \sqrt{A}$ and the closed subsets of affine varieties, we now get a correspondence between *all* ideals A and closed subpreschemes.

Theorem 3. *Let* $X = \mathrm{Spec}\ (R)$, *and let* $Y \subset X$ *be a closed subprescheme,* $f : Y \to X$ *the inclusion. Let* $\mathcal{Q}$ *be the* $\underline{o}_X$*-ideal defining* Y, $A = \Gamma(X, \mathcal{Q})$. *Then* Y *is canonically isomorphic to* Spec (R/A), *i.e., there is a commutative diagram:*

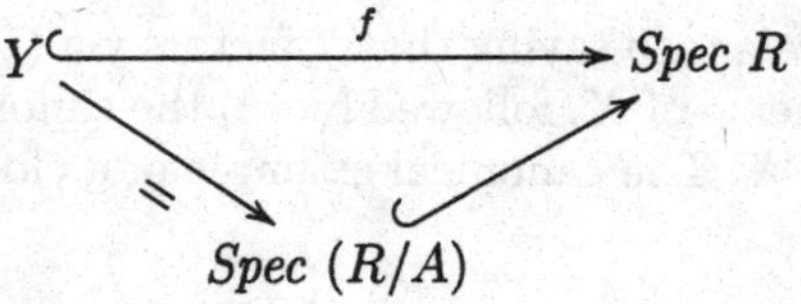

Proof. First of all, note that f factors through Spec (R/A). For, $f^* : R \to \Gamma(Y, \underline{o}_Y)$ factors through R/A by definition of A; hence f factors through Spec (R/A), by Th. 1, §2. So we may assume $A = (0)$, i.e., $f^* : R \to \Gamma(Y, \underline{o}_Y)$ is injective.

Y is quasi-compact, since it is a closed subspace of Spec (R). Hence it is covered by a finite number of open affines, say Spec (S_i).

We now claim f is surjective. Since $f(Y)$ is closed in Spec (R), $f(Y) = V(B)$ for some ideal B in R. If $s \in B$, $s(x) = 0$ for all $x \in V(B)$, so $f^*s = 0$ at all points of Y. In particular, $\mathrm{res}_{Y,\mathrm{Spec}\ (S_i)}(f^*s)$ is given by an element $\sigma_i \in S_i$ which is 0 at every point of Spec (S_i), and hence is nilpotent: suppose $\sigma_i^{n_i} = 0$. There are only finitely many S_i, so $\sigma_i^n = 0$ for all i if n is big enough. Then $(f^*s)^n$ is an element which restricts to 0 on each open piece Spec (S_i), so it is 0. As f^* is injective, $s^n = 0$. Thus $B \subset \sqrt{(0)}$, hence $V(B) = B((0)) = \mathrm{Spec}\ (R)$, and f is indeed surjective.

It follows that f is a homeomorphism, since by definition it is closed and injective. That is, we can suppose that we have a single topological space and two sheaves of rings $\underline{o}_X$, $\underline{o}_Y$ on it, together with a surjective homomorphism $\underline{o}_X \to \underline{o}_Y$; we must show that it is injective.

Say $y \in Y$, and $f(y) = [P]$; we have a map $\underline{o}_{f(y),X} = R_P \to \underline{o}_{y,Y}$ which we must show injective. Suppose not: then there is an element $a \in R$ in the kernel. We want to show $a = 0$ in R_P, i.e., $ab = 0$ for some $b \in R - P$.

We know f^*a goes to 0 in $\underline{o}_{y,Y}$. Set $U = \{y' \in Y \mid f^*a \text{ goes to } 0 \text{ in } \underline{o}_{y',Y}\}$; U is open, since an element goes to 0 in a stalk if and only if it is 0 in a neighbourhood. As f is a homeomorphism, $f(U)$ is an open neighbourhood of $f(y) = [P]$. Hence $f(U)$ contains a set $(\text{Spec } R)_t$ for some $t \in R - P$, i.e., f^*a is 0 in the open set $Y_{f^*(t)} = \{y' \mid f^*t(y') \neq 0\}$.

Let $\sigma_i = \text{res}_{Y,\text{Spec}(S_i)}(f^*t)$. We have then $Y_{f^*t} \cap \text{Spec}(S_i) = (\text{Spec}(S_i))_{\sigma_i}$. Now $\Gamma\left((\text{Spec } S_i)_{\sigma_i}, \underline{o}_Y\right) = (S_i)_{\sigma_i}$, and f^*a is 0 in $(S_i)_{\sigma_i}$. If $\alpha_i \in S_i$ gives f^*a on Spec S_i, this tells us $\alpha_i \cdot \sigma_i^{n_i} = 0$ in S_i for some n_i. Hence, as there are only finitely many S_i's, $\alpha_i \sigma_i^n = 0$ for all i and big enough n. Thus $(f^*a) \cdot (f^*t)^n$ restricted to each Spec (S_i) is 0, so $f^*(at^n) = 0$ in $\Gamma(Y, \underline{o}_Y)$. Since $f^* : R \to \Gamma(Y, \underline{o}_Y)$ is objective, $at^n = 0$ in R. As $t \notin P$, a does go to 0 in R_P. □

Corollary 1. *Let $f : Y \to X$ be a morphism of preschemes. Then the following are equivalent:*

(i) f is a closed immersion,
(ii) for all affine open sets $U \subset X$, $f^{-1}(U)$ is affine and the map

$$\Gamma(U, \underline{o}_X) \longrightarrow \Gamma\left(f^{-1}(U), \underline{o}_Y\right)$$

is surjective,
(iii) there exists an affine open covering U_i of X such that $f^{-1}(U_i)$ is affine and the map

$$\Gamma(U_i, \underline{o}_X) \longrightarrow \Gamma\left(f^{-1}(U_i), \underline{o}_Y\right)$$

is surjective for all i.

Proof. (i) $\Longrightarrow$ (ii) by the theorem; (ii) $\Longrightarrow$ (iii) obviously; (iii) $\Longrightarrow$ (i) by Prop. 2. □

Corollary 2. *Let X be a prescheme and let $\mathcal{Q}$ be an $\underline{o}_X$-ideal. Let $Y = \{y \in X \mid 1 \notin \mathcal{Q}_y\}$ – a closed subset of X. The cokernel of $O \to \mathcal{Q} \to \underline{o}_X$ is O outside of X, so consider it as a sheaf of rings $\underline{o}_Y$ on Y. Then $(Y, \underline{o}_Y)$ is a subprescheme of $(X, \underline{o}_X)$ if and only if*

(*) *For all $y \in X$, there is a neighbourhood U of y and sections $\{s_\alpha\}$ in $\Gamma(U, \mathcal{Q})$ such that*

$$\mathcal{Q}_x = \sum \text{res}(s_\alpha) \cdot \underline{o}_{x,X}$$

for all $x \in U$.

Proof. If $(Y, \underline{o}_Y)$ is a subprescheme, and $y \in X$, then by the theorem (*) holds if you take U to be any *affine* open neighbourhood of y, and s_α to be generators of $\Gamma(U, \mathcal{Q})$ over $\Gamma(U, \underline{o}_X)$. Conversely, if (*) holds, and U is taken to be affine, say $U = \text{Spec}(R)$, then

$$(U \cap Y, \underline{o}_Y|_U) \cong \text{Spec}(R/\Sigma s_\alpha \cdot R) .$$

□

(*) is the condition for $\mathcal{Q}$ to be *quasi-coherent*. We shall investigate this fully in Ch. III, §1.

Example I bis. Closed subschemes of Spec $(k[t])$, k algebraically closed. Since $k[t]$ is a PID, all non-zero ideals are of the form

$$A = \left(\prod_{i=1}^{n} (t - a_i)^{r_i} \right) .$$

The corresponding subscheme Y of $A^1 = \text{Spec}\,(k[t])$ is supported by the n points $a_1, \ldots, a_n$, and at a_i its structure sheaf is

$$\underline{o}_{a_i,y} = \underline{o}_{a_i,\mathbb{A}^1} \Big/ m_i^{r_i} \quad ,$$

where $m_i = m_{a_i,\mathbb{A}^1} = (t - a_i)$. Y is the union of the a_i's "with multiplicity r_i". The real significance of the multiplicity is that if you restrict a function f on $\mathbb{A}^1$ to this subscheme, the restriction can tell you not only the value $f(a_i)$ but the first $r_i - 1$-derivatives:

$$\frac{d^i f}{dt^i}(a_i) , \qquad i \leq r_i - 1 \ .$$

In other words, Y contains the $(r_i - 1)^{\text{st}}$ order normal neighbourhood of $\{a_i\}$ in $\mathbb{A}^1$.

Consider all possible subschemes supported by $\{0\}$. These are the schemes

$$Y_n = \text{Spec}\ \left(k[t]/\,(t^n)\right) \quad .$$

Y_1 is just the point as a reduced scheme, but the rest are not reduced. Corresponding to the fact that the defining ideals are included in each other:

$$(t) \supset (t^2) \supset (t^3) \supset \ldots \supset (t^n) \supset \ldots \supset (0) \ ,$$

the various schemes are subschemes of each other:

$$Y_1 \subset Y_2 \subset Y_3 \subset \ldots \subset Y_n \subset \ldots \subset \mathbb{A}^1 \ .$$

Example P. Closed subschemes of Spec $(k[x,y])$, k algebraically closed. Every ideal $A \subset k[x,y]$ is of the form:

$$(f) \cap Q$$

for some $f \in k[x,y]$ and Q of finite codimension (to check this use noetherian decomposition and the fact that prime ideals are either maximal or principal). Let $Y = \text{Spec}\,([x,y]/A)$ be the corresponding subscheme of $\mathbb{A}^2$. First, suppose $A = (f)$. If $f = \prod_{i=1}^{n} f_i^{r_i}$, with f_i irreducible, then the subscheme Y is the union of the irreducible curves $f_i = 0$, "with multiplicity r_i". As before, if g is a function in $\mathbb{A}^2$, then one can compute solely from the restriction of g to Y the first $r_i - 1$ normal derivatives of g to the curve $f_i = 0$. Second, look at the case A of finite codimension. Then

$$A = Q_1 \cap \ldots \cap Q_t$$

where $\sqrt{Q_i}$ is the maximal ideal $(x - a_i, y - b_i)$. Therefore, the support of Y is the finite set of points (a_i, b_i), and the stalk of Y at (a_i, b_i) is the finite dimensional algebra $k[x,y]/Q_i$. For simplicity, look at the case $A = Q_1$, $\sqrt{Q_1} = (x, y)$. The lattice of such ideals A is much more complicated than the one-dimensional case. Consider, for example, the ideals:

$$(x,y) \supset (\alpha x + \beta y, x^2, xy, y^2) \supset (x^2, xy, y^2) \supset (x^2, y^2) \supset (0) .$$

These define subschemes:

$$\left\{ \begin{array}{l} (0,0) \text{ with} \\ \text{reduced structure} \end{array} \right\} \subset Y_{\alpha,\beta} \subset Y_2 \subset Y_3 \subset \mathbb{A}^2 .$$

Since $(\alpha x + \beta y, x^2, xy, y^2) \supset (\alpha x + \beta y)$, $Y_{\alpha,\beta}$ is a subscheme of the reduced line $\ell_{\alpha,\beta}$ defined by $\alpha x + \beta y = 0$: $Y_{\alpha,\beta}$ is the point and *one* normal direction. But Y_2 is not a subscheme of any reduced line: it is the full double point and is invariant under rotations. Y_3 is even bigger, is *not* invariant under rotations, but still does not contain the 2nd order neighbourhood of $(0,0)$ along any line. If g is a function on $\mathbb{A}^2$, $g|_{Y_{\alpha,\beta}}$ determines *one* directional derivative of g at $(0,0)$, $g|_{Y_2}$ determines both partial derivatives of g at $(0,0)$ and $g|_{Y_3}$ even determines the mixed partial $\frac{\partial^2 g}{\partial x \partial y}(0,0)$.

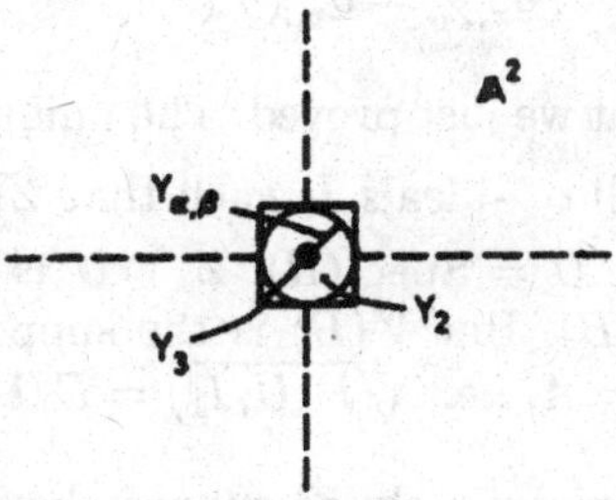

As an example of the general case, look at $A = (x^2, xy)$. Then $A = (x) \cap (x^2, xy, y^2)$. Since $\sqrt{A} = (x)$, the support of Y is the y-axis. The stalk $\underline{o}_{z,y}$ that has no nilpotents in it except when $z = (0,0)$. This is an "embedded point", and if a function g on $\mathbb{A}^2$ is cut down to Y, the restriction determined both partials of g at $(0,0)$, but only $\frac{\partial}{\partial y}$ at other points :

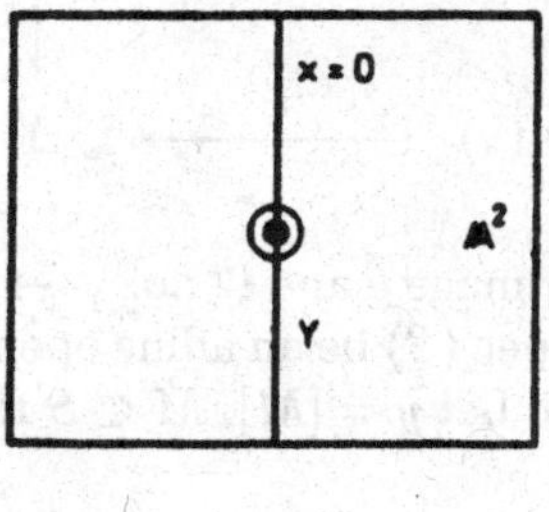

Proposition 4. *Let X be a prescheme and $Z \subset X$ a closed subset. Consider the various closed subschemes $Z_1 = (Z, \underline{o}_X/I_1)$ that can be defined with support Z. Among all these, there is a unique reduced subscheme – call it $Z_0 = (Z, \underline{o}_X/I_0)$. Furthermore, if $Z_1 = (Z, \underline{o}_X/I_1)$ is any other, then $I_1 \subset I_0$ (so Z_0 is a subscheme of Z_1) and $I_0 = \sqrt{I_1}$ (i.e., $\Gamma(U, I_0) = \sqrt{\Gamma(U, I_1)}$ for U affine).*

Proof. Define I_0 by

$$\Gamma(U, I_0) = \{s \in \Gamma(U, \underline{o}_X) \mid s(x) = 0 \text{ for all } x \in U \cap Z\} \ .$$

Now suppose $U = \text{Spec}(R)$, and $Z \cap U = V(A)$ where $A \subset R$ is an ideal such that $A = \sqrt{A}$. Then I claim $I_{0,x} = A \cdot \underline{o}_x$, all $x \in U$, hence I_0 is quasi-coherent and defines a closed subscheme Z_0. First of all, $A \cdot \underline{o}_x \subset I_{0,x}$ since all elements of A vanish on $Z \cap U$. Conversely, suppose $s \in (I_0)_x$; then s is the restriction of some $t \in \Gamma(U_f, I_0)$, where $x \in U_f$, since the U_f are a basis in U. But inside $U_f = \text{Spec}(R_f)$,

$$Z \cap U_f = V(A \cdot R_f)$$

and moreover $A \cdot R_f = \sqrt{A \cdot R_f}$. And $t(x) = 0$, all $x \in Z \cap U_f$ means that t is in all prime ideals P of R_f containing $A \cdot R_f$, hence $t \in A \cdot R_f$. Therefore $s \in A \cdot \underline{o}_x$.

Z_0 is reduced since for all $x \in Z$,

$$\underline{o}_{x,Z_0} = \underline{o}_{x,X} / (I_0)_x$$

and $(I_0)_x = \sqrt{(I_0)_x}$ by what we just proved. The uniqueness assertion will follow if we prove $I_0 = \sqrt{I_1}$ for all $\underline{o}_X$-ideals I_1 such that $Z_1 = (Z, \underline{o}_X/I_1)$ is a closed subscheme. Again, suppose $U = \text{Spec}(R)$. $Z_1 \cap U$ is a closed subscheme of U, hence $Z_1 \cap U = \text{Spec}(R/B)$. But $V(B)$ is the support of $\text{Spec}(R/B)$, hence $V(B) = V(A)$, hence $\sqrt{B} = A$, i.e., $\sqrt{\Gamma(U, I_1)} = \Gamma(U, I_0)$. □

A very important example of a closed subprescheme are the *fibres of a morphism.* Let $X \xrightarrow{f} Y$ be a morphism and let $y \in Y$ be a closed point. We shall put a subscheme structure on the fibre $f^{-1}(y)$ over y. Let $k = \mathbf{k}(y)$. Consider the fibre product X_y in the diagram:

$$\begin{array}{ccc} X_y & \overset{j}{\dashrightarrow} & X \\ {\scriptstyle g}\,\vdots & & \downarrow{\scriptstyle f} \\ \text{Spec}(k) & \xrightarrow{\ i\ } & Y \end{array}$$

where i is the morphism with image y and $i^* : \underline{o}_{y,Y} \to k$ the canonical map of $\underline{o}_{y,Y}$ to its residue field. Let $V = \text{Spec}(S)$ be an affine open neighbourhood of y and let $U_i = \text{Spec}(R_i)$ cover $f^{-1}(V)$. Let $y = [M]$, $M \subset S$ maximal. Then by Theorems

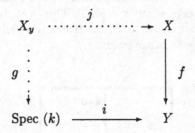

3a, b, §2, X_y is covered by open affine pieces $X_y^{(i)} = \text{Spec } (R_i \otimes_S k)$. But $k \cong S/M$, hence $X_y^{(i)} \cong \text{Spec } (R_i/M \cdot R_i)$. Therefore j is a *closed immersion.* In other words, j is an isomorphism of X_y with a closed subscheme of X with support $f^{-1}(y)$ defined by the sheaf of ideals I, where

$$\Gamma(U, I) = M \cdot \Gamma(U, \underline{o}_X)$$

(for all affine open $U \subset f^{-1}(V)$).

I mentioned at the beginning of this chapter that one of the reasons why it is necessary to include schemes with nilpotents in our set-up is that they occur as fibres of morphisms between very nice varieties. To see this, let

$$X = \text{Spec } k[X_1, \ldots X_n]/(f_1, \ldots, f_m)$$

be an arbitrary affine scheme of finite type over k. Define

$$\phi : \mathbb{A}_k^n \longrightarrow \mathbb{A}_k^m$$

by the homomorphism:

$$\begin{array}{ccc} k[X_1, \ldots, X_n] & \xleftarrow{\phi^*} & k[Y_1, \ldots, Y_m] \\ f_i & \longleftarrow\!\!\!\mid & Y_i \, . \end{array}$$

Then X is exactly the fibre of ϕ over the origin!

When the residue field $\Bbbk(y)$ of a point $y \in Y$ is not algebraically closed, it is sometimes more important to embed $\Bbbk(y)$ in an algebraically closed field k, and to take the fibre product as above, but via the morphism $i : \text{Spec } (k) \to Y$, defined by Image $(i) = y$, and $i^* : \underline{o}_{y,Y} \to \Bbbk(y) \to k$. The prescheme X_y obtained in this way is called a *geometric fibre* of f.

Problem. Show that the closed subpreschemes of $\mathbb{P}_k^n$ correspond bijectively with the homogeneous ideals

$$A \subset k[X_0, \ldots, X_n]$$

such that

$$A : (X_0, \ldots, X_n) = A$$

(i.e., if $f \in k[X_0, \ldots, X_n]$, $X_i \cdot f \in A$ for all i, then $f \in A$. This condition says that A is biggest among those ideals that induce the same ideals in each affine piece of $\mathbb{P}^n$.)

§6. The functor of points of a prescheme

We have had several indications that the underlying point set of a scheme is peculiar from a geometric point of view. Non-closed points are odd for one thing. A serious difficulty is that the point set of a fibre product $X \times_S Y$ does not map injectively into the set-theoretic product of X and Y. For example:

$$\begin{aligned} \text{Spec}\,(\mathbb{C}) \times_{\text{Spec}\,(\mathbb{R})} \text{Spec}\,(\mathbb{C}) &\cong \text{Spec}\,(\mathbb{C} \times_{\mathbb{R}} \mathbb{C}) \\ &\cong \text{Spec}\,(\mathbb{C}) \amalg \text{Spec}\,(\mathbb{C}) \end{aligned}$$

($\amalg$ denotes open disjoint union). As we will see, this prevents the point set of a group variety from being an abstract group! The explanation of these confusing facts is that there are really *two* concepts of "point" in the language of preschemes. To see this in its proper setting, look at some examples in other categories:

Example Q. Let C = category of differentiable manifolds. Let z be the manifold with *one* point. Then for any manifold Z,

$$\text{hom}_C(z, X) \cong X \quad \text{as a point set}\ .$$

Example R. Let C = category of groups. Let $z = \mathbb{Z}$. Then for any group G,

$$\text{hom}_C(Z, G) \cong H \quad \text{as a point set}\ .$$

Example S. Let C = category of rings with 1 (and homomorphisms f such that $f(1) = 1$). Let $z = \mathbb{Z}[X]$. Then for any ring R,

$$\text{hom}_C(z, R) \cong R \quad \text{as a point set}\ .$$

This indicates that if C is any category, whose objects may not be point sets to begin with, and z is an object, one can try to conceive of $\text{hom}_C(z, X)$ as the underlying set of points of the object X. In fact,

$$X \longmapsto \text{hom}_C(z, X)$$

extends to a functor from the category C to the category (Sets), of sets. But it is not satisfactory to call $\text{hom}_C(z, X)$ the set of points of X unless this functor is *faithful*, i.e., unless a morphism f from X_1 to X_2 is determined by the map of sets:

$$\widetilde{f} : \text{hom}_C(z, X_1) \longrightarrow \text{hom}_C(z, X_2)\ .$$

Example T. Let (Hot) be the category of CW-complexes, where $\text{hom}(X, Y)$ is the set of homotopy-classes of continuous maps from X to Y. If z = the 1 point complex, then

$$\text{hom}_{(\text{Hot})}(z, X) = \pi_0(X), \qquad \text{(the set of components of } X)$$

and this does *not* give a faithful functor.

Example U. Let C = category of pre-schemes. Taking the lead from Examples 1 and 4, take for z the *final* object of the category $C : z = \text{Spec}\ (\mathbb{Z})$. Now

$$\text{hom}_C(\text{Spec}\ (\mathbb{Z}), X)$$

is absurdly small, and does not give a faithful functor.

Grothendieck's ingenious idea is to remedy this defect by considering (for arbitrary categories C) not *one* z, but *all* z: attach to X the whole set:

$$\bigcup_z \text{hom}_C(z, X) .$$

In a natural way, this always gives a faithful functor from the category C to the category (Sets). Even more than that, the "extra structure" on the set $\bigcup_z \text{hom}_C(z, X)$ which characterizes the object X, can be determined. It consists in:

i) the decomposition of $\bigcup_z \text{hom}_C(z, X)$ into subsets $S_z = \text{hom}_C(z, X)$, one for each z,
ii) the natural maps from one set S_z to another $S_{z'}$, given for each morphism $g : z' \to z$ in the category.

Putting this formally, it comes out like this:

Attach to each X in C, the *functor* h_X (contravariant, from C itself to (Sets)) via

$$(*) \quad h_X(z) = \text{hom}_C(z, X), \quad z \text{ an object in } C .$$

$$(**) \quad h_X(g) = \left\{ \begin{array}{l} \text{induced map from } \text{hom}_C(z, X) \\ \text{to } \text{hom}_X(z', X) \end{array} \right\} , \quad \begin{array}{l} g : z' \to z \\ \text{a morphism in } C \end{array} .$$

Now the functor h_X is an object in a category too: viz.,

$$\mathit{Funct}\ (C^0, (\text{Sets})) ,$$

(where *Funct* stands for functors, C^0 for C with arrows reversed). It is also clear that if $g : X_1 \to X_2$ is a morphism in C, then one obtains a morphism of functors $h_g : h_{X_1} \to h_{X_2}$. All this amounts to one big functor:

$$h : C \longrightarrow \mathit{Funct}\ (C^0, (\text{Sets})) .$$

Proposition 1. *h is fully faithful, i.e., if X_1, X_2 are objects of C, then, under h,*

$$\text{hom}_C\,(X_1, X_2) \xrightarrow{\sim} \text{hom}_{\mathit{Funct}}\,(h_{X_1}, h_{X_2}) .$$

Proof. Easy.

The conclusion, heuristically, is that an object X of C can be *identified* with the functor h_X, which is basically just a structured set.

Return to algebraic geometry! What we have said motivates I hope:

Definition 1. If X and K are preschemes, a *K-valued point of X* is a morphism $f: K \to X$; if $K = \text{Spec}(R)$, we call this an *R-valued point of X*. If X and K are preschemes over a third prescheme S, i.e., we are given morphisms $p_X : X \to S$, $p_K : K \to S$, then f is a *K-valued point of X/S* if $p_X \cdot f = p_K$; if $K = \text{Spec}(R)$, we call this an *R-valued point of X/S*. The set of all R-valued points of a prescheme X, or of X/S, is denoted $X(R)$.

"Points" in this sense are compatible with products. That is to say, if K, X, and Y are preschemes over S, then the set of K-valued points of $X \times_S Y/S$ is just the (set-theoretic) product of the set of K-valued points of X/S and the set of K-valued points of Y/S. This is just the definition of the fibre product. In particular K-valued points of a group prescheme X/K – to be defined in Ch. IV – will actually form a group!

The concept of an R-valued point generalizes the notion of a solution of a set of diophantine equations in the ring R. In fact, let:

$$\left\| \begin{array}{l} f_1, \ldots, f_m \in \mathbb{Z}[X_1, \ldots, X_n] \\ X = \text{Spec}\ (\mathbb{Z}[X_1, \ldots, X_n] / (f_1, \ldots, f_M)) \end{array} \right. .$$

I claim an R-valued point of X is the "same thing" as an n-tuple $a_1, \ldots, a_n \in R$ such that

$$f_1(a_1, \ldots, a_n) = \ldots = f_m(a_1, \ldots, a_n) = 0.$$

But in fact a morphism

$$\text{Spec}\ (R) \xrightarrow{g} \text{Spec}\ (\mathbb{Z}[X_1, \ldots, X_n] / (f_1, \ldots, f_m))$$

is determined by the n-tuple $a_i = g^*(X_i)$, $1 \le i \le n$, and those n-tuples that occur are exactly those such that $h \mapsto h(a_1, \ldots, a_n)$ defines a homomorphism

$$R \xleftarrow{g^*} \mathbb{Z}[X_1, \ldots, X_n] / (f_1, \ldots, f_m) \quad ,$$

i.e., solutions of $f_1, \ldots, f_m$.

An interesting point is that a prescheme is actually determined by the functor of its R-valued points as well as by the larger functor of its K-valued points. To state this precisely, let X be a prescheme, and let $h_X^{(0)}$ be the *covariant* functor from the category (Rings) of commutative rings with 1 to the category (Sets) defined by:

$$h_X^{(0)}(R) = h_X(\text{Spec}\ (R)) = \text{hom}[\text{Spec}\ (R), X] .$$

Regarding $h_X^{(0)}$ as a functor in X in a natural way, one has:

Proposition 2. *For any two preschemes X_1, X_2,*

$$\text{hom}(X_1, X_2) \xrightarrow{\sim} \text{hom}\left(h_{X_1}^{(0)}, h_{X_2}^{(0)}\right) \quad .$$

Hence $h^{(0)}$ is a fully faithful functor from the category of preschemes to

$$\textit{Funct}\ ((\text{Rings}),(\text{Sets})) \quad .$$

This result is more readily checked privately than proven formally, but it may be instructive to sketch how a morphism $F : h_{X_1}^{(0)} \to h_{X_2}^{(0)}$ will induce a morphism $f : X_1 \to X_2$. One chooses an affine open covering $U_i \cong \text{Spec}\,(A_i)$ of X_1; let

$$I_i : \text{Spec}\,(A_i) \cong U_i \to X_1$$

be the inclusion. Then I_i is an A_i-valued point of X_1. Therefore $F(I_i) = f_i$ is an A_i-valued point of X_2, i.e., f_i defines

$$U_i \cong \text{Spec}\,(A_i) \longrightarrow X_2 \ .$$

Modulo a verification that these f_i patch together on $U_i \cap U_j$, these f_i give the morphism f via

$$\begin{array}{ccc} U_i & \xrightarrow{\ f_i\ } & X_2 \\ \cap & \nearrow_{f} & \\ X_1 & & \end{array} \ .$$

Grothendieck's existence problem comes up when one asks why not *identify* a prescheme X with its corresponding functor $h_X^{(0)}$, and try to define preschemes as suitable functors:

$$F : (\text{Rings}) \longrightarrow (\text{Sets}) \ .$$

The problem is to find "natural" conditions on the functor F to ensure that it is isomorphic to a functor of the type $h_X^{(0)}$. For example, let me mention one property of all the functors $h_X^{(0)}$ which was discovered by Grothendieck (Compatibility with faithfully flat descent):

Let $q : A \to B$ be a homomorphism of rings making B into a faithfully flat A-algebra, i.e.,

$$(*) \qquad \begin{cases} \forall \text{ ideals } I \subset A, \\ I \otimes_A B \xrightarrow{\sim} I \cdot B, \text{ and } q^{-1}(I \cdot B) = I \end{cases} .$$

Then if $p_1, p_2 : B \to B \otimes_A B$ are the homomorphisms $\beta \mapsto \beta \otimes 1$ and $\beta \mapsto 1 \otimes \beta$, the induced diagram of sets:

$$F(A) \xrightarrow{F(q)} F(B) \underset{F(p_2)}{\overset{F(p_1)}{\rightrightarrows}} F(B \otimes_A B)$$

is exact, (i.e., $F(g)$ injective, and Im $F(g) = \{x \mid F(p_1)x = F(p_2)x\}$).

In terms of preschemes, this means the following: Let A, B be as above and let X be a prescheme. Let $f : \text{Spec}\,(B) \to X$ be a morphism and consider the diagram:

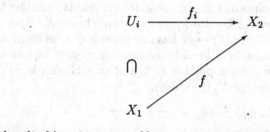

$$\begin{array}{ccc}
\mathrm{Spec}(B)\times_{\mathrm{Spec}(A)}\mathrm{Spec}(B) & & \\
p_1 \downarrow\downarrow p_2 & & \\
\mathrm{Spec}(B) & \xrightarrow{\quad f \quad} & X \\
\downarrow q & \nearrow_{g} & \\
\mathrm{Spec}(A) & &
\end{array}$$

(where q denotes the morphism of schemes induced by the homomorphism q). Then there is a morphism $q : \mathrm{Spec}\ (A) \to X$ such that $f = g \cdot q$ if and only if $f \cdot p_1 = f \cdot p_2$, and if g exists at all, it is unique.

To tie these R-valued points in with our usual idea of points, consider the case where $R = k$, a field. What is a k-valued point of a prescheme X? Spec (k) has just one point, so a map $f : \mathrm{Spec}\ (k) \to X$ has one point $x \in X$ as its image. The ring map $f^* : \underline{o}_{x,X} \to k$ must be a local homomorphism, so it factors through $\Bbbk(x)$. Conversely, if we are given a point $x \in X$ and an inclusion $\Bbbk(x) \subset k$, we get a k-valued point with image X, if we define f^* by:

$$\Gamma(U, \underline{o}_X) \xrightarrow{\text{res}} \underline{o}_{x,X} \longrightarrow \Bbbk(x) \hookrightarrow k\ ,$$

all $U \subset X$ open, $x \in U$. Thus:

$$\left\{ \begin{array}{l} \text{set of } k\text{-valued} \\ \text{points of } x \end{array} \right\} \cong \left\{ \begin{array}{l} \text{set of points } x \in X, \text{ plus} \\ \text{injections } \Bbbk(x) \hookrightarrow k \end{array} \right\} .$$

For example, for each point $x \in X$, there is a canonical morphism

$$i_x : \mathrm{Spec}\ (\Bbbk(x)) \longrightarrow X$$

with image x. If X is a prescheme over k_0 to start with and $k \supset k_0$, then the k-valued points of X *over* k_0 correspond to those $x \in X$ and injections $\Bbbk(x) \hookrightarrow k$ which reduce to the identity on the subfield k_0:

$$\begin{array}{ccc}
\Bbbk(x) & \hookrightarrow & k \\
\cup & & \subset \\
 & k_o . &
\end{array}$$

For example, suppose $k = k_0$ is algebraically closed and X is of finite type over k. Then

$$\left\{ \begin{array}{l} \text{set of } k\text{-valued points} \\ \text{of } X/k \end{array} \right\} \cong \left\{ \begin{array}{l} \text{set of points } x \in X \\ \text{such that } \ k \xrightarrow{\sim} \Bbbk(x) \end{array} \right\}$$

$$\Vert$$

$$\left\{ \begin{array}{l} \text{set of closed points} \\ x \in X \end{array} \right\} .$$

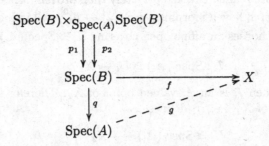

Here is another example: let k_0 be any field, but assume k algebraically closed. Let X_0 be a prescheme of finite type over k_0, and let $X = X_0 \times_{\text{Spec }(k_0)} \text{Spec }(k)$. Then:

$$\left\{ \begin{array}{l} \text{set of } k\text{-valued points} \\ \text{of } X_0/k_0 \end{array} \right\} \cong \left\{ \begin{array}{l} \text{set of } k\text{-valued} \\ \text{points of } X/k \end{array} \right\}$$
$$\cong \left\{ \begin{array}{l} \text{set of closed points} \\ \text{of } X \end{array} \right\}.$$

Thus if X_0 is a prevariety/k_0, we recover the underlying set of the old variety over k associated to X. For this reason, when k is an algebraically closed overfield of k_0, k-valued points of X_0/k_0 are often called *geometric points of* X_0. Among prevarieties X_0/k_0, $X_0 \to h_{X_0/k_0}(\text{Spec }(k))$ is a *faithful* functor into (Sets); but this is false if X is allowed to have nilpotent functions on it.

We can describe very simply the set of R-valued points of a scheme X when R is any *local ring.* In fact, if $\mathcal{M}$ is the maximal ideal of R, then Spec (R) has just one closed point, $[\mathcal{M}]$. If $f : \text{Spec }(R) \to X$ is a morphism, then (by simple topology) any open set in X containing $f([\mathcal{M}])$ contains the whole image of f. In particular the image always lies in *one* affine piece of X. (Cf. "Illustration" of Spec (R) in §1.)

Proposition 3. *Let $x \in X$. Then there is a bijection between the set of R-valued points of X such that $f([\mathcal{M}]) = x$ and the set of local homomorphisms $q : \underline{o}_{x,X} \to R$.*

Proof. Notice that $\underline{o}_{[\mathcal{M}],\text{Spec } R} = R$, so from f we get a local homomorphism g. To show we have a bijection we may replace X by an affine neighbourhood Spec (A) of x; suppose $x = [P]$. We have a commutative diagram:

$$\begin{array}{ccc} \left\{ \begin{array}{l} \text{set of } f = \text{Spec }(R) \to X \\ \text{such that } f([\mathcal{M}]) = x \end{array} \right\} & \longrightarrow & \left\{ \begin{array}{l} \text{set of local homomorphisms} \\ g : A_P = \underline{o}_{x,X} \to R \end{array} \right\} \\ \nwarrow & & \swarrow \\ & \left\{ \begin{array}{l} \text{set of homomorphisms } \varphi : A \to R \\ \text{such that } \varphi^{-1}(\mathcal{M}) = P \end{array} \right\}. & \end{array}$$

So all arrows are bijections. □

The final topic that we want to consider is the Hausdorff condition in the category of preschemes. Let X be a prescheme. The basic idea is to study X by taking a test prescheme K and considering two K-valued points of X:

$$K \underset{g}{\overset{f}{\rightrightarrows}} X.$$

If $x \in K$, when should we say that f and g are "equal" at x? Among varieties we just asked that $f(x) = g(x)$. For preschemes, the approximate definition is:

Definition 2. $f(x) \equiv g(x)$ if $f \cdot i_x = g \cdot i_x$, where $i_x : \text{Spec}\,(\mathbb{k}(x)) \to K$ is the canonical morphism. Equivalently, this means that $f(x) = g(x)$, and that the 2 maps $f_x^*, g_x^* : \mathbb{k}(f(x)) \to \mathbb{k}(x)$ are equal.

It is easy to check that if K is reduced, then $f = g$ if and only if $f(x) \equiv g(x)$, all $x \in K$.

Proposition 4. *For all $f, g : K \to X$,*

$$\{x \in K \mid f(x) \equiv g(x)\}$$

is locally closed.

Proof. Call this set Z; assume $x \in Z$. Let $y = f(x) = g(x)$, and let $U_1 = \text{Spec}\,(R_1)$ be an affine open neighbourhood of y in X. Let $U_2 = \text{Spec}\,(R_2)$ be an affine open neighbourhood of x in K such that

$$U_2 \subset f^{-1}(U_1) \cap g^{-1}(U_1)\,.$$

Then f and g induce homomorphisms:

$$f^*, g^* : R_1 \longrightarrow R_2\,.$$

Let A be the R_2-ideal generated by the elements $f^*(\alpha) - g^*(\alpha)$, $\alpha \in R_1$. Then I claim $Z \cap U_2 = V(A)$. In fact, if $[P_2] \in U_2$, $P_2 \subset R_2$ being prime, then

$$\begin{aligned} f([P_2]) \equiv g([P_2]) \iff & {f^*}^{-1}(P_2) = {g^*}^{-1}(P_2) = P_1 \text{ say, and the} \\ & \text{homomorphisms } \overline{f}^*, \overline{g}^* : R_1/P_1 \to R_2/P_2 \text{ are equal} \\ \iff & f^*(\alpha) - g^*(\alpha) \in P_2, \quad \text{all } \alpha \in R_1 \\ \iff & P_2 \supset A. \end{aligned}$$

□

Definition 3. A prescheme X is a *scheme* if for all preschemes K and all K-valued points f, g of X, $\{x \in K \mid f(x) \equiv g(x)\}$ is closed.

Proposition 5. *If X is a prescheme over a ring R, then the criterion for X to be a scheme is satisfied for all K, f, g if it is satisfied in the case:*

$$\begin{aligned} K &= X \times_{\text{Spec}\,(R)} X \\ f &= p_1 \\ g &= p_2\,. \end{aligned}$$

Proof. Let $f, g : K \to X$ be given, K arbitrary. Let $\pi : X \to \text{Spec}\,(R)$ be the given morphism. Let

$$\begin{aligned} Z_1 &= \{x \in K \mid f(x) \equiv g(x)\} \\ Z_2 &= \{x \in K \mid \pi(f(x)) \equiv \pi((x))\}\,, \end{aligned}$$

and let $\mathcal{Q}$ be the $\underline{o}_K$-ideal generated by the functions $f^*\pi^*\alpha - g^*\pi^*\alpha$, all $\alpha \in R$. Then $\mathcal{Q}$ is quasi-coherent and $1 \notin \mathcal{Q}_x \Leftrightarrow x \in Z_2$. Therefore $(Z_2, \underline{o}_K/\mathcal{Q})$ is a closed subscheme $\mathcal{Z}$ of K. Moreover, $Z_1 \subset \mathcal{Z}$, so Z_1 is closed in K if and only if it is closed in $\mathcal{Z}$. But the restrictions of $\pi \cdot f$ and $\pi \cdot g$ to $\mathcal{Z}$ are equal, since the homomorphisms

$$(\pi \cdot f)^*, \quad (\pi \cdot g)^* : R \to \Gamma(\mathcal{Z}, \underline{o}_{\mathcal{Z}})$$

are equal. Therefore f and g induce a morphism

$$h = (f, g) : \mathcal{Z} \to X \times_{\mathrm{Spec}\,(R)} X .$$

But for all $x \in \mathcal{Z}$, let $y = h(x)$, and then:

$$f(x) \equiv g(x) \Longleftrightarrow p_1(y) \equiv p_2(y).$$

Therefore

$$Z_1 = h^{-1}\left(\{y \in X \times_{\mathrm{Spec}\,(R)} X \mid p_1(y) \equiv p_2(y)\}\right) ,$$

hence if the set in braces is closed, so is T_1. □

Corollary 1. *If k is an algebraically closed field, then a prevariety over k is a variety in the sense of Ch. I if and only if it is a scheme.*

Proof. Let $Z \subset X \times_k X$ be the set of points z such that $p_1(z) \equiv p_2(z)$. If z is a closed point then $z \in Z$ if and only if $p_1(z) = p_2(z)$. But the locally closed set Z is closed if and only if its intersection with the set of closed points of $X \times_k X$ is closed in the induced topology, i.e., if and only if X is a variety in the old sense. □

In Ch. I, §5, we asserted that varieties X also satisfied the "local criterion":

> If $x, y \in X$, $x \neq y$, there is no local ring $\mathcal{O} \subset k(X)$ such that $\mathcal{O} > \underline{o}_x$ and $\mathcal{O} > \underline{o}_y$.

We can now prove this. If such an $\mathcal{O}$ existed, then by Proposition 3 we can define 2 $\mathcal{O}$-valued points of X:

$$\mathrm{Spec}\,(\mathcal{O}) \underset{g}{\overset{f}{\rightrightarrows}} X$$

such that if x_0 is the closed point of Spec $(\mathcal{O})$, $f(x_0) = x$, $g(x_0) = y$. But if x_1 is the generic point of Spec $(\mathcal{O})$, then $f(x_1) = g(x_1) =$ the generic point of X, and $f^*_{x_1}, g^*_{x_1}$ both give the identity map from $k(X)$ to the quotient field of $\mathcal{O}$. This shows that

$$x_1 \in \{z \mid f(z) \equiv g(z)\}$$
$$x_0 \notin \{z \mid f(z) \equiv g(z)\} .$$

Therefore $\{z \mid f(z) \equiv g(z)\}$ is not closed.

Proposition 6. *If X is a scheme and $U, V \subset X$ are open affine sets, then $U \cap V$ is affine and the canonical homomorphism:*

$$\Gamma(U, \underline{o}_X) \otimes \Gamma(V, \underline{o}_X) \longrightarrow \Gamma(U \cap V, \underline{o}_X)$$

is surjective.

Proof. Let $\Delta : X \to X \times X$ be the diagonal ($X \times X$ being the absolute product, i.e., over Spec $(\mathbb{Z})$). First, $\Delta(X)$ is closed since X is a scheme. And if you cover X by open affines $U_i = \text{Spec}\,(R_i)$, then $\Delta(X)$ is covered by the affines $U_i \times U_i$. But $U_i = \Delta^{-1}(U_i \times U_i)$ and Δ^* is the canonical surjection:

$$\begin{array}{ccc} \Gamma(U_i, \underline{o}_X) & \xleftarrow{\;\Delta^*\;} & \Gamma(U_i \times U_i, \underline{o}_{X\times X}) \\ \| & & \| \\ R_i & & R_i \otimes R_i \,. \end{array}$$

This shows that Δ is a closed immersion (cf. Cor. 1 of Th. 3, §5). But $U \times V$ is an open affine in $X \times X$ with ring $\Gamma(U, \underline{o}_X) \otimes \Gamma(V, \underline{o}_X)$. Therefore $U \cap V = \Delta^{-1}(U \times V)$ is also affine and its ring is a quotient of $\Gamma(U, \underline{o}_X) \otimes \Gamma(V, \underline{o}_X)$. □

Problem

1) Check that a fibre product of schemes is a scheme.
2) Prove that $\mathbb{P}^n_{\mathbb{Z}}$ is a scheme (cf. Example J).
3) Given a diagram of schemes:

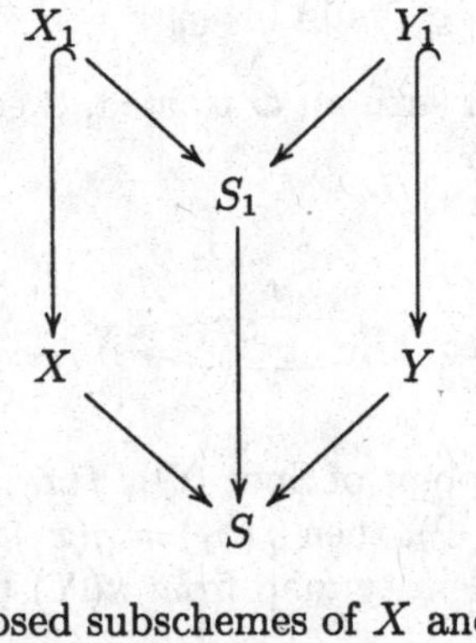

where X_1 and Y_1 are closed subschemes of X and Y, show that $X_1 \times_{S_1} Y_1$ is a closed subscheme of $X \times_S Y$.
(*Hint:* You must use the fact that S_1 is a scheme.)

From now on, all our preschemes will be assumed to be schemes.

§7. Proper morphisms and finite morphisms

In this section, we will generalize the results of Ch. I, §8.

Definition 1. Let $f : X \to Y$ be a morphism. f is *of finite type* if for all open affines $U \subset Y$, $f^{-1}(U)$ is of finite type over $\Gamma(U, \underline{o}_Y)$, i.e., $f^{-1}(U)$ is quasi-compact and $\Gamma(V, \underline{o}_X)$ is finitely generated over $\Gamma(U, \underline{o}_Y)$ for all open affine $V \subset f^{-1}(U)$.

Proposition 1. *Let $f : X \to Y$ be a morphism. To prove that f is of finite type, it suffices to check the defining property for the open affine sets $U_i \subset Y$ of one covering $\{U_i\}$ of Y.*

Proof. Let $U_i = \text{Spec }(R_i)$ and assume $f^{-1}(U_i)$ of finite type$/R_i$. Note that for all $f \in R_i$, $f^{-1}(U_{i,f})$ is of finite type over $(R_i)_f$. Let $U = \text{Spec }(R)$ be any open affine in Y. Then U admits a finite covering by open sets of the type $(U_i)_f$, $f \in R_i$, since it is quasi-compact. Therefore we can assume $U = U_1 \cup \ldots \cup U_n$, U_i affine and $f^{-1}(U_i)$ of finite type over $\Gamma(U_i, \underline{o}_Y)$, and hence of finite type over R. Therefore by Prop. 1, §3, U is of finite type over R. □

Definition 2. Let $f : X \to Y$ be a morphism. Then f is *proper* if f is of finite type and for all morphisms $g : K \to Y$, K any prescheme, the projection

$$p_2 : X \times_Y K \longrightarrow K$$

is a closed map of topological spaces.

Since p_2 being closed is a local property on K, it certainly suffices, in this definition, to look at *affine* K's. The following functorial properties of properness follow immediately from the definition:

i) closed immersions are proper,
ii) the composition of proper morphisms is proper,
iii) if $f : X \to Y$ is proper and $g : K \to Y$ is any morphism, then $p_2 : X \times_Y K \to K$ is proper.
iv) Suppose $X \xrightarrow{f} Y \xrightarrow{g} Z$ are morphisms and that g is of finite type. f is surjective and $g \cdot f$ is proper. Then g is proper.

We certainly should check that a variety X over k is complete if and only if $\pi : X \to \text{Spec }(k)$ is proper. This follows easily from Chow's lemma and the next theorem:

Theorem 2. *The morphism $f : \mathbb{P}^n_{\mathbb{Z}} \longrightarrow \text{Spec }(\mathbb{Z})$ is proper.*

Proof. We must show that the projection

$$p : \mathbb{P}^n_R \longrightarrow \text{Spec }(R)$$

is closed, for every R (cf. Ex. J bis, §2). We proved in Ch. I, §8, that p is closed when R is a finitely generated integral domain over an algebraically closed field

k. However, with small modifications, the identical proof works in general. Let's follow this proof. Start with $Z \subset \mathbb{P}^n_R$, a closed subset. As before, Z can be described as a homogeneous ideal

$$A \subset R[X_0, X_1, \ldots, X_n] .$$

In fact, essentially the same argument shows

Lemma. *For all closed subschemes $Z \subset \mathbb{P}^n_R$, for all i and all $g \in I(Z \cap U_i)$, there is a homogeneous polynomial $G \in R[X_0, \ldots, X_n]$ of degree r such that $G/X_j^r \in I(Z \cap U_j)$, all j and $G/X_i^r \in I(Z \cap U_j)$, all j and $G/X_i^r = g$.*

Next take a point $y \in \text{Spec}(R) - p(Z)$, and assume $y = [P]$, where $P \subset R$ is a prime ideal (no longer necessarily maximal). One can still conclude that

$$1 \in I(Z \cap U_i) + P \cdot R_P \left[\frac{X_0}{X_i}, \ldots, \frac{X_n}{X_i}\right] .$$

From this one gets

(*)' $$(S_N/A_N) \otimes_R (R_P/P \cdot R_P) = (0) ,$$

hence by Nakayama's lemma

(**)' $$(S_N/A_N) \otimes_R R_P = (0) ,$$

Since S_N is a finitely generated R-module, this implies that there is an element $f \in R_P$ such that $f \cdot S_N \subset A_N$. As in Ch. I, this implies that Z is disjoint from $\mathbb{P}^n_{\mathbb{Z}} \times \text{Spec}(R_f)$, hence $p(Z) \subset V((f))$. This shows that $p(Z)$ is closed. □

In particular, if Y is any scheme and X is a closed subscheme of $\mathbb{P}^n_{\mathbb{Z}} \times Y$, then $p_2 : X \to Y$ is proper.

Proposition 3. *Suppose $f : X \to Y$ is proper and R is a valuation ring, with quotient field K. Suppose we are given morphisms ϕ, ψ_0:*

$$\begin{array}{ccc} \text{Spec}(K) & \xrightarrow{\psi_0} & X \\ {\scriptstyle i}\downarrow & \nearrow & \downarrow{\scriptstyle f} \\ \text{Spec}(R) & \xrightarrow[\phi]{} & Y. \end{array}$$

Then there is one and only one morphism ψ : Spec $(R) \to X$ *such that ψ extends to ψ_0 and $f \cdot \psi = \phi$.*

Proof. If ψ_0 had 2 extensions ψ', ψ'' : Spec $(R) \to X$, then apply the definition of scheme to ψ' and ψ'' and it follows that $\psi' = \psi''$. This takes care of uniqueness. Next apply the definition of proper to ϕ : Spec $(R) \to Y$ and to the closed subset of $X \times_Y$ Spec (R) obtained by taking the closure Z of (ψ_0, i) [Spec K]. Give Z the structure of reduced closed subschemes. Then $p_2(Z)$ is a closed subset of Spec (R) containing its generic point, hence equal to Spec (R). I claim that $p_2 : Z \to$ Spec (R) is an isomorphism, hence Z is the graph of the ψ we are looking for.

Everything follows from:

Lemma. *With R and K as above, let $f : Z \to$ Spec (R) be a surjective birational morphism, where Z is a reduced and irrational scheme. Then f is an isomorphism.*

Proof. Let $x \in Z$ be a point over the closed point of Spec (R), and let y be the generic point of Z. We get a diagram:

$$\begin{array}{ccc} \underline{o}_{y,Z} & \hookleftarrow & \underline{o}_{x,Z} \\ f_y^* \uparrow & \approx & \uparrow f_x^* \\ K & \hookleftarrow & R \end{array}$$

Since f_x^* is a local homomorphism, and R is a valuation ring, $R = \underline{o}_{x,Z}$. But then the local homomorphism

$$(f_x^*)^{-1} : \underline{o}_{x,Z} \longrightarrow R$$

defines a morphism g : Spec $(R) \to Z$ by Prop. 3, §6. It is clear that $f \cdot g = 1_{\text{Spec }(R)}$, and $g \cdot f = 1_Z$ since $g \cdot f(y) \equiv 1_Z(y)$ and Z is a scheme. □

Corollary. *Let X be a complete variety over a field k. Then for all valuation rings $R \subset K$, such that $k \subset R$ and $K =$ quotient field of R, $R \supset \underline{o}_x$ for some $x \in X$.*

The *valuation criterion* asserts that the property in Prop. 3 implies conversely that f is proper. We will not use this; for a proof, cf. EGA, Ch. II.

Proposition 4. *Let $\phi : R \to S$ be a homomorphism of rings such that S is integrally dependent on $\phi(R)$. Then the corresponding morphism Φ :* Spec $(S) \to$ Spec (R) *is a closed map.*

Proof. This is exactly the going-up theorem of Cohen-Seidenberg. In fact, if $V(A)$ is a closed subset of Spec (S), then I claim $\Phi(V(A)) = V\left(\phi^{-1}(A)\right)$. To see this, let $[P] \in V\left(\phi^{-1}(A)\right)$. Then P is a prime ideal in R containing $\phi^{-1}(A)$. By the going-up theorem applied to S/A and the subring $R/\phi^{-1}(A)$, there is a prime ideal $P' \subset S$ such that $\phi^{-1}(P') = P$ and $P' \supset A$. Therefore $[P'] \in V(A)$ and $\Phi([P']) = [P]$. □

Corollary. *Let $\phi : R \to S$ be a homomorphism of rings such that S is a finite R-module. Then the corresponding morphism Φ :* Spec $(S) \to$ Spec (R) *is proper.*

It is very useful to have a class of morphisms $f : X \to Y$ which over every affine in Y look like those in this corollary, but where the image itself is not necessarily affine.

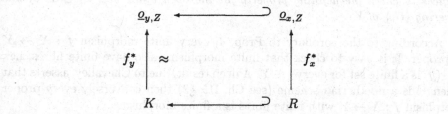

Definition 3. A morphism $f : X \to Y$ is affine if for all open affine sets $U \subset Y$, $f^{-1}(U)$ is affine. f is *finite* if it is affine and if for all open affine sets $U \subset Y$, $\Gamma\left(f^{-1}(U), \underline{o}_X\right)$ is a finite module over $\Gamma(U, \underline{o}_Y)$.

We will prove the following key point in Ch. III, §1, when we have the appropriate machinery.

Proposition 5. *To prove that a morphism $f : X \to Y$ is affine or finite, it suffices to check the defining property for the open affine sets $U_i \subset Y$ of one covering $\{U_i\}$ of Y.*

According to the corollary to Prop. 4, every finite morphism $f : X \to Y$ is proper. It is easy to check that finite morphisms also have finite fibres, i.e., $f^{-1}(y)$ is a finite set for every $y \in Y$. A deep result, due to Chevalley, asserts that when Y is a noetherian scheme (see Ch. III, §2) then conversely every proper morphism $f : X \to Y$ with finite fibres is a finite morphism.

As an example of how the definition of "proper" works, consider the projection

$$p : \mathbb{A}_k^2 \longrightarrow \mathbb{A}_k^1$$

taking (x, y) to x. If $x \subset \mathbb{A}_k^2$ is the closed subset $x \cdot y = 1$, then $p(X) = \mathbb{A}_k^1 - (0)$ which is not closed. Therefore p is not proper. Looking back to Chapter I, Examples N and O are examples of finite bijective morphisms which are not isomorphisms. But Example *P* is clearly not a finite morphism and it is worthwhile seeing that it is not even proper. We started with a finite morphism

$$f : \mathbb{A}^1 \longrightarrow C\,,$$

C an affine plane curve. Letting $D_0 = \mathbb{A}^1 - (1)$, and $f' =$ restriction of f to D_0, we obtained a bijection between D_0 and C. Now f' itself is certainly a closed map since the topology of a one-dimensional variety is so trivial. Therefore to show that f' is not proper, we have to take a fibre product. Let $D_1 = \mathbb{A}^1 - (-1)$. Consider the fibre product:

$$\begin{array}{ccccc}
D_0 \times_C D_1 & & \xrightarrow{\quad p_1 \quad} & & D_o \\
\downarrow{\scriptstyle p_2} & \underset{\text{open}}{\cap} & & & \cap\ \text{open} \\
 & & \mathbb{A}^1 \times_C \mathbb{A}^1 & \xrightarrow{\quad p_1 \quad} & \mathbb{A}^1 \\
 & & \downarrow{\scriptstyle p_2} & & \downarrow{\scriptstyle f} \\
D_1 & \underset{\text{open}}{\subset} & \mathbb{A}^1 & \xrightarrow{\quad f \quad} & C.
\end{array}$$

Then $p_2 : D_0 \times_C D_1 \to D_1$ is *not* closed. To see this, look at the structure of $\mathbb{A}^1 \otimes_C \mathbb{A}^1$. It is a closed subscheme of $\mathbb{A}^1 \times_k \mathbb{A}^1$ whose closed points are the pairs (x_1, x_2) such that $f(x_1) = f(x_1)$. Therefore it is the union of (1) the diagonal $\{(x, x)\}$, and (2) the 2 isolated points $(1, -1)$ and $(-1, 1)$. $D_0 \times_C D_1$ consists in the open subset U of diagonal points $\{(x,x) \mid x \neq 1, -1\}$ and the open subset V containing the single point $(-1,1)$. Therefore U and V are closed too. But $p_2(U) = D_1 - \{1\}$ is not closed in D_1.

A natural way in which finite morphisms occur is via ***projections***. To fix notation, let $\mathbb{P}^n_R$ be the union of $(n+1)$-affines

$$U_i = \text{Spec } R\left[\frac{X_0}{X_i}, \ldots, \frac{X_n}{X_i}\right], \quad 0 \le i \le n$$

with the usual patching. Let

$$\ell_i = \sum_{j=0}^{n} a_{ij}x_j, \qquad 0 \le i \le r$$

be any linear forms, with $a_{ij} \in R$. Recall that the open set W_i given by $\ell_i \neq 0$ is the affine scheme

$$\text{Spec } R\left[\frac{X_0}{\ell_i}, \ldots, \frac{X_n}{\ell_i}\right]$$

(cf. Ex. J bis). Let L be the "linear" subset $\ell_0 = \ldots = \ell_r = 0$. Now let $\mathbb{P}^r_R$ be the union of the $(r+1)$-affines:

$$V_i = \text{Spec } R\left[\frac{Y_0}{Y_i}, \ldots, \frac{Y_r}{Y_i}\right], \qquad 0 \le i \le r.$$

Define the projection:

$$\pi_1 : \mathbb{P}^n_R - L \longrightarrow \mathbb{P}^r_R$$

by requiring that $\pi_1(W_i) \subset V_i$ and that $\pi_1^*(Y_k/Y_i) = \ell_k/\ell_i$, for all $0 \le i, k \le r$. Then clearly $\pi_1^{-1}(V_i) = W_i$, so π_1 is an affine morphism of finite type.

Proposition 6. *Let Z be a closed subscheme of $\mathbb{P}^n_R$ disjoint from L. Let π be the restriction of π_1 to Z. Then*

$$\pi : Z \longrightarrow \mathbb{P}^r_R$$

is a finite morphism.

Proof. π is certainly affine since it is the restriction of π_1 to a closed subscheme. For all k such that $0 \le k < n$, $Z \cap U_k$ and $V\left(\left(\frac{\ell_0}{X_k}, \ldots, \frac{\ell_r}{X_k}\right)\right)$ are disjoint closed sets in U_k. Therefore

$$1 \in I(Z \cap U_k) + \left(\frac{\ell_0}{X_i}, \ldots, \frac{\ell_r}{X_k}\right).$$

Suppose in fact that

$$(*) \qquad 1 = f_k + \sum_{i=0}^{r} \frac{\ell_i}{X_k} \cdot g_{ik},$$

where $f_k \in I(Z \cap U_k)$ and $g_{ik} \in R\left[\frac{X_0}{X_k}, \dots, \frac{X_n}{X_k}\right]$. By the lemma in Th. 2, there is a homogeneous polynomial $F_k \in R[X_0, \dots, X - n]$ of degree d such that

$$\begin{aligned} &\text{a)} \quad F_k/X_k^d = f_k \\ &\text{b)} \quad F_k/X_i^d \in I(Z \cap U_i), \quad \text{all } i. \end{aligned}$$

Equation $(*)$ shows that F_k must have the form

$$F_k = X_k^d + \sum_{i=0}^{r} \ell_i \cdot G_{ik}(X_0, \dots, X_n),$$

degree $(G_{ik}) = d - 1$.

Now consider the ring extension given by π^*:

$$\begin{array}{ccc} \Gamma\left(\pi^{-1}(V_i), \underline{o}_Z\right) & \xleftarrow{\quad \pi^* \quad} & \Gamma\left(V_i, \underline{o}_{\mathbb{P}^r_R}\right) \\ \| & & \| \\ \Gamma(Z \cap W_i, \underline{o}_Z) & & R\left[\frac{Y_0}{Y_i}, \dots, \frac{Y_r}{Y_i}\right] \\ \| & & \\ R\left[\frac{X_0}{\ell_i}, \dots, \frac{X_n}{\ell_i}\right] \Big/ I(Z \cap W_i) & & \end{array}$$

But because $I(Z \cap W_i)$ contains $F_0/\ell_i^d, \dots, F_n/\ell_i^d$, it follows that the monomials

$$\prod_{j=0}^{n} (X_j/\ell_i)^{a_j}, \qquad 0 \le a_j < d$$

form a module basis of $R\left[\frac{X_0}{\ell_i}, \dots, \frac{X_n}{\ell_i}\right] \Big/ I(Z \cap U_i)$ over $R[\ell_0/\ell_i, \dots, \ell_r/\ell_i]$. Therefore π is finite. □

At this point, it is almost trivial to give the classical proof of Noether's normalization lemma so as to illustrate the power of this method (cf. Ch. I, §6). Let k be an algebraically closed field and let X be an affine variety over k. If $X \subset \mathbb{A}_k^n$, embed $\mathbb{A}_k^n \subset \mathbb{P}_k^n$ as the complement of the hyperplane $X_0 = 0$ and let $\overline{X}$ be the closure of X in $\mathbb{P}_k^n$. Let L be a maximal linear subspace of $\mathbb{P}^n - \mathbb{A}^n$ disjoint from $\overline{X}$. Let $\ell_0 = X_0, \ell_1, \dots, \ell_r$ be a set of independent linear equations defining L. Use these to define the projection:

$$\begin{array}{ccc} \mathbb{P}^n - L & \xrightarrow{\pi_1} & \mathbb{P}^r \\ \cup & \nearrow_{\pi} & \cup \\ \overline{X} & & V_0 \cong \mathbb{A}^r \\ \cup & \nearrow_{\pi'} & \\ X & & \end{array}$$

π is finite by the Proposition. Since $\pi^{-1}(V_0) = \overline{X} \cap \{\ell_0 \neq 0\} = X$, the restriction π' of π to a morphism from X to V_0 is finite. Since $V_0 \cong \mathbb{A}^r$, this gives the normalization lemma, provided that π (and hence π') is surjective. But if $\pi(\overline{X})$ were a proper subset of $\mathbb{P}^r$, it would be irreducible and it could not contain completely the hyperplane $\mathbb{P}^r - \mathbb{A}^r$ either. Then choose a closed point $x \in \mathbb{P}^r - \pi(\overline{X}) - \mathbb{A}^r$: it follows that $L \cup \pi_1^{-1}(x)$ is a linear subspace of $\mathbb{P}^n - \mathbb{A}^n$ disjoint from $\overline{X}$ and bigger than L. This contradiction proves that π' is surjective, so Noether's lemma comes out.

§8. Specialization

Let k be an algebraically closed field and let $R \subset k$ be a valuation ring. Then the residue field L of R is also algebraically closed. Let $\pi : R \to L$ denote the canonical map, and let $M = \ker(\pi)$. Let $\mathbb{P}^n(k)$ and $\mathbb{P}^n(L)$ denote the set of closed points of $\mathbb{P}^n_k$ and $\mathbb{P}^n_L$ respectively (= the set of k-valued and L-valued points). Then there is a remarkable map:

$$\mathbb{P}^n(k) \xrightarrow{\rho} \mathbb{P}^n(L)$$

defined as follows: let $(\alpha_0, \alpha_1, \ldots, \alpha_n)$ be homogeneous coordinates of $x \in \mathbb{P}^n(k)$. Then for some $\lambda \in k$, all the elements $\lambda\alpha_0, \ldots, \lambda\alpha_n$ will be in R, and not all of them M. Set

$$\rho((\alpha_0, \ldots, \alpha_n)) = (\pi(\lambda\alpha_0), \ldots, \pi(\lambda\alpha_n)).$$

Since λ is unique up to a unit in R, $(\pi(\lambda\alpha_0), \ldots, \pi(\lambda\alpha_n))$ is unique up to multiplication by a non-zero element of L, so ρ makes sense. Note that ρ is surjective.

Definition 1. For all closed points $x \in \mathbb{P}^n_k$, $\rho(x)$ is the *specialization of* x with respect to R.

For example, look at the case $n = 1$, and consider $k \cong \mathbb{A}^1(k) = \mathbb{P}^1(k) - \{\infty\}$. Then $\rho : k \to \mathbb{P}^1(L) = L \cup \{\infty\}$ is just the place associated to R.

Although ρ does not extend to a continuous map of the whole scheme $\mathbb{P}^n_k$ to $\mathbb{P}^n_L$, (Reader: why not?) it has the following remarkable property:

Theorem 1. *For all closed subsets $Z \subset \mathbb{P}^n_k$, there is a unique closed subset $W \subset \mathbb{P}^n_L$ such that*

$$\rho(Z(k)) = W(L).$$

In fact, W can be constructed as follows. Consider the diagram of fibre products:

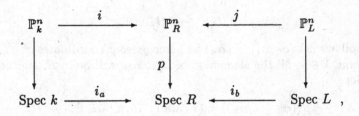

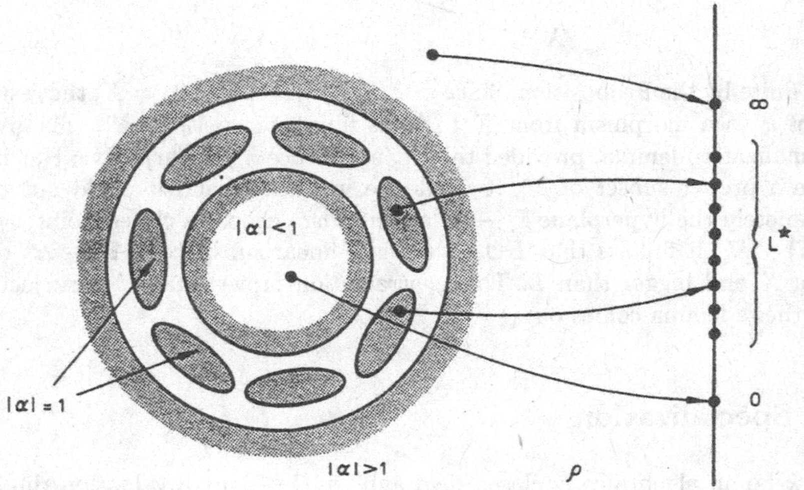

case of real valuation

$$\begin{array}{ccccc} \mathbb{P}^n_k & \xrightarrow{\;i\;} & \mathbb{P}^n_R & \xleftarrow{\;j\;} & \mathbb{P}^n_L \\ \big\downarrow & & \big\downarrow p & & \big\downarrow \\ \operatorname{Spec} k & \xrightarrow{\;i_a\;} & \operatorname{Spec} R & \xleftarrow{\;i_b\;} & \operatorname{Spec} L \end{array},$$

where a and b are the generic and closed points of Spec (R). Then j is a closed immersion, so we will identify $\mathbb{P}^n_L$ with the closed subscheme $p^{-1}(b)$ in $\mathbb{P}^n_R$. Moreover i is an isomorphism of $\mathbb{P}^n_k$ with the subset $p^{-1}(a)$ of $\mathbb{P}^n_R$, plus the restriction of $\underline{O}_{\mathbb{P}^n_R}$ to $p^{-1}(a)$. In other words, i is a homeomorphism, and for all $x \in \mathbb{P}^n_k$, $i^* : \underline{O}_{i(x),\mathbb{P}^n_R} \to \underline{O}_{x,\mathbb{P}^n_k}$ is an isomorphism.

Theorem 1 bis. $W = \mathbb{P}^n_K \cap \overline{i(Z)}$.

To see why this is reasonable, let's check that for all closed points $x \in \mathbb{P}^n_k$,

$$\{\rho(x)\} = \mathbb{P}^n_L \cap \overline{\{i(x)\}}\,.$$

In fact, if $x = (\alpha_0, \ldots, \alpha_n)$ where we assume $\alpha_i \in R$ and $\alpha_0 \notin M$ to simplify notations, then $\rho(x) = (\pi\alpha_0, \ldots, \pi\alpha_n)$. Cover $\mathbb{P}^n_R$ by open affines:

$$U_i = \operatorname{Spec} R\left[\frac{X_0}{X_i}, \ldots, \frac{X_n}{X_i}\right].$$

Consider the subset

$$\Sigma = V\left(\left(\frac{X_1}{X_0} - \frac{\alpha_1}{\alpha_0}, \ldots, \frac{X_n}{X_0} - \frac{\alpha_n}{\alpha_0}\right)\right) \subset U_0 .$$

Clearly $i(x) \in \Sigma$, and $\Sigma \cap \mathbb{P}^n_L = \{\rho(x)\}$. Via p, $\Sigma \xrightarrow{\sim} \operatorname{Spec}(R)$, so Σ is the graph of a morphism $f : \operatorname{Spec}(R) \to \mathbb{P}^n_R$ which is a section of p. Therefore Σ is closed in $\mathbb{P}^n_R$ and the point $i(x)$, which corresponds to the generic point $a \in \operatorname{Spec}(R)$, is dense in Σ. Therefore $\Sigma = \overline{\{i(x)\}}$. (Compare Prop. 3, §7).

It follows that $\rho(Z(k)) \subset \mathbb{P}^n_L \cap \overline{i(Z)}$. The main point is the lifting problem: to show that every L-valued point of $\mathbb{P}^n_L \cap \overline{i(Z)}$ is the intersection of $\mathbb{P}^n_L$ with an R-valued point of $\overline{i(Z)}$. We shall do this via the Going-Down Theorem of Cohen-Seidenberg:

Going-Down Theorem. *Let $f : X \to Y$ be a finite morphism. Assume that Y is an irreducible normal scheme, i.e., all of its local rings $\underline{o}_y$ are domains, integrally closed in its quotient field. Assume that for all $x \in X$, no non-zero element of $\underline{o}_{f(x),y}$ is a 0-divisor in $\underline{o}_x$*[7]. *Then for every pair of points $x_1 \in X$, $y_0 \in Y$ such that $f(x_1) \in \overline{\{y_0\}}$, there is a point $x_0 \in f^{-1}(y_0)$ such that $x_1 \in \overline{\{x_0\}}$.*

This is very nearly the same as saying that f is an *open* map. In fact, if f is open, the conclusion of the Theorem has to hold; and conversely, whenever you know that under f constructable sets go to constructable sets, then the conclusion of the theorem implies f is open.

We also need a new form of Noether's Normalization lemma:

Normalization lemma over R. *Let $Z \subset \mathbb{P}^n_k$ be an (irreducible closed subset of dimension r and let $\mathcal{Z} = \overline{i(Z)}$. There exist $(r+1)$ linear forms $\ell_0(X), \ldots, \ell_r(X)$ with coefficients in R such that the subset $L \subset \mathbb{P}^n_R$ defined by $\ell_0 = \ldots = \ell_r = 0$ is disjoint from $\mathcal{Z}$. Let τ be the projection*

$$\tau : \mathbb{P}^n_R - L \longrightarrow \mathbb{P}^r_R .$$

Then giving $\mathcal{Z}$ any structure of closed subscheme, $\tau : \mathcal{Z} \to \mathbb{P}^r_R$ is a finite surjection morphism.

Proof. Once we find $\ell_0, \ldots, \ell_r$ such that $L \cap \mathcal{Z} = \emptyset$, the last part is easy: τ is finite by Prop. 6, §7. Therefore, it is surjective if the image contains the generic point of $\mathbb{P}^r_R$. But restricted to $\mathbb{P}^n_k$, τ defines a finite morphism from the

[7] In the language to be introduced in Ch. 3: all associated points of $\underline{o}_X$ lie over the generic point of Y.

r-dimensional variety Z to $\mathbb{P}_k^r$. If this were not dominating, its fibres would be positive dimensional by the results of Ch. I, §7. Therefore τ is surjective.

To construct the ℓ's, let s be the *smallest* integer such that $\mathcal{Z}$ is disjoint from $\ell_0 = \ldots = \ell_s = 0$ for some set of $s+1$ linear forms. Since $\mathcal{Z}$ is disjoint from the empty set defined by $X_0 = \ldots = X_n = 0$, therefore $s \leq n$. Let L' be $\ell_0 = \ldots = \ell_s = 0$ and let $\tau' : \mathbb{P}_R^n - L' \to \mathbb{P}_R^s$ be the projection. We want to show $s \leq r$ (in fact, $s < r$ is easily seen to be impossible by the first argument). Assume $s > R$, and let $\mathcal{Z}' = \tau'(\mathcal{Z})$. $\mathcal{Z}'$ is closed by Prop. 6, §7, and if $Z' = \tau'(Z)$ in $\mathbb{P}_k^s$, then $\mathcal{Z}'$ must be just the closure of $i(Z')$. Since $s > r$, $i(Z')$ is still a proper closed subset of $\mathbb{P}_k^s$. But then $i(Z')$ is contained in some hypersurface $F = 0$ of $\mathbb{P}_k^s$. Normalize the coefficients of F so that they all lie in R, but not all in M. Then $F = 0$ defines a subset $H \subset \mathbb{P}_R^s$: i.e., let

$$H \cap U_i = V\left((F/X_i^d)\right), \qquad d = \text{degree } F.$$

Since $H \supset i(Z')$, it follows that $H \supset \mathcal{Z}'$ too. Furthermore $H \not\supset \mathbb{P}_L^s$ since the equation $\pi(F)$, reduced mod M, is not identically 0. Choose a closed point $(\overline{\alpha}_0, \ldots, \overline{\alpha}_s)$ in $\mathbb{P}_L^s - H$: assume for simplicity that $\overline{\alpha}_0 \neq 0$. Let $\alpha_i \in R$ be elements such that $\pi(\alpha_i) = \overline{\alpha}_i$ and let $\Sigma \subset \mathbb{P}_R^s$ be the section of p defined by

$$X_1 = \frac{\alpha_1}{\alpha_0} X_0, \ldots, X_s = \frac{\alpha_s}{\alpha_0} X_0 \,.$$

Then $\Sigma \cap \mathcal{Z}' = \emptyset$, since $\{\overline{\alpha}_0, \ldots, \overline{\alpha}_s)$ is the only closed point of Σ and it is not in $\mathcal{Z}'$. On the other hand, back in $\mathbb{P}_R^n$,

$$L' \cup \tau'^{-1}(\Sigma) = \left\{ \begin{array}{l} \text{locus of zeroes of} \\ \ell_1 - \frac{\alpha_1}{\alpha_0}\ell_0 = \ldots = \ell_s - \frac{\alpha_s}{\alpha_0}\ell_0 = 0 \end{array} \right\}.$$

Call this L''. Then L'' is disjoint from $\mathcal{Z}$ and is defined by only $s-1$ equations. This shows that s was *not* as small as possible, and the lemma is proven. □

The theorem now follows easily: suppose $Z \subset \mathbb{P}_k^n$ is given. First of all, we may assume that Z is irreducible, since the general case reduces to this one right away. Let $\mathcal{Z} = \overline{i(R)}$ and choose a finite surjective morphism:

$$\begin{array}{ccc} \mathcal{Z} & \xrightarrow{\quad\tau\quad} & \mathbb{P}_R^r \\ & {\scriptstyle p}\searrow \quad \swarrow{\scriptstyle q} & \\ & \operatorname{Spec}(R) & \end{array}$$

($r = \dim_k Z$). Let $W = \mathcal{Z} \cap \mathbb{P}_L^n$, let y be any closed point of W, and let $y_0 = \tau(y)$. If $y_0 = (\overline{\alpha}_0, \ldots, \overline{\alpha}_r)$, then choose $\alpha_i \in R$ such that $\pi(\alpha_i) = \overline{\alpha_i}$. The $\underline{\alpha\text{'s define}}$ a closed point $(\alpha_0, \ldots, \alpha_r) \in \mathbb{P}_k^r$. Let $x_0 = i\left((\alpha_0, \ldots, \alpha_r)\right)$. Then $y_0 \in \overline{\{x_0\}}$. By the Going-Down Theorem, choose a point $x \in \tau^{-1}(x_0)$ such that $y \in \overline{\{x\}}$. Since x_0 lies over the generic point of Spec (R), so does x. Therefore $x = i(\tilde{x})$ for some point $\tilde{x} \in Z$. τ restricts to a finite morphism:

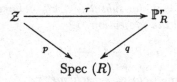

$$\begin{array}{ccc} Z & \xrightarrow{\tau_k} & \mathbb{P}^r_k \\ & \searrow \quad \swarrow & \\ & \mathrm{Spec}\,(k). & \end{array}$$

Then since $\tau_k(\tilde{x})$ is closed, $\tilde{x}$ is a closed point too. In other words, $\tilde{x} \in Z(k)$ and y is the point $\overline{\{i(\tilde{x})\}} \cap \mathbb{P}^n_L$, i.e., $\rho(\tilde{x})$. This proves that $y \in \rho(Z(k))$, hence $\rho(Z(k)) = W(L)$. □

Definition 2. The subset W of $\mathbb{P}^n_L$ is called the *specialization of* Z *with respect to* R.

Corollary (of proof). *If* Z *is irreducible and* $\dim Z = r$, *then all components of its specialization* W *are* r*-dimensional.*

Proof. Assuming that Z is irreducible and r-dimensional, we constructed a finite surjective morphism $\tau : \overline{i(Z)} \to \mathbb{P}^r_R$. Intersecting with $\mathbb{P}^n_L$ this gives a finite surjective morphism

$$\tau_L : W \longrightarrow \mathbb{P}^r_L .$$

We would like to know that every component of W dominates $\mathbb{P}^r_L$, which would show that it is r-dimensional. Suppose this did not happen: let $w_0 \in W$ be the generic point of a component of W such that $\tau_L(w_0) \neq y$, y the generic point of $\mathbb{P}^r_R$. The going-down theorem applies to $\tau : \overline{i(Z)} \to \mathbb{P}^r_R$. Since $\tau_L(w_0) \in \overline{\{y\}}$, there is a point $w \in \tau^{-1}(y)$ such that $w_0 \in \overline{\{w\}}$. But then w must be a point of W over the generic point of $\mathbb{P}^r_L$ and $w_0 \in \overline{\{w\}}$ implies that w_0 is not generic. □

To actually determine W, here is what you have to do: describe Z by its homogeneous ideal

$$A \subset k[X_0, \ldots, X_n] .$$

Then take the intersection $A \cap R[X_0, \ldots, X_n]$ and take the reductions mod M of all the polynomials in this intersection:

$$B = \pi\{A \cap R[X_0, \ldots, X_n]\} .$$

Then W is the closed subset $V(B)$.

Proof. In fact, for all $F \in A \cap R[X_0, \ldots, X_n]$, $F = 0$ defines a closed subset H of $\mathbb{P}^n_R$. Since $F \in A$, $H \supset i(Z)$. Therefore $H \supset \overline{i(Z)}$ and $H \cap \mathbb{P}^n_L \supset W$. But $H \cap \mathbb{P}^n_L$ is the locus $\pi F = 0$. Therefore $W \subset V(B)$.

Conversely, suppose $x \in \mathbb{P}^n_L - W$. Then $x \in \mathbb{P}^n_R - \overline{i(Z)}$, and by the lemma in Th. 2, §7, there is a homogeneous polynomial $F \in R[X_0, \ldots, X_n]$ such that $F = 0$ on $\overline{i(Z)}$, $F \neq 0$ at x. Therefore $F \in A \cap R[X_0, \ldots, X_n]$ and $\pi F \in B$. Since $\pi F(x) \neq 0$, $x \notin V(B)$. □

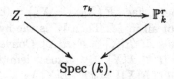

Example V. Let Z be a hypersurface $F(X_0, \ldots, X_n) = 0$. Normalize the coefficients so they all lie in R, not all in M. Then W is the hypersurface $\pi F = 0$.

(In fact, $A = F \cdot k[X]$. Thus $A \cap R[X] \supset F \cdot R[X]$. Conversely, if $G \in A \cap R[X]$, then $G = F \cdot H$ for some $H \in k[X]$. By Gauss' lemma, $H \in R[X]$, hence $G \in F \cdot R[X]$. Therefore $\pi\{A \cap R[X]\} = \pi F \cdot L[X]$.)

More specifically, take Z to be the hyperbola $X_1 X_2 = aX_0^2$ in $\mathbb{P}_k^2$, where $a \in M$. Then W is given by $X_1 X_2 = 0$, i.e., it is the union of 2 lines. Thus W can be reducible even when Z is irreducible. As a matter of fact, the only other elementary property that W does have when Z is irreducible is that it is connected. More generally, the *connectedness theorem* of Zariski states that if $Z \subset \mathbb{P}_k^n$ is any closed connected set, then its specialization over R is still connected.

Example W. One of the reasons why specialization is so important is that char (L) may be finite when char $(k) = 0$. For example, take k to be the field of algebraic numbers, and let R be a valuation ring such that $R \cap \mathbb{Q} = \mathbb{Z}_{(p)}$. Then L is an algebraic closure of $\mathbb{Z}/p\mathbb{Z}$. Take $n = 2$ and let Z be the Fermat curve:

$$X_1^k + X_2^k = X_0^k .$$

Its reduction mod R is again the plane curve $X_1^k + X_2^k = K_0^k$. This equation is irreducible, however, only when $p \not| k$. If $p^\nu \mid k$, $p^{\nu+1} \not| k$, then let $k = p^\nu \cdot k_0$. We find:

$$X_1^k + X_2^k - X_0^k = \left(X_1^{k_0} + X_2^{k_0} - X_0^{k_0}\right)^{p^\nu} ,$$

and W is a plane curve of lower order k_0, "taken with multiplicity p^ν".

This example makes it clear that the process of specialization should be further refined to take this multiplicity into account:

Definition 3. Let Z be a closed *subscheme* of $\mathbb{P}_k^n$ defined by a homogeneous ideal A. Let $B = \pi\{A \cap R[X_0, \ldots, X_n]\}$, and let B define the closed subscheme W of $\mathbb{P}_L^n$. Then W is the *specialization of* Z with respect to R.

Proposition 2. *Let Z be a closed subscheme of $\mathbb{P}_k^n$. Then there is one and only one closed subscheme $\mathcal{Z} \subset \mathbb{P}_R^n$ such that:*

1) For all $x \in \mathcal{Z}$, the ring $\underline{o}_{x,\mathcal{Z}}$ is a torsion-free R-module,
2) $Z = \mathcal{Z} \times_{\mathbb{P}_R^n} \mathbb{P}_k^n = \mathcal{Z} \times_{\mathrm{Spec}\,(R)} \mathrm{Spec}\,(k)$.

Moreover, the specialization of Z is the closed fibre of $\mathcal{Z}$ over b.

Proof. Suppose that a subscheme $\mathcal{Z}$ in $\mathbb{P}_R^n$ is defined by a collection of ideals

$$A_i \subset R\left[\frac{X_0}{X_i}, \ldots, \frac{X_n}{X_i}\right] .$$

Then condition (2) on $\mathcal{Z}$ means that the ideal $A_i \cdot K \subset K\left[\frac{X_0}{X_i}, \ldots, \frac{X_n}{X_i}\right]$ should be the ideal $I(Z \cap U_i)$. And condition (1) is equivalent to requiring that

$$R\left[X_0/X_i,\ldots,X_n/X_i\right]/A_i$$

be torsion-free for all i; or that for all $\alpha \in R$, $\alpha \neq 0$, and $f \in R\left[X_0/X_i,\ldots,X_n/X_i\right]$, $\alpha \cdot f \in A_i$ implies $f \in A_i$. But once the ideal $A_i^* = I\left(Z \cap U_i\right)$ is given, I claim the only ideal A_i with both these properties is precisely the ideal

$$A_i = A_i^* \cap R\left[X_0/X_i,\ldots,X_n/X_i\right] .$$

In fact, say B_i has both properties. Then $A_i^* = B_i \cdot K$ implies $B_i \subset A_i^* \cap R\left[X_0/X_i,\ldots,X_n/X_i\right]$ and that for all $f \in A_i^*$, $\alpha \cdot f \in B_i$ for some non-zero $\alpha \in R$; and if $f \in A_i^* \cap R\left[X_0/X_i,\ldots,X_n/X_i\right]$ so that $\alpha \cdot f \in B_i$, some α, then the torsion-freeness implies that f itself is in B_i. This shows that $\mathcal{Z}$ is unique. Now define $\mathcal{Z}$ by these ideals A_i: we leave it to the reader to check that on $U_i \cap U_j$, these define the same subscheme. The fibre of $\mathcal{Z}$ over b is defined by the affine ideals

$$\overline{A}_i = \pi\left\{A_i^* \cap R\left[X_n/X_i,\ldots,X_n/X_i\right]\right\} \subset L\left[X_0/X_i,\ldots,X_n/X_i\right] .$$

But W is given by the ideals

$$\begin{aligned} I\left(W \cap U_i\right) &= \left\{ f/X_i^d \,\middle|\, \begin{array}{l} f \in \pi A \cap R\left[X_0,\ldots,X_n\right] \\ f \text{ homogeneous, degree } d \end{array} \right\} \\ &= \left\{ \pi\left(f/X_i^d\right) \,\middle|\, \begin{array}{l} f \in A \cap R\left[X_0,\ldots,X_n\right] \\ f \text{ homogeneous, degree } d \end{array} \right\} \\ &= \pi\left\{I\left(Z \cap U_i\right) \cap R\left[X_0/X_i,\ldots,X_n/X_i\right]\right\} \\ &= \overline{A_i} . \end{aligned}$$

□

Example X. In $\mathbb{P}_k^2$, let Z be the union of the 3 points $z = y = 0$; $x = 0$, $y = \alpha$; $x = \alpha$, $y = 0$, where $x = X_1/X_0$, $y = X_2/X_0$ are affine coordinates and $\alpha \in M$. Then the homogeneous ideal of Z is generated by:

$$\begin{aligned} &X_1X_2 \\ &X_1\left(X_1 - \alpha X_0\right) \\ &X_2\left(X_2 - \alpha X_0\right) . \end{aligned}$$

Reducing this ideal mod M, we get $\left(X_1^2, X_1X_2, X_2^2\right)$: therefore W is the single point $x = y = 0$ in $\mathbb{P}_L^2$ with structure sheaf $k[x,y]/\left(x^2, xy, y^2\right)$ (cf. Ex. P, §5). It is interesting to work out other examples, where n points x_i are given in $\mathbb{P}_k^2$, such that $\rho(x_1) = \ldots = \rho(x_n) = (0,0)$, and see which ideal $A \subset k[x,y]$ of codimension n comes out.

Example Y (Hironaka). In $\mathbb{P}_k^3$, let Z be the twisted cubic which is the image of

$$\begin{aligned} X_0 &= 1 \\ X_1 &= t^3 \\ X_2 &= t^2 \\ X_3 &= \alpha t , \end{aligned}$$

$\alpha \in M$. The ideal A of Z is generated by:

$$\alpha X_0 X_1 - X_2 X_3, \quad \alpha^2 X_2 X_0 - X_3^2, \quad \alpha X_2^2 - X_1 X_3 \ .$$

It also contains the element $X_1^2 X_0 - X_2^3$. Therefore $\pi\{A \cap R[X_0, X_1, X_2, X_3]\}$ contains:

$$X_3^2, X_2X_3, X_1X_3, \quad \text{and} \quad X_1^2 X_0 - X_2^3 \ .$$

These elements define, as a set, the plane cubic curve C: $X_3 = 0$, $X_1^2 X_0 = X_2^3$. Also, giving t a value in R, one sees that $\rho(Z(k))$ contains all the points $(1, \beta^3, \beta^2, 0)$, $\beta \in L$. Since these are dense in C, it follows that C, *as a set*, is the specialization of Z over R. But the ideal $B = (X_3^2, X_2X_3, X_1X_3, X_1^2X_0 - X_2^3)$ is itself the intersection:

$$B = (X_3, X_1^2 X_0 - X_2^3) \cap (X_1, X_2, X_3)^2 \ .$$

It can be checked that $B = \pi\{A \cap R[X_0, \ldots, X_3]\}$ so that the scheme specialization of Z has nilpotent elements in its structure sheaf at the *one* point $P = (1,0,0,0)$. The reason for this can be seen in the affine coordinates $x = X_1/X_0$, $y = X_2/X_0$, $z = X_3/X_0$. Then Z, like C, contains $(1,0,0,0)$. After specialization, the whole curve Z is squashed into the horizontal plane $z = 0$. But at P, and only at P, Z has a vertical tangent. For this reason, it resists being squashed there and gets an embedded component in the end:

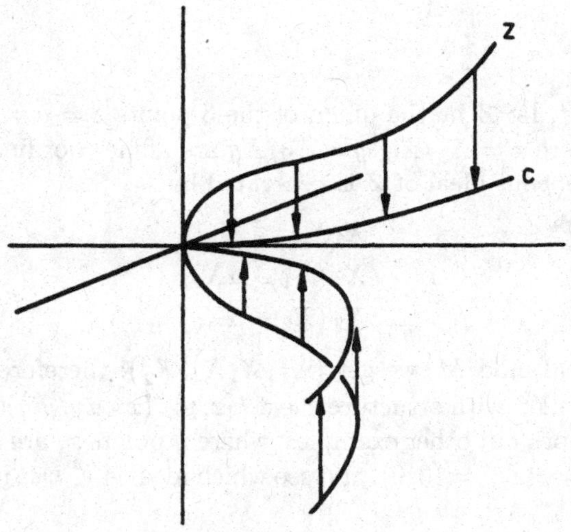

More concretely, suppose f is a polynomial in x, y, z, over R, that vanishes on Z. Then not only is $f(0,0,0) = 0$, but $\frac{\partial f}{\partial z}(0,0,0) = 0$. Therefore the same must hold for polynomials over L that vanish on the scheme-specialization of Z. In particular, z itself cannot vanish on the specialized scheme.

In practice, specializations usually arise like this: one is given a Dedekind domain D and a subscheme $\mathcal{Z} \subset \mathbb{P}^n_D$. For example, a set of homogeneous Diophantine equations define a subscheme $\mathcal{Z} \subset \mathbb{P}^n_{\mathbb{Z}}$. Or, in a geometric situation over an algebraically closed field L, one might have a subvariety $\mathcal{Z} \subset \mathbb{P}^n_L \times \mathbb{A}^1_L$. Then suppose you want to compare the generic fibre of $\mathcal{Z}$ over D with some closed fibre of $\mathcal{Z}$ over D. Let k_0 be the quotient field of D and $\pi : D \to L_0$ a map of D onto D/M, some maximal ideal M. Then by fibre product, we get

$$\begin{array}{ccccc} Z_{k_0} & \longrightarrow & \mathcal{Z} & \longleftarrow & Z_{L_0} \\ \cap & & \cap & & \cap \\ \mathbb{P}^n_{k_0} & \longrightarrow & \mathbb{P}^n_D & \longleftarrow & \mathbb{P}^n_{L_0} \\ \downarrow & & \downarrow & & \downarrow \\ \mathrm{Spec}\,(k_0) & \longrightarrow & \mathrm{Spec}\,(D) & \longleftarrow & \mathrm{Spec}\,(L_0) \end{array}$$

Then $\mathcal{Z}$ sets up a link between Z_{k_0} and Z_{L_0} which is very interesting. The point is that just as in Prop. 2, if $\underline{o}_{\mathcal{Z}}$ is a sheaf of torsion-free D-modules, then Z_k determines $\mathcal{Z}$, and therefore determines Z_{L_0}. This is just the non-algebraically closed generalization of the concept of specialization. It can be reduced to the previous case by taking fibre products. In fact, let k be an algebraic closure of k_0; using the place extension theorem, let $R \subset k$ be a valuation ring, quotient field k, dominating D_M. The residue field L over R is an algebraic closure of L_0. We obtain

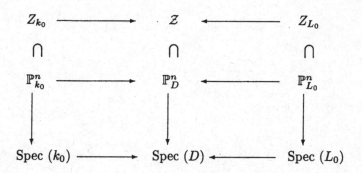

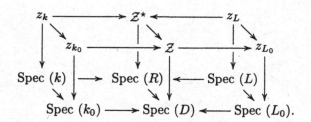

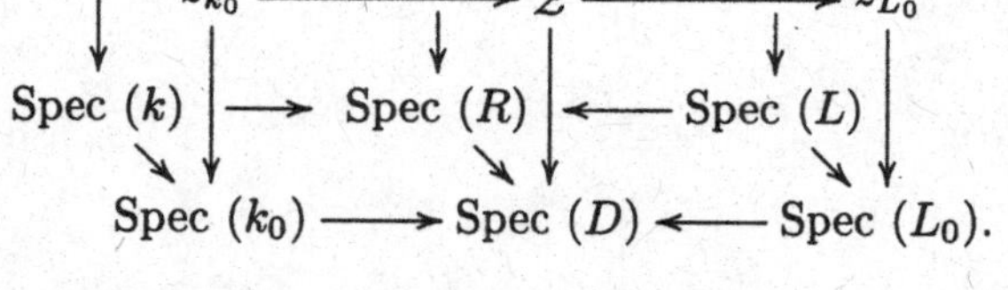

Assuming for a minute that $\mathcal{Z}^*$ is torsion-free over R, it follows from Prop. 2 that the subscheme $Z_L \subset \mathbb{P}^n_L$ is the specialization over R of the subscheme $Z_k \subset \mathbb{P}^n_k$. Z_{L_0} is just an L_0-structure on Z_L and Z_{k_0} is a k_0-subscheme on Z_k; so the added touch here is that having D around lets us specialize a k_0-structure to an L_0-structure as well. That $\mathcal{Z}^*$ is torsion-free over R follows from the following (by taking $R_2 = D_M$, $R_1 = R$, $M = \underline{o}_{x,\mathcal{Z}}$):

Lemma. *Let R_2 be a valuation ring, and R_1 an R_2-algebra. Let M be a torsion-free R_2-module. Then $M \otimes_{R_2} R_1$ is a torsion-free R_1-algebra.*

Proof. Recall that a finitely generated torsion-free module over a valuation ring is free. Therefore M, as a torsion-free module, is the direct limit of free R_2-modules. Therefore $M \otimes_{R_2} R_1$ is a direct limit of free R_1-modules. Therefore $M \otimes_{R_2} R_1$ is torsion-free. □

III. Local Properties of Schemes

So far, besides mere definitions, the only properties of varieties that we have studied are i) their dimensions, and ii) whether or not they are complete. In this chapter, we will analyze 2 local concepts: i) when is a variety "manifold-like" at a given point, and ii) if it is not manifold-like, how many "branches" does it have at a given point. The main problem here is to carry over our intuitive differential-geometric and topological ideas into a purely algebraic setting. In some cases, as with the module of differentials, this involves setting up an algebraic theory in close analogy to the geometric one. In other cases, as with the concepts of normality and flatness, an idea is introduced into the geometry primarily because of the simplicity and naturality of the algebra. In these cases, the geometric meaning is hidden and often subtle, but the first aim of the theory is to elucidate this meaning: one should expect a true synthesis of algebraic and geometric ideas in cases like these.

Many different techniques have been used to develop the local theory of schemes. For example, there are:

i) *projective methods.* These involve reducing the question to graded rings – either by adding a new variable to make your polynomials homogeneous, or by passing from a filtered object to its graded object. This enables you to use global projective methods in a local situation.
ii) *power series methods.* These involve taking completions of our rings in I-adic topologies and a close comparison of the original ring with its completion. The Weierstrass preparation theorem is a key tool here.
iii) *homological methods.* These involve the study of syzygies, i.e., projective resolutions of modules, the dual notion of injective resolutions, and are adapted to concepts like that of 0-divisor which have good exact-sequence translations.

These methods often overlap (e.g., (i) and (iii) in concept of Hilbert polynomial). I will try to give a sampling of all of them, although we prefer (i) as being closer to the spirit of this book. To do this, however, I have had to assume many more standard results in Commutative Algebra than in previous chapters. My advice to the reader, if he has not seen these results before, is to skip back and forth between this book and books on Commutative Algebra, using the geometric examples to enrich the algebra, and the clean-cut algebraic techniques to clarify the geometry.

§1. Quasi-coherent modules

Almost all local theory involves modules as well as rings and ideals. The purpose of this section is to put modules in their convenient geometric setting. This is really just a technical digression before we get on to the more interesting geometry.

Definition 1. Let $(X, \underline{o}_X)$ be a scheme. An *$\underline{o}_X$-module* is a sheaf F of abelian groups on X plus, for all $U \subset X$ open, a $\Gamma(U, \underline{o}_X)$-module structure on $\Gamma(U, F)$, such that if $V \subset U$, the diagram:

$$\begin{array}{ccccc} \Gamma(U, \underline{o}_X) & \times & \Gamma(U, F) & \longrightarrow & \Gamma(U, F) \\ \downarrow & & \downarrow & & \downarrow \\ \Gamma(V, \underline{o}_X) & \times & \Gamma(V, F) & \longrightarrow & \Gamma(V, F) \end{array}$$

commutes. A *homomorphism* of $\underline{o}_X$-modules is a sheaf homomorphism that preserves the module structure on every open set.

Basic Example. Let $X = \operatorname{Spec} R$, and let M be an R-module. We define an associated $\underline{o}_X$-module $\widetilde{M}$ in the same way in which we defined $\underline{o}_X$ itself. To the distinguished open set X_d we assign the module M_f, checking that when $X_f \subset X_g$ we get a natural map $M_h \to M_f$. We check that $\varinjlim_{[P] \in X_f} M_f = M_P$, and that our assignment gives a "sheaf on the distinguished open sets", i.e., satisfies Lemmas 1 and 2 of §1, Ch. 2. (The proofs are almost word-for-word the same as in the construction of $\underline{o}_X$.) We then can extend to a sheaf on X in the canonical way, letting $\Gamma(U, \widetilde{M})$ be the set of all $s \in \prod_{[P] \in U} M_P$ which are given locally by sections over distinguished open sets. We put an $\underline{o}_X$-module structure on this set by restricting to $\Gamma(U, \underline{o}_X) \times \Gamma(U, \widetilde{M})$ the natural map $\prod R_P \times \prod M_P \to \prod M_P$ and observing that it lands us in $\Gamma(U, \widetilde{M})$; clearly, this commutes with restriction, so we do have an $\underline{o}_X$-module.

Proposition 1. *Let M, N be R-modules. The natural map* $\hom_{\underline{o}_X}(\widetilde{M}, \widetilde{N}) \to \hom_R(M, N)$ *gotten by taking global sections is a bijection.*

Proof. Given a homomorphism $\varphi : M \to N$ we get induced homomorphisms $\varphi_f : M_f \to N_f$ for every f. This gives us homomorphisms $\Gamma(U, \widetilde{M}) \to \Gamma(U, \widetilde{N})$ on every distinguished open set U, compatible with restriction; that necessarily induces homomorphisms for all open sets U. It is easily seen that this process gives an inverse to $\hom(\widetilde{M}, \widetilde{N}) \to \hom(M, N)$. □

Corollary. *The category of R-modules is equivalent to the category of $\underline{o}_X$-modules of the form $\widetilde{M}$.*

Proposition 2. *The sequence $M \to N \to P$ is exact if and only if the sequence $\widetilde{M} \to \widetilde{N} \to \widetilde{P}$ is exact.*

Proof. The sequence of sheaves is exact if and only if it is exact at all stalks, i.e., for all primes P, $M_P \to N_P \to P_P$ is exact. This is equivalent to exactness of $M \to N \to P$ by a standard theorem of commutative algebra (cf. Bourbaki). □

Corollary. *If $\widetilde{M} \to \widetilde{N}$ is a homomorphism, then its kernel, cokernel, and image are of the form $\widetilde{K}$ for some R-module K.*

Theorem-Definition 3. *Let X be a scheme, $\mathcal{F}$ an $\underline{o}_X$-module. The following are equivalent:*

(1) For all $U \subset X$ affine open, $\mathcal{F}|_U \cong \widetilde{M}$ for some $\Gamma(U, \underline{o}_X)$-module M.
(2) There is an open affine cover $\{U_i\}$ of X such that for all i, $\mathcal{F}|_{U_i} \cong \widetilde{M}_i$ for some $\Gamma(U_i, \underline{o}_X)$-module M_i.
(3) For all $x \in X$, $\exists$ a neighbourhood U of x and an exact sequence of $\underline{o}_X|_U$-modules

$$\underline{o}_X|_U^{(I)} \longrightarrow \underline{o}_X|_U^{(J)} \longrightarrow \mathcal{F}|_U \longrightarrow 0$$

(here the exponent in parentheses denotes direct sum).
(4) For all $V \subset U$ open affines, the canonical map

$$\Gamma(U, \mathcal{F}) \otimes_{\Gamma(U, \underline{o}_X)} \Gamma(X, \underline{o}_X) \longrightarrow \Gamma(X, \mathcal{F})$$

is an isomorphism.

An $\underline{o}_X$-module with these properties is called *quasi-coherent.*

Proof. Let's check

$$(4) \Longrightarrow (3) \Longrightarrow (2) \Longrightarrow (1) \Longrightarrow (4).$$

(4) $\Longrightarrow$ (3): Before we prove this, we really ought to define a direct sum of $\underline{o}_X$-modules. In fact, if $\{F_\alpha\}$ is a collection of $\underline{o}_X$-modules, then $\oplus F_\alpha$ is the sheaf associated to the presheaf:

$$U \longmapsto \bigoplus_\alpha \Gamma(U, F_\alpha).$$

For all affine U, check that:

$$\oplus \widetilde{M}_\alpha \cong \widetilde{\oplus M_\alpha}.$$

Now assume (4) and let $x \in X$. Take any open affine neighbourhood U of x and let $R = \Gamma(U, \underline{o}_X)$. The R-module $\Gamma(U, \mathcal{F})$ has a presentation:

$$R^{(I)} \longrightarrow R^{(J)} \longrightarrow \Gamma(U, \mathcal{F}) \longrightarrow 0.$$

This induces the top sequence of homomorphisms:

$$\begin{array}{ccccccc}
\widetilde{R^{(I)}} & \longrightarrow & \widetilde{R^{(J)}} & \longrightarrow & \widetilde{\Gamma(U,\mathcal{F})} & \longrightarrow & 0 \\
\| & & \| & & \downarrow \alpha & & \\
\underline{o}_X\big|_U^{(I)} & \longrightarrow & \underline{o}_X\big|_U^{(J)} & \longrightarrow & \mathcal{F}\big|_U & \longrightarrow & 0\,.
\end{array}$$

There is a canonical homomorphism $\alpha : \widetilde{\Gamma(U,\mathcal{F})} \to \mathcal{F}\big|_U$ that then gives the bottom sequence. To prove (3), it suffices to check that α is an isomorphism. But α is given, on distinguished open subsets U_f of U by:

$$\begin{array}{ccc}
\Gamma\left(U_f, \widetilde{\Gamma(U,\mathcal{F})}\right) & \longrightarrow & \Gamma(U_f,\mathcal{F}) \\
\| & & \\
\Gamma(U,\mathcal{F}) \underset{R}{\otimes} R_f & &
\end{array}$$

which is an isomorphism by (4).

(3) $\Longrightarrow$ (2): Cover X by open affines $U_i = \operatorname{Spec}(R_i)$ on which there are exact sequences:

$$\underline{o}_X\big|_{U_i}^{(I_i)} \xrightarrow{\phi_i} \underline{o}_X\big|_{U_i}^{(J_i)} \longrightarrow \mathcal{F}\big|_{U_i} \longrightarrow 0\,.$$

Since $\underline{o}_X\big|_{U_i}^{(I_i)} = \widetilde{R_i^{(I_i)}}$, $\underline{o}_X\big|_{U_i}^{(J_i)} = \widetilde{R^{(J_i)}}$, the cokernel of ϕ_i is a module of type $\widetilde{K}$, by the Cor. of Prop. 2 and this is (2).

(2) $\Longrightarrow$ (1): Notice first that if U is an open affine set such that $\mathcal{F}\big|_U \cong \widetilde{M}$ for some $\Gamma(U,\underline{o}_X)$-module M, then for all $f \in \Gamma(U,\underline{o}_X)$, $\mathcal{F}\big|_{U_f} \cong \widetilde{M_f}$.

Therefore, starting with condition (2), we deduce that there is a *basis* $\{U_i\}$ for the topology of X consisting of open affines such that $\mathcal{F}\big|_{U_i} \cong \widetilde{M_i}$. Now if U is any open affine set and $R = \Gamma(U,\underline{o}_X)$, we can cover each of these U_i's by smaller open affines of the type U_g, $g \in R$. Since $U_g = (U_i)_g$, $\mathcal{F}\big|_{U_g}$ is still of the form $\widetilde{(M_i)_g}$. In other words, we get a finite covering of U by affines U_{g_i} such that $\mathcal{F}\big|_{U_{g_i}} \cong \widetilde{N_i}$, N_i an R_{g_i}-module.

For every open set $V \subset U$, the sequence

$$0 \longrightarrow \Gamma(V,\mathcal{F}) \longrightarrow \prod_i \Gamma(V \cap U_{g_i},\mathcal{F}) \longrightarrow \prod_{i,j} \Gamma(V \cap U_{g_i} \cap U_{g_j},\mathcal{F})$$

is exact. Define new sheaves $\mathcal{F}_i{*}$ and $\mathcal{F}^*_{i,j}$ by:

$$\begin{aligned}\Gamma(V,\mathcal{F}_i^*) &= \Gamma(V\cap U_{g_i},\mathcal{F})\\ \Gamma(V,\mathcal{F}_{i,j}^*) &= \Gamma(V\cap U_{g_i}\cap U_{g_j},\mathcal{F})\ .\end{aligned}$$

Then the sequence of sheaves

$$0\longrightarrow\mathcal{F}\longrightarrow\prod_i\mathcal{F}_i^*\longrightarrow\prod_{i,j}\mathcal{F}_{i,j}^*$$

is exact, so to prove that $\mathcal{F}$ is of the form $\widetilde{M}$, it suffices to prove this for $\mathcal{F}_i^*$ and $\mathcal{F}_{i,j}^*$. But if M_i^0 is M_i viewed as an R-module, then $\mathcal{F}_i^*\cong\widetilde{M_i^0}$. In fact, for all distinguished open sets U_g,

$$\begin{aligned}\Gamma(U_g,\mathcal{F}_i^*) &= \Gamma(U_g\cap U_{g_i},\mathcal{F})\\ &= \Gamma\left(\left(U_{g_i}\right)_g,\mathcal{F}\big|_{U_{g_i}}\right)\\ &= (M_i)_g\\ &= \Gamma\left(U_g,\widetilde{M_i^0}\right).\end{aligned}$$

The same argument works for the $\mathcal{F}_{i,j}^*$'s.

(1) $\Longrightarrow$ (4): Let $U=\text{Spec}\ (R)$, $V=\text{Spec}\ (S)$ be open affines such that $U\supset V$. Let $\mathcal{F}\big|_U=\widetilde{M}$. Present the R-module M via:

$$R^{(I)}\xrightarrow{\ \alpha\ }R^{(J)}\longrightarrow M\longrightarrow 0\ .$$

This gives us:

$$\underline{o}_X\big|_U^{(I)}\xrightarrow{\ \widetilde{\alpha}\ }\underline{o}_X\big|_U^{(J)}\longrightarrow\mathcal{F}\big|_U\longrightarrow 0\ .$$

By restriction, we get an exact sequence on V:

$$\underline{o}_X\big|_V^{(I)}\xrightarrow{\ \widetilde{\beta}\ }\underline{o}_X\big|_V^{(J)}\longrightarrow\mathcal{F}\big|_V\longrightarrow 0\ .$$

But $\mathcal{F}\big|_V$ is an $\underline{o}_X$-module of type $\widetilde{N}$, so we get an exact sequence of S-modules:

$$S^{(I)}\xrightarrow{\ \beta\ }S^{(J)}\longrightarrow N\longrightarrow 0\ .$$

On the other hand, the first exact sequence gives, by tensor product, the exact sequence:

$$\begin{array}{ccccccc}R^{(I)}\otimes_R S & \xrightarrow{\ \alpha'\ } & R^{(J)}\otimes_R S & \longrightarrow & M\otimes_R S & \longrightarrow & 0\\ \Big\| & & \Big\| & & & & \\ S^{(I)} & & S^{(J)}\ . & & & & \end{array}$$

But β and α' are the same homomorphisms. Hence N and $M \otimes_R S$ are cokernels of the same map, hence they are isomorphic. □

Example A. Let $X = \mathrm{Spec}\,(R)$, where R is a principal valuation ring (= discrete, and rank 1). Spec (R) has one open and one closed point: x_1, x_0. Let U be the open set $\{x_1\}$. Then a sheaf $\mathcal{F}$ on X just consists in 2 sets

$$\begin{aligned} M_0 &= \Gamma(X, \mathcal{F}) \\ M_1 &= \Gamma(U, \mathcal{F}) \end{aligned}$$

and a restriction map res : $M_0 \to M_1$. If K is the quotient field of R, and the $\underset{\sim}{o}_X$-module $\mathcal{F}$ consists in an R-module M_0, a K-module M_1, and an R-linear restriction map res : $M_0 \to M_1$. By condition (4) of the theorem, it is quasi-coherent if and only if the induced map

$$M_0 \otimes_R K \longrightarrow M_1$$

is an isomorphism. In this case, $\mathcal{F} = \widetilde{M_0}$.

$\mathcal{F}$ could fail to be quasi-coherent in two different ways. There might not be enough global sections, e.g., $M_1 = K$, $M_0 = (0)$; on the other hand, all the sections might be supported only on the closed point, e.g., $M_1 = (0)$, $M_0 = R$. In a quasi-coherent sheaf, the closed point can support only the "torsion" part, i.e., the kernel of $M_0 \to M_0 \otimes_R K$.

Note that an $\underset{\sim}{o}_X$-ideal $\mathcal{Q}$ is quasi-coherent in this definition if and only if it is quasi-coherent in our former sense (Ch. 2, §5). Moreover, kernels, cokernels, and images of maps between quasi-coherent modules are quasi-coherent. For by the theorem quasi-coherence is a local condition, and we know the result for affine schemes. Yet another way of combining quasi-coherent $\underset{\sim}{o}_X$-modules is this: Let $\mathcal{Q}$ be a quasi-coherent $\underset{\sim}{o}_X$-ideal, $\mathcal{F}$ a quasi-coherent $\underset{\sim}{o}_X$-module. Let $\mathcal{Q}\cdot\mathcal{F}$ be the least submodule of $\mathcal{F}$ containing all elements $\alpha \cdot s$ ($\alpha \in \Gamma(U, \mathcal{Q})$, $s \in \Gamma(U, \mathcal{F})$). Then $\mathcal{Q}\cdot\mathcal{F}$ is also quasi-coherent, and if U is any open affine, $\mathcal{Q}\cdot\mathcal{F}\big|_U = \widetilde{I \cdot M}$ where $I = \Gamma(U, \mathcal{Q})$, $M = \Gamma(U, \mathcal{F})$. [The first statement follows from the second, whoose proof is left as an exercise.]

As a less trivial example, we will define for every morphism $X \xrightarrow{f} Y$ a very important quasi-coherent $\underset{\sim}{o}_X$-module $\Omega_{X/Y}$, called the *sheaf of relative differentials*. First, we recall the commutative algebra involved.

Let $A \to B$ be a homomorphism of rings. We define a B-module $\underline{\Omega_{B/A}}$ by taking a free B-module on the symbols $\{d\beta \mid \beta \in B\}$ and dividing out by the relations

a) $d(\beta_1 + \beta_2) = d\beta_1 + d\beta_2$
b) For $\alpha, \beta \in B$, $\alpha d\beta + \beta d\alpha = d(\alpha\beta)$
c) $d\beta = 0$ if β comes from A.

Notice that for all B-modules C,

$$\hom_B(\Omega_{B/A}, C) \cong \left\{ \begin{array}{l} \text{module of } A\text{-derivations} \\ D : B \to C \end{array} \right\}.$$

In fact, if $\tau : \Omega_{B/A} \to C$ is a B-homomorphism, then $\beta \to \tau(d\beta)$ is an A-derivation from B to C. And if $D : B \to C$ is an A-derivation, define τ by

$$\tau\left(\sum \beta\alpha_i d\beta_i\right) = \sum \alpha_i \cdot D(\beta_i) .$$

Let $B \otimes_A B \xrightarrow{\delta} B$ be the map $\beta_1 \otimes \beta_2 \to \beta_1\beta_2$, and let $I = \ker \delta$. I is a $B \otimes_A B$ module, and I/I^2 is a $B \otimes_A B/I$-module, that is, a B-module (since multiplication by $\beta \otimes 1$ and by $1 \otimes \beta$ give the same result in I/I^2).

Theorem 4. *I/I^2 is canonically isomorphic to $\Omega_{B/A}$.*

Proof. To map $\Omega_{B/A}$ to I/I^2, we let $d\beta$ go to $1 \otimes \beta - \beta \otimes 1$ in I. Clearly a) and c) are satisfied. And $d(\alpha\beta)$ goes to:

$$\begin{aligned} 1 \otimes \alpha\beta - \alpha\beta \otimes 1 &= 1 \otimes \alpha\beta - \alpha \otimes \beta + \alpha \otimes \beta - \alpha\beta \otimes 1 \\ &= (1 \otimes \beta)(1 \otimes \alpha - \alpha \otimes 1) + (\alpha \otimes 1)(1 \otimes \beta - \beta \otimes 1) . \end{aligned}$$

This is to be the image of $\beta d\alpha + \alpha d\beta$, so the relations b) are satisfied too. So we get a map $\Omega_{B/A} \to I/I^2$.

To get a map backwards, we define a ring $R = B \oplus \Omega_{B/A}$, where B acts on $\Omega_{B/A}$ through the module action and elements of $\Omega_{B/A}$ have square zero. Define a map $B \times B \to E$ by $(\beta_1, \beta_2) \longmapsto (\beta_1\beta_2, \beta_1 d\beta_2)$. This is clearly biadditive; it is A-bilinear, for if α comes from A,

$$\begin{aligned} (\alpha\beta_1, \beta_2) &\longmapsto (\alpha\beta_1\beta_2, \alpha\beta_1 d\beta_2) \\ (\beta_1, \alpha\beta_2) &\longmapsto (\alpha\beta_1\beta_2, \beta_1 d(\alpha\beta_2)) \end{aligned}$$

and $d(\alpha\beta_2) = \alpha d\beta_2$. Hence this induces a map $B \otimes_A B \to E$. It is easily checked that this is a ring homomorphism. Moreover, I goes into $\Omega_{B/A}$ (by definition, I is the set of those elements whose images have first coordinate 0; and all squares are 0 in $\Omega_{B/A}$, so the map must factor via $I/I^2 \to \Omega_{B/A}$).

It is easy to see that our two maps are inverses. □

In some books, the module of differentials is defined as the dual of the module of derivations, i.e., the double dual of $\Omega_{B/A}$. Double dualizing will in general lose information, so the definition here, due to Kähler, is preferable.

Example B. Let $A = k$, $B = k[X_1, \ldots, X_n]$. Then $\Omega_{B/A}$ is a free B-module with generators $dX_1, \ldots, dX_n$, and

$$dg = \sum_{i=1}^{n} \frac{\partial g}{\partial X_i} \cdot dX_i, \qquad \text{all } g \in B .$$

More generally, if

$$B = k[X_1, \ldots, X_n] / (f_1, \ldots, f_m) ,$$

then $\Omega_{B/A}$ is generated, as B-module, by $dX_1, \ldots, dX_n$, but with relations:

$$df_i = \sum_{j=1}^{n} \frac{\partial f_i}{\partial X_j} \cdot dX_j = 0 .$$

Example C. Let $A = k$, and let $B = K \supset k$ be an extension field. Then the dual of $\Omega_{K/k}$ (over K) is the vector space of k-derivations from K to K. In particular, if K is finitely generated and separable over k, the dimension of both of these vector spaces is the transcendence degree of K over k (cf. Zariski-Samuel, vol. 1). If $n = \text{tr.d.}(K/k)$ and $f_1, \ldots, f_n \in K$, then

$$\left\{\begin{array}{l} df_1, \ldots, df_n \text{ are a basis} \\ \text{of } \Omega_{K/k} \end{array}\right\} \Longleftrightarrow \left\{\begin{array}{l} \text{all } k\text{-derivations of } K \text{ that} \\ \text{annihilate } f_1, \ldots, f_n \text{ are } 0 \end{array}\right\}$$
$$\Longleftrightarrow \left\{\begin{array}{l} f_1, \ldots, f_n \text{ are a separating} \\ \text{transcendence basis of } K/k \end{array}\right\}.$$

On the other hand, if char $k = p$ and $K = k(b)$ where $b^p = a \in k$, then $\Omega_{B/A}$ is the free K-module generated by db.

Example D. Let $A = k$, $B = k[X,Y]/(X \cdot Y)$. Then by Ex. B, dX and dY generate $\Omega_{B/A}$ with the one relation $XdY + YdX = 0$.

Consider the element $\omega = XdY = -YdX$. Then $X\omega = Y\omega = 0$, so the submodule M generated by ω is $k\omega$, a one-dimensional k-space. On the other hand, in Ω/M, we have $XdY = YdX = 0$, so $\Omega/M \cong B \cdot dX \oplus B \cdot dY$. Note that $B \cdot dX \cong B/(Y) \cong \Omega_{B_X/k}$, where $B_X = B/(Y) \cong k[X]$; likewise, $B \cdot dY \cong \Omega_{B_Y/k}$. That is, the module of differentials on B (which looks like $+$) is the module of differentials on the horizontal and vertical lines separately *extended* by a torsion module. (One can even check that the extension is non-trivial, i.e., does not split.)

Problem. Let R be a finitely generated k-algebra. Show that $\Omega_{R/k} = (0)$ if and only if R is a direct sum of finite separable extension fields over k.

Hint: Let P be a prime ideal in R and let $L =$ quotient field of R/P. Show $\Omega_{L(k)} = (0)$, and hence L is finite separable over k. Thus show that R satisfies the d.c.c. Finally show that nilpotents can't occur either.

All this is easy to globalize. Let a morphism $f : X \to Y$ be given. The closed immersion

$$\Delta : X \longrightarrow X \times_Y X$$

globalizes the homomorphism $\delta : B \otimes_A B \to B$. In fact, if $U = \text{Spec}\,(B) \subset X$ and $V = \text{Spec}\,(A) \subset Y$ are open affines such that $f(U) \subset V$, then $U \times_V U$ is an open affine in $X \times_Y X$, $\Delta^{-1}\,(U \times_V U) = U$, and we get a commutative diagram:

$$\begin{array}{ccc} \Gamma\,(U, \underline{o}_X) & \xleftarrow{\ \Delta^*\ } & \Gamma\,(U \times_V U,\ \underline{o}_{X \times_Y X}) \\ \| & & \| \\ B & \longleftarrow & B \otimes_A B\ . \end{array}$$

Let $\mathcal{Q}$ be the quasi-coherent $\underline{o}_{X\times_Y X}$-ideal defining the closed subscheme $Z = \Delta(X)$. Then $\mathcal{Q}^2$ is also a quasi-coherent $\underline{o}_{X\times_Y X}$-ideal, and $\mathcal{Q}/\mathcal{Q}^2$ is a quasi-coherent $\underline{o}_{X\times_Y X}$-module. It is also a module over $\underline{o}_{X\times_Y X}/\mathcal{Q}$, which is $\underline{o}_Z$ extended by zero. As its stalks are all 0 off Z, $\mathcal{Q}/\mathcal{Q}^2$ is actually a $(Z, \underline{o}_Z)$-module. It is quasi-coherent by virtue of the nearly tautologous:

Lemma. *If $S \subset T$ are a schema and a closed subscheme, and if $\mathcal{F}$ is an $\underline{o}_S$-module, then $\mathcal{F}$ is a quasi-coherent $\underline{o}_S$-module on S if and only if $\mathcal{F}$, extended by O on $T - S$, is a quasi-coherent $\underline{o}_T$-module on T.*

Definition 2. $\Omega_{X/Y}$ is the quasi-coherent $\underline{o}_X$-module obtained by carrying $\mathcal{Q}/\mathcal{Q}^2$ back to X by the isomorphism $\Delta : X \xrightarrow{\sim} Z$.

Clearly, for all $U = \text{Spec}\,(B) \subset X$ and $V = \text{Spec}\,(A) \subset Y$ such that $f(U) \subset V$, the restriction of $\Omega_{X/Y}$ to U is just $\Omega_{B/A}$. Therefore, we have globalized our affine construction.

There are many unsolved problems concerning $\Omega_{X/Y}$. For example, if $Y = \text{Spec}\,(k)$ and X is a variety over the field k, when is $\Omega_{X/Y}$ torsion-free?

We can now fill in the gap in Chapter 2, §7:

Proposition 5. *To prove that a morphism $f : X \to Y$ is affine or finite, it suffices to check the defining property for the open sets $U_i \subset Y$ of one affine covering.*

Proof. Let $Y = \cup U_i$, $U_i = \text{Spec}\,(R_i)$ and assume $f^{-1}(U_i)$ is affine for all i. First of all, note that the sheaf $f_*(\underline{o}_X)$ is a quasi-coherent sheaf of $\underline{o}_Y$-algebras such that, if $f^{-1}(U_i) = \text{Spec}\,(S_i)$, then regarding S_i as an R_i-module,

$$f_*(\underline{o}_X)\big|_{U_i} \cong \widetilde{S}_i\,.$$

To see this, notice that we have natural maps:

$$\Gamma\left((U_i)_g, \widetilde{S}_i\right) = (S_i)_g \longrightarrow \Gamma\left(f^{-1}(U_i)_g, \underline{o}_X\right) = \Gamma\left((U_i)_g, f_*(\underline{o}_X)\right)$$

for all $g \in R_i$. And since $f^{-1}(U_i)_g \cong f^{-1}(U_i) \times_{U_i} (U_i)_g$,

$$\Gamma\left(f^{-1}(U_i)_g, \underline{o}_X\right) \cong \Gamma\left(f^{-1}(U_i), \underline{o}_X\right) \otimes_{R_i} (R_i)_g \cong (S_i)_g$$

therefore all these maps are isomorphisms.

Now let $U = \text{Spec}\,(R) \subset Y$ be any open affine subset. Let $S = \Gamma(U, f_*(\underline{o}_X))$. I claim $f^{-1}(U) \cong \text{Spec}\,(S)$, which will prove that f is affine. To begin with, the homomorphism

$$S = \Gamma(U, f_*(\underline{o}_X)) \xrightarrow{\sim} \Gamma(f^{-1}(U), \underline{o}_X)$$

defines a morphism of schemes over R:

$$\begin{array}{ccc} f^{-1}(U) & \xrightarrow{\Psi} & \mathrm{Spec}\,(S) \\ & \searrow \quad \swarrow & \\ & U = \mathrm{Spec}(R). & \end{array}$$

Look at the restriction ψ_i of ψ to the parts of $f^{-1}(U)$ and Spec (S) over the open subset $U \cap U_i$ of U. First of all, $f^{-1}(U) \cap f^{-1}(U_i)$, as a subset of $f^{-1}(U_i)$, is isomorphic to

$$f^{-1}(U_i) \times_{U_i} (U \cap U_i)$$

and since all of these are affine schemes, it is affine. If $R_i' = \Gamma\,(U \cap U_i, \underline{o}_Y)$, its coordinate ring is just $S_i \otimes_{R_i} R_i'$, which (by (4), Theorem 3) is $\Gamma\,(U \cap U_i, f_*(\underline{o}_X))$. On the other hand, again by (4), Theorem 3,

$$\Gamma\,(U \cap U_i, f_*(\underline{o}_X)) \cong \Gamma\,(U, f_*(\underline{o}_X)) \otimes_R R_i' = S \otimes_R R_i'$$

which is the ring of Spec $(S) \times_U (U \cap U_i)$, i.e., the inverse image of $U \cap U_i$ in Spec (S). Putting all this together means that ψ_i is a morphism on affine schemes such that ψ_i^* is an isomorphism. Therefore ψ_i and hence ψ are isomorphisms.

Finally, assume S_i is a finite R_i-module. Then using the above notation, $S_i \otimes_{R_i} R_i'$ is a finite R_i'-module, hence so is $S \otimes_R R_i'$. Now U is covered by a finite number of these subsets $U \cap U_i$: say by $U \cap U_i, \ldots, U \cap U_n$. Then build up a finitely generated submodule S^* of S by throwing in enough elements to generate $S \otimes_R R_i'$ over R_i' for $1 \leq i \leq n$. Then the submodule S^* of S is big enough so that for all prime ideals $P \subset R$, the induced map $S_P^* \to S_P$ is surjective (i.e., localize in 2 stages, from R to R_i' and then R_P, for a suitable i). Therefore $S^* = S$ (cf. Bourbaki), so S was a finite R-module anyway. □

Problem. Suppose $f : X \to Y$ is a morphism of schemes such that for all $U \subset Y$ open and affine, $f^{-1}(U)$ admits a *finite* open affine covering (i.e., $f^{-1}(U)$ is quasi-coherent). Let $\mathcal{F}$ be any quasi-coherent $\underline{o}_X$-module. Prove that $f_*(\mathcal{F})$ is a quasi-coherent $\underline{o}_Y$-module.

§2. Coherent modules

For one more section, we must continue to study modules for their own sakes, this time with finiteness assumptions.

Definition 1. A scheme X is *noetherian* if for all open sets $U \subset X$, the partially ordered set of closed subschemes of U satisfies the descending chain condition.

This implies first of all that for all open sets U in X the closed sub*sets* of U satisfy the d.c.c., hence U is quasi-compact. In other words, X is a noetherian topological space. As in Ch. I, this implies that every closed subset $Z \subset X$ can be written in exactly one way as a finite union of irreducible closed subsets

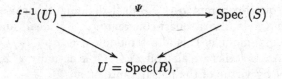

$Z_1, \ldots, Z_n$ such that $Z_i \not\supset Z_j$. These Z_i are called, as before, the *components* of Z. Secondly, if $U = \text{Spec }(R)$ is an *affine* open set, then in view of the order-reversing bijection of R, X being noetherian implies that the ring R is noetherian. Moreover, conversely,

Proposition 1. *Let X be a scheme. If X can be covered by a finite number of open affine sets $U_i = \text{Spec }(R_i)$ such that each R_i is a noetherian ring, then X is a noetherian scheme.*

Proof. Let U be any open subset of X, and let $Z_1 \supset Z_2 \supset Z_3 \supset \ldots$ be a chain of closed subschemes. The open subset $U \cap U_i$ of U_i can be covered by a finite number of distinguished open sets $(U_i)_{g_{ij}}$, $g_{ij} \in R_i$, since R_i is noetherian. Therefore U itself is covered by a *finite* number of open affines $(U_i)_{g_{ij}}$ which are the Spec's of the *noetherian* rings $(R_i)_{g_{ij}}$. By the noetherianness of $(R_i)_{g_{ij}}$ the chain of subschemes of $(U_i)_{g_{ij}}$:

$$\left\{ Z_k \cap (U_i)_{g_{ij}} \right\}_{k=1,2,3,\ldots}$$

is stationary for each i, j. By the finiteness of the covering, this implies that the chain $\{Z_k\}$ itself is stationary. □

Definition 2. Let X be a noetherian scheme, $\mathcal{F}$ a quasi-coherent $\underline{o}_X$-module. Then $\mathcal{F}$ is *coherent* if for all open affines $U \subset X$, $\Gamma(U, \mathcal{F})$ is a finite $\Gamma(U, \underline{o}_X)$-module.

As usual, to check that a quasi-coherent $\mathcal{F}$ is coherent, it suffices to check that $\Gamma(U_i, \mathcal{F})$ is a finite $\Gamma(U_i, \underline{o}_X)$-module for the open sets U_i of *one* affine covering $\{U_i\}$ of X.

[In fact, if this is so, then by (4), Theorem 3, §1, $\Gamma(V, \mathcal{F})$ is a finite $\Gamma(V, \underline{o}_X)$-module for all affine subsets V of any U_i. Then if U is any affine, $M = \Gamma(U, \mathcal{F})$, and $M_i = \Gamma(U \cap U_i, \mathcal{F})$. M_i is a finite $\Gamma(U \cap U_i, \underline{o}_X)$-module. Cover U by a *finite* set of $U \cap U_i$'s: say $U \cap U_1, \ldots, U \cap U_n$. Then to generate M over $\Gamma(U, \underline{o}_X)$, it suffices to take enough elements of M so that their images in each M_i $(1 \leq i \leq n)$ generate M_i over $\Gamma(U \cap U_i, \underline{o}_X)$.]

Note that all quasi-coherent $\underline{o}_X$-ideals are coherent. More generally quasi-coherent sub and quotient modules of coherent $\underline{o}_X$-modules are coherent. If $f : X \to Y$ is a morphism of finite type, then $\Omega_{X/Y}$ is a coherent $\underline{o}_X$-module. If f is affine, then $f_*(\underline{o}_X)$ is coherent if and only if f is a finite morphism.

Definition 3. Let $\mathcal{F}$ be a coherent $\underline{o}_X$-module on the noetherian scheme X, $x \in X$ is an ***associated point*** of $\mathcal{F}$ if $\exists$ an open neighbourhood U of x and an element $a \in \Gamma(U, \mathcal{F})$ whose support is the closure of x, i.e.,

$$s_y \neq 0 \iff y \in \overline{\{x\}}, \qquad \text{all } y \in U .$$

In other words, for all open $U \subset X$ and all $s \in \Gamma(U, \mathcal{F})$, look at the support of $s : \{y \in U \mid s_y \neq 0\}$. Call this W. It is a closed subset of U since, by definition of the stalk $\mathcal{F}_y$, $s_y = 0$ only if s is 0 already in an open neighbourhood of y. Then the generic points of the components of these W's are the associated points of $\mathcal{F}$.

Notice that if $X = \text{Spec}\,(R)$, $\mathcal{F} = \widetilde{M}$, then for all prime ideals $P \subset R$, $[P]$ is an associated point of $\mathcal{F}$ if and only if P is an associated prime ideal of M. (For the definition and theory of these, see Bourbaki, Ch. 4, §1; or Zariski-Samuel, vol. 1, pp. 252–3, where these are called the prime ideals "associated to the 0 submodule of M".) The only problem in proving this is to check that, when X is *affine*, to find the associated points of $\mathcal{F}$ it suffices to look at the supports of *global* sections $s \in \Gamma(X, \mathcal{F})$; we leave this point to the reader to check. One of the main facts in the theory of noetherian decompositions is that a finite R-module has only a finite number of associated prime ideals. Since any noetherian scheme can be covered by a finite set of open affines, this implies

Proposition 2. *Let $\mathcal{F}$ be a coherent $\underline{o}_X$-module on a noetherian scheme X. Then $\mathcal{F}$ has only a finite number of associated points.*

The associated points of $\underline{o}_X$ itself are obviously very important:

Proposition 3. *Let X be noetherian. The generic points of the components of X – for short, the generic points of X – are associated points of $\underline{o}_X$. On the other hand, if $U \subset X$ is open, $s \in \Gamma(U, \underline{o}_X)$ and $Z = \{y \in U \mid s_y \neq 0\}$ is the support of s, and if Z does contain any generic points of X, then s is nilpotent. In particular, if X is reduced, the generic points of X are the only associated points of $\underline{o}_X$.*

Proof. Since X itself is the support of $1 \in \Gamma(X, \underline{o}_X)$, the generic points of X are associated points of $\underline{o}_X$. Now suppose $s \in \Gamma(U, \underline{o}_X)$. If $s(x) \neq 0$ for some $x \in U$, then U_s is non-empty. But U_s is open, so it will contain some generic point of X. On the other hand, $U_s \subset Z$ since $s(y) \neq 0 \Rightarrow s_y \neq 0$. Therefore Z contains a generic point of X. Therefore conversely if we assume that Z does *not* contain any generic point of X, it follows that $s(x) = 0$, all $x \in U$, and therefore s is nilpotent. □

When $\underline{o}_X$ does have non-generic associated points, their closures are called the *embedded components* of X. The simplest case is in Example P, Ch. II:

$$X = \text{Spec}\ \left(k[x,y]/\left(x^2, xy\right)\right),$$

with the origin as an embedded component, since it is the support of the nilpotent x.

Among coherent modules, the most important are the locally free ones:

Definition 4. Let X be a scheme. An $\underline{o}_X$-module $\mathcal{F}$ is *locally free of rank r* if there is an open covering $\{U_i\}$ of X such that

$$\mathcal{F}|_{U_i} \cong \underline{o}_X^r|_{U_i}.$$

If $\mathcal{F}$ is locally free of rank 1, it is called *invertible*[8].

Locally free modules are the most convenient algebraic form of the concept of vector bundles familiar in topology and differential geometry. And invertible sheaves are the algebraic analogs of line bundles. To see this clearly, suppose we mimic literally the usual definition of vector bundle in scheme language[9]. We can then check the sort of object we get is the equivalent of a locally free module.

Definition 5. Let X be a scheme. A *vector bundle of rank r on X with atlas* is

1) a scheme $\mathbb{E}$ and a morphism

$$\pi : \mathbb{E} \longrightarrow X$$

2) a covering $\{U_i\}$ of X, and
3) isomorphisms of schemes$/U_i$:

$$\begin{array}{ccc} \pi^{-1}(U_i) & \xrightarrow{\quad\phi_i\quad} & \mathbb{A}^r \times U_i \\ & {\scriptstyle\pi}\searrow \quad \swarrow{\scriptstyle p_2} & \\ & U_i & \end{array}$$

such that for all i, j, if we restrict ϕ_i and ϕ_j to $U_i \cap U_j$:

$$\begin{array}{ccccc} \mathbb{A}^r \times (U_i \cap U_j) & \xleftarrow{\text{res of } \Phi_i} & \pi^{-1}(U_i) \cap \pi^{-1}(U_j) & \xrightarrow{\text{res of } \Phi_j} & \mathbb{A}^r \times (U_i \cap U_j) \\ & {\scriptstyle p_2}\searrow & \downarrow{\scriptstyle\pi} & \swarrow{\scriptstyle p_2} & \\ & & U_i \cap U_j & & \end{array}$$

so as to get a morphism (after composing p_1):

$$\mathbb{A}^r \times (U_i \cap U_j) \xrightarrow{\psi_{i,j}} \mathbb{A}^r$$

then the dual homomorphism $\psi^*_{i,j}$ takes the coordinates $X_1, \ldots, X_r$ on $\mathbb{A}^r$ into *linear forms* in the X's:

$$\psi^*_{i,j}(X_k) = \sum_{\ell=1}^{r} a^{(i,j)}_{k,\ell} \cdot X_\ell$$

where $a^{(i,j)}_{k,\ell} \in \Gamma(U_i \cap U_j, \underline{o}_X)$.

---

8 This terminology stems from the fact that if R is a ring, M an R-module, then M is locally free of rank 1 if and only if there is an R-module N such that $M \otimes_R N \cong R$ (cf. Bourbaki).

9 The reader who is not familiar with vector bundles can skip the discussion that follows if he wants.

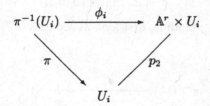

(Cf. Atiyah, *K-theory*, p. 1; Auslander and Mackenzie, *Introduction to Differentiable Manifolds*, Ch. 9).

Thus a vector bundle is just a scheme over X locally isomorphic to $\mathbb{A}^r \times X$. The simplest way to define a vector bundle, without distinguishing one " atlas", is to simply add the extra condition that the atlas is *maximal*, i.e., every possible open $V \subset X$ and isomorphism $\phi : \pi^{-1}(V) \xrightarrow{\sim} \mathbb{A}^r \times V$ compatible with the given $\{U_i, \phi_i\}$ is already there. This makes the indexing set of i's unaesthetically large, but this is unimportant.

The classical way of obtaining an $\underline{o}_X$-module from a vector bundle $\mathbb{E}/X$ is by taking the sheaf $\mathcal{E}$ of sections of $\mathbb{E}$:

For all $U \subset X$,

$$\Gamma(U, \mathcal{E}) = \left\{ \begin{array}{l} \text{set of morphisms } s : U \to \mathbb{E} \\ \text{such that } \pi \cdot s = 1_U \end{array} \right\} .$$

To make $\mathcal{E}$ into an $\underline{o}_X$-module, first look at sections of $\mathbb{E}$ over subsets V of a set U_i in the atlas:

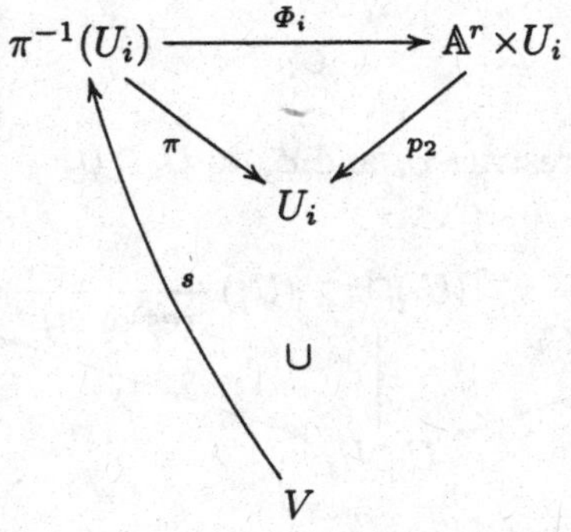

Then via ϕ_i sections of $\pi^{-1}(U_i)$ over V and sections of $\mathbb{A}^r \times U_i$ over V correspond to one another. But sections s of $\mathbb{A}^r \times U_i$ over V are given by r-tuples of functions $f_1 = s_1^*(X_1), \ldots, f_r = s_1^*(X_r)$ in $\Gamma(V, \underline{o}_X)$. In other words, we get isomorphisms:

$$\left\{ \begin{array}{l} \text{sections of the scheme} \\ \Gamma^{-1}(U_i) \text{ over } V \end{array} \right\} \cong \left\{ \begin{array}{l} \text{sections of the} \\ \text{sheaf } \underline{o}_X^r \text{ of } V \end{array} \right\}$$

hence

$$\mathcal{E}\big|_{U_i} \cong \underline{o}_X^r\big|_{U_i} .$$

This induces an $\underline{o}_X$-module structure in $\mathcal{E}\big|_{U_i}$. The reader can check that the compatibility demanded between ϕ_i and ϕ_j over $U_i \cap U_j$ is *exactly* what is needed to insure that the 2 $\underline{o}_X$-module structures that we get on $\mathcal{E}\big|_{U_i \cap U_j}$ are the same. The main theorem in this direction is that every locally free $\underline{o}_X$-module arises as the sheaf of sections of a unique vector bundle (up to isomorphism).

The point here is this: both $\mathbb{E}$ and $\mathcal{E}$ are structures which are locally "trivial", i.e., isomorphic to $\mathbb{A}^r \times U$ or $\underline{o}_X^r|_U$. Therefore, to get either $\mathbb{E}$ or $\mathcal{E}$ globally is only a matter of taking a collection of pieces $\pi^{-1}(U_i)$ or $\mathcal{E}|_{U_i}$, each "trivial" as before, and patching them together.

To prove that the two sets of global objects that we get in this way are isomorphic, it suffices to show that the patching data required to put together pieces of the type $\mathbb{A}^r \times U_i$ or of the type $\underline{o}_X^r|_{U_i}$ are exactly the same. In the first case, what is needed is an isomorphism of the two open subsets:

$$\begin{array}{ccccc} & & \mathbb{A}^r\times U_i & & \\ & \subset & & \supset & \\ \mathbb{A}^r\times(U_i\cap U_j) & & & & \\ \cong\Big\downarrow \text{suitable isomorphism} & & & & E \\ \mathbb{A}^r\times(U_i\cap U_j) & & & \subset & \\ & \subset & & & \\ & & \mathbb{A}^r\times U_i & & \end{array}$$

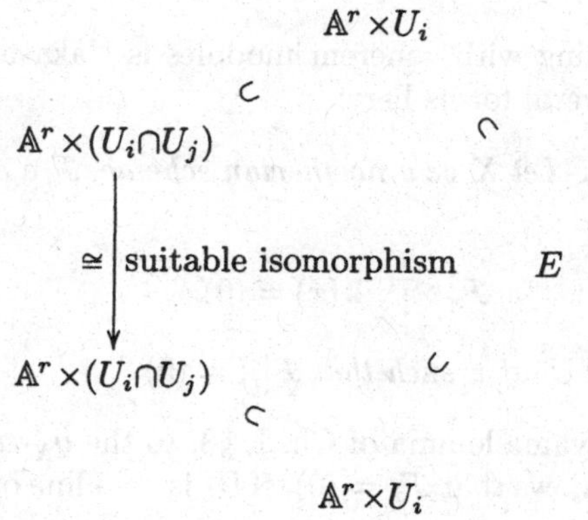

In the second case, since $\mathcal{E}|_{U_i\cap U_j}$ is isomorphic in two different ways with $\underline{o}_X^r|_{U_i\cap U_j}$ (i.e., by restricting the isomorphism over U_i and over U_j), we need to know the *automorphism* of $\underline{o}_X^r|_{U_i\cap U_j}$ by which these 2 isomorphisms differ.

Now the first isomorphism is required to be linear and therefore it is given by an $r\times r$ matrix $a_{k,\ell}^{(i,j)} \in \Gamma(U_i\cap U_j, \underline{o}_X)$, $1\le k,\ell\le r$. To be an *isomorphism* means exactly that the determinant of $a_{k,\ell}$ is invertible in $\Gamma(U_i\cap U_j,\underline{o}_X)$. On the other hand, an $\underline{o}_X$-module homomorphism

$$\chi : \underline{o}_X^r|_{U_i\cap U_j} \longrightarrow \underline{o}_X^r|_{U_i\cap U_j}$$

is also determined by the matrix of components of the sections $\chi((0,\ldots,0,1,0,\ldots,0))$. Again χ is an isomorphism if and only if the determinant of this matrix is invertible. The situation is even nicer: if $\mathcal{E}$ is the sheaf of sections of $\mathbb{E}/X$, then the *same* $r\times r$ matrices give the induced automorphisms of $\mathbb{A}^r\times(U_i\cap U_j)$ and of $\underline{o}_X^r|_{U_i\cap U_j}$, whenever you have corresponding "trivializations" of $\mathbb{E}$ and $\mathcal{E}$ over U_i and over U_j. From these considerations, it follows easily that a vector bundle $\mathbb{E}$ is determined by its sheaf of sections $\mathcal{E}$, and that every locally free $\underline{o}_X$-module $\mathcal{E}$ arises in this way.

For most purposes, the sheaf $\mathcal{E}$ turns out to be more convenient than the bundle $\mathbb{E}$. A word of warning though: Grothendieck has introduced a *dual* method of going back and forth between $\mathbb{E}$ and $\mathcal{E}$. In his approach, the module $\mathcal{E}$ associated to the bundle $\mathbb{E}$ is defined by:

$$\Gamma(U,\mathcal{E}) = \left\{ s \in \Gamma\left(\pi^{-1}(U), \mathbb{E}\right) \;\middle|\; \begin{array}{l} s \text{ is a linear function} \\ \text{on } \mathbb{E} \end{array} \right\}$$

where a linear function is one such that, over $U \cap U_i$, the induced element $\phi_i^*(s) \in \Gamma\left(\mathbb{A}^r \times (U \cap U_i), \underline{o}_{\mathbb{A}^r \times U_i}\right)$ is of the form:

$$\sum_{k=1}^{r} a_k \cdot X_k, \qquad a_k \in \Gamma\left(U \cap U_i, \underline{o}_X\right) \ .$$

His method has a good generalization to arbitrary coherent $\underline{o}_X$-modules (cf. EGA, Ch. II).

A basic tool in dealing with coherent modules is Nakayama's lemma which we want to recall in several forms here:

Nakayama's Lemma. *Let X be a noetherian scheme, $\mathcal{F}$ a coherent $\underline{o}_X$-module and $x \in X$. If*

$$\mathcal{F}_x \otimes_{\underline{o}_X} \mathbb{k}(x) = (0) \quad ,$$

then $\exists$ a neighbourhood U of x such that $\mathcal{F}\big|_U = (0)$.

Proof. Apply the Nakayama lemma of Ch. I, §8, to the $\underline{o}_X$-module $\mathcal{F}_x$, and the maximal ideal m_x. It follows that $\mathcal{F}_x = (0)$. If U_1 is an affine open neighbourhood of x, then $\mathcal{F}\big|_{U_1} = \widetilde{M}$. Let $a_1, \ldots, a_n$ be generators of M as $\Gamma\left(U_1, \underline{o}_X\right)$-module. Since $a_i \longmapsto 0$ in $\mathcal{F}_x$, it follows that $a_i \longmapsto 0$ in $\Gamma(U, \mathcal{F})$ for some neighbourhood U of x. Since the a_i's generate $\mathcal{F}$, this implies that $\mathcal{F}\big|_U = (0)$. □

Souped-up version I. *Let $X, \mathcal{F}, x$ be as above. If U is a neighbourhood of x and $a_1, \ldots, a_n \in \Gamma(U, \mathcal{F})$ have the property:*

(*) *the images $\overline{a_1}, \ldots, \overline{a_n}$ generate $\mathcal{F}x \otimes_{\underline{o}_x} \mathbb{k}(x)$*

then $\exists$ a neighbourhood $U_0 \subset U$ of x such that $a_1, \ldots, a_n$ generate $\mathcal{F}\big|_{U_0}$.

Proof. Define a module homomorphism:

$$\phi : \underline{o}_X^n\big|_U \longrightarrow \mathcal{F}\big|_U$$

by $\phi\left((b_1, \ldots, b_n)\right) = \sum a_i b_i$. Let $\mathcal{K}$ be the cokernel of ϕ, a coherent module on U. Then $\mathcal{K}_x \otimes_{\underline{o}_x} \mathbb{k}(x) = (0)$, so by Nakayama's lemma $\mathcal{K}\big|_{U_0} = (0)$, some $U_0 \subset U$ containing x. Therefore ϕ is surjective on U_0. □

Souped-up version II. *Let $X, \mathcal{F}$ be as above. Define*

$$e(x) = \dim_{\mathbb{k}(x)} \left[\mathcal{F}_x \otimes_{\underline{o}_x} \mathbb{k}(x)\right]$$

all $x \in X$. Then e is upper semi-continuous, i.e., $\{x \mid e(x) \le r\}$ is open, for all r. Assume further that X is reduced. Then for all $x \in X$, $\mathcal{F}$ is a free $\underline{o}_X$-module in some neighbourhood of x if and only if e is constant near x.

Proof. Let $x_1 \in X$ and $r_1 = e(x_1)$. Then there are r_1 elements $a_1, \ldots, a_{r_1} \in \mathcal{F}_{x_1}$ whose images generate $\mathcal{F}_{x_1} \otimes \mathbb{k}(x_1)$. Lift the a_i to sections of $\mathcal{F}$ in some open U_1 containing x_1. By version I, the a_i generate $\mathcal{F}$ in some open $U_2 \subset U_1$ still containing x_1. But then $\overline{a_1}, \ldots, \overline{a_{r_1}}$ generate $\mathcal{F}_y \otimes \mathbb{k}(y)$, all $y \in U_2$, i.e., $e(y) \leq r_1$ if $y \in U_2$. To prove the second statement, note that if $\mathcal{F}$ is a free $\underline{o}_X$-module of rank r in a neighbourhood U_1 of x, then $e(y) = r$, all $y \in U_1$. Conversely, assume $e(y) = r$, all $y \in U_1$. As in the first part, construct a surjective homomorphism

$$\underline{o}_X^r\big|_{U_2} \xrightarrow{\psi} \mathcal{F}\big|_{U_2} \longrightarrow 0$$

in some (possibly smaller) affine neighbourhood U_2 of x. Let $\mathcal{K}$ be the kernel of ψ. If $\mathcal{K} \neq (0)$, $\mathcal{K}$ has a non-0 section s. Since, $\mathcal{K} \subset \underline{o}_X^r\big|_{U_2}$, s is an r-tuple of elements of $\Gamma(U_2, \underline{o}_X)$. Now since X is *reduced*, s will be non-zero at some generic point y of U_2. Moreover, since y is generic, the stalk $\underline{o}_y$ of $\underline{o}_X$ at y is a field. Looking at stalks at y, we get an exact sequence of vector spaces over $\underline{o}_y$:

$$0 \longrightarrow \mathcal{K}_y \longrightarrow \underline{o}_y^r \longrightarrow \mathcal{F}_y \longrightarrow 0\,.$$

$$\in$$

$$s_y \neq 0$$

But $e(y) = \dim_{\underline{o}_y} \mathcal{F}_y = r$, by assumption, so this is a contradiction. This proves that ψ is an isomorphism, hence $\mathcal{F}\big|_{U_2}$ is free. □

Problem. Let X be an irreducible noetherian scheme all of whose stalks $\underline{o}_x$ at closed points are principal valuation rings. If $\mathcal{F}$ is a coherent $\underline{o}_X$-module, show that

$$\mathcal{F} \cong \mathcal{F}_1 \oplus \mathcal{F}_2\,,$$

where $\mathcal{F}_1$ is a locally free $\underline{o}_X$-module, and $\mathcal{F}_2$ has support at a finite number of closed points $x_1, \ldots, x_n$, hence

$$\mathcal{F}_2 \cong \bigoplus_{i=1}^{n} \underline{o}_{x_i}/m_{x_i}^{r_i}$$

(here $\underline{o}_x/m_x^r$ should be considered an $\underline{o}_x$-module by being extended by 0 outside $\{x\}$).

§3. Tangent cones

We are ready to go back to geometry. Let k be an algebraically closed field, let X be a scheme of finite type over k, and let x be a closed point of X. The scheme X has a *tangent cone* at x defined as follows:

1. Let $U \subset X$ be an affine open neighbourhood of x.

2. Let $i : U \to \mathbb{A}^n$ be a closed immersion, making U isomorphic with the subscheme Spec $(k[X_1,\ldots,X_n]/A)$ of $\mathbb{A}^n$. By adding suitable constants to the X_i's, we can assume that $X_1,\ldots,X_n \in I(\{x\})$, or equivalently that $i(x) = \mathbb{O}$, the origin in $\mathbb{A}^n$.
3. For all polynomials $f \in k[X_1,\ldots,X_n]$, let f^* be their "leading form", i.e., if

$$\begin{aligned} f &= \sum_{i=r}^{N} f_i, \qquad f_i \text{ homogeneous of degree } i \\ f_r &\neq 0 \end{aligned}$$

then $f^* = f_r$. Let A^* be the ideal of all polynomials f^*, for all $f \in \mathbb{A}$.

Provisional Definition 1. Spec $(k[X_1,\ldots,X_n]/A^*)$ = the *tangent cone* to X at x.

A priori, it might look as though this definition depended on the particular embedding of a neighbourhood of x in affine space. We can easily make it intrinsic through:

Definition 2. Let $\mathcal{O}$ be a local ring, with maximal ideal m. Then

$$\mathrm{gr}(\mathcal{O}) = \sum_{n=0}^{\infty} m^n/m^{n+1} \quad .$$

If $k = \mathcal{O}/m$, then $\mathrm{gr}(\mathcal{O})$ is seen to be a graded k-algebra, generated by m/m^2, the elements of degree 1. In fact, if $x_1,\ldots,x_n$ generate m/m^2 over k, then

$$\mathrm{gr}(\mathcal{O}) \cong k[X_1,\ldots,X_n] \Big/ \left(\begin{matrix} \text{some homogenenous} \\ \text{ideal} \end{matrix} \right)$$

$$x_i \longleftrightarrow X_i \quad .$$

Final Definition. If $x \in X$ is a closed point, Spec $[\mathrm{gr}(\underline{o}_x)]$ is the *tangent cone* to X at x.

Why are these the same? With notation as above, let

$$\begin{aligned} R &= k[X_1,\ldots,X_n]/A = \Gamma(U, \underline{o}_X) \\ M &= (X_1,\ldots,X_n)/A = I(\{x\}) \end{aligned}$$

so $R_M = \underline{o}_x$.
Then

$$\begin{aligned} \mathrm{gr}(\underline{o}_x) &= \sum_k M^k/M^{k+1} \\ &\cong \sum_k (X_1,\ldots,X_n)^k \big/ (X_1,\ldots,X_n)^{k+1} + A \cap (X_1,\ldots,X_n)^k \\ &\cong \sum_k (X_1,\ldots,X_n) \big/ (X_1,\ldots,X_n)^{k+1} + A_k^* \end{aligned}$$

if $A_k^* =$ homogeneous piece of A^* of degree k. But

$$(X_1,\ldots,X_n)^k \,/\, (X_1,\ldots,X_n)^{k+1} + A_k^* \quad = \quad k^{\text{th}}\text{-graded piece of } k\,[X_1,\ldots,X_n]\,/A^* \,.$$

Therefore

$$\operatorname{gr}(\underline{o}_x) \cong k\,[X_1,\ldots,X_n]\,/A^* \,.$$

Example E. Let $f(x,y)$ be an irreducible polynomial such that $f(0,0) = 0$. Then $X = V((f))$ is an affine plane curve through the origin. In this case, A^* is generated by the leading form f^* of f, and since f^* is a homogeneous form in 2 variables, f^* will factor into a product of linear forms ℓ_i:

$$f^* = \prod_i \ell_i^{r_i} \,.$$

Therefore, the tangent cone is a union of lines through the origin taken with multiplicities.

Case i). f has a non-zero linear term:
$$f = \alpha x + \beta y + f_2 + f_3 + \ldots$$

where α or $\beta \neq 0$, f_k is homogeneous of degree k. Then the tangent cone is the one line

$$\alpha x + \beta y = 0 \,,$$

and the origin is called a non-singular point of X.

Case ii). f begins with a quadratic term
$$f = f_2 + f_3 + \ldots$$

where $f_2 \neq 0$ and f_2 is the product of *distinct* linear factors $\ell(x,y)$, $m(x,y)$. Then the tangent cone consists of 2 distinct lines with multiplicity 1, and X is said to have a ***node*** at the origin.

Case iii). f begins with a square quadratic term:
$$f = (\alpha x + \beta y)^2 + f_3 + \ldots$$

where α or $\beta \neq 0$. The tangent cone is a double line. There are several subcases depending on the power of t dividing $f(\beta t, -\alpha t)$. If $f_3(\beta, -\alpha) \neq 0$, this is divisible only by t^3, and t is said to have a ***cusp*** at the origin.

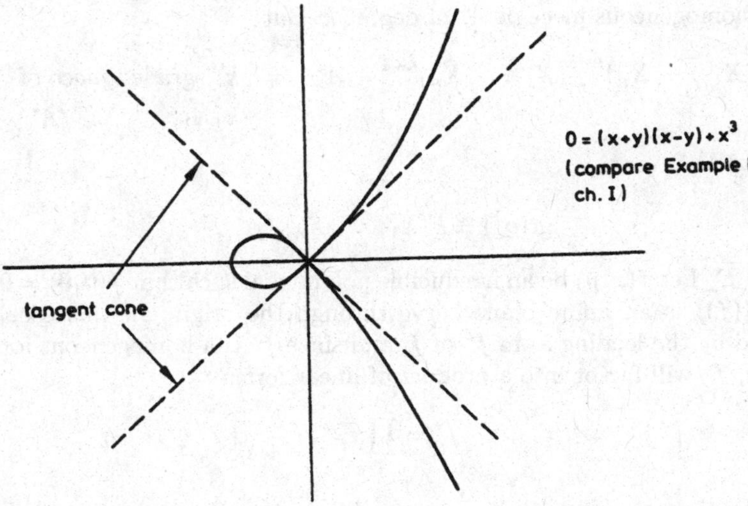

Figure to Case ii): X is said to have a node at the origin

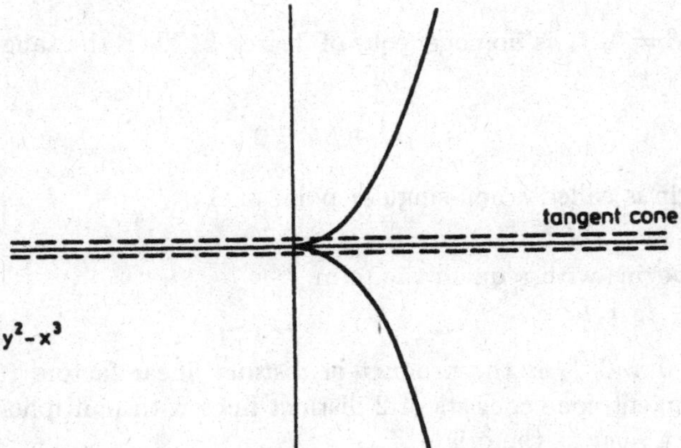

Figure to Case iii): f is said to have a cusp at the origin

Example F. Let $f(X_1,\ldots,X_n)$ be an irreducible polynomial such that $f(0,\ldots,0) = 0$, and let $X = V((f))$. Again A^* is generated by the leading form f^* of f. If f^* is linear, then the tangent cone to X at $\mathbb{O}$ is a hyperplane and we say that X has a non-singular point at the origin. If f is a non-degenerate quadratic form, then the tangent cone is the simplest type of quadratic cone and we say that X has an ordinary double point at the origin.

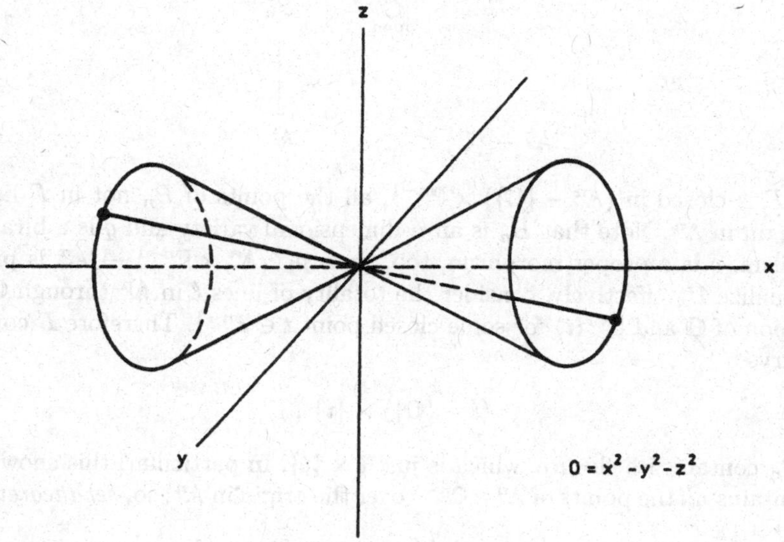

Warning: If X is an affine variety of higher codimension, defined by $f_1 = \ldots = f_r = 0$, then its tangent cone may *not* be the locus $f_1^* = \ldots = f_r^* = 0$. One may need the leading forms of some more of the polynomials $\sum g_i f_i$.

Since the tangent cone is defined by a homogeneous ideal A^*, it is natural to projectivize it: i.e., look at the locus of roots $(a_1, \ldots, a_n)$ of the polynomials in A^* as the set of *homogeneous* coordinates of a subset $T \subset \mathbb{P}^{n-1}$. More precisely, A^* defines a subscheme T in $\mathbb{P}^{n-1}$, which we call the projectivized tangent cone. An amazing fact is that there is a natural way to put $X - \{x\}$ and T together into a new scheme $B_x(X)$ in such a way that locally on $B_x(X)$, T is the subscheme defined by the vanishing of a single function f (which is not a 0-divisor). This is known as *blowing-up* x because $\{x\}$ is replaced in the process by T which is a picture of the infinitesimal behaviour of X at x.

We first define the variety B_n obtained by blowing up the origin $\mathbb{O}$ in $\mathbb{A}^n$. Let

$$p : \mathbb{A}^n - \{\mathbb{O}\} \longrightarrow \mathbb{P}^{n-1}$$

be the projection morphism taking a closed point with affine coordinates $(a_1, \ldots, a_n)$ into the closed point with homogeneous coordinates $(a_1, \ldots, a_n)$. Look at the graph Γ of p:

$$\Gamma \subset (\mathbb{A}^n - \{\mathbb{O}\}) \times \mathbb{P}^{n-1} .$$

Define B_n to be the closure of Γ in $\mathbb{A}^n \times \mathbb{P}^{n-1}$ (as a subvariety as well as a subset). By restricting the projection $p_1 : \mathbb{A}^n \times \mathbb{P}^{n-1} \to \mathbb{A}^n$ to B_n, we get a diagram:

$$\begin{array}{ccc} \Gamma & \underset{\text{open}}{\subset} & B_n \\ \downarrow & & \downarrow q \\ \mathbb{A}^n - \{\mathbb{O}\} & \subset & \mathbb{A}^n \end{array} .$$

Since Γ is closed in $(\mathbb{A}^n - \{\mathbb{O}\}) \times \mathbb{P}^{n-1}$, all the points of B_n not in Γ lie over the origin in $\mathbb{A}^n$. Note that B_n is an n-dimensional variety and q is a birational morphism. q is a proper morphism, too, since $p_1 : \mathbb{A}^n \times \mathbb{P}^{n-1} \to \mathbb{A}^n$ is proper. To visualize B_n effectively, consider the totality of lines ℓ in $\mathbb{A}^n$ through $\mathbb{O}$. ℓ is the union of $\mathbb{O}$ and $p^{-1}(t)$ for some closed point $t \in \mathbb{P}^{n-1}$. Therefore Γ contains the curve

$$(\ell - \{\mathbb{O}\}) \times \{t\} ,$$

and B_n contains its closure, which is just $\ell \times \{t\}$. In particular, this shows that B_n contains *all* the points of $\mathbb{A}^n \times \mathbb{P}^{n-1}$ over the origin in $\mathbb{A}^n$: so, *set-theoretically*,

$$B_n = \Gamma \cup \left(\{\mathbb{O}\} \times \mathbb{P}^{n-1}\right) .$$

More than that, this shows that whereas $\mathbb{A}^n$ is the union of all these lines ℓ, with their common origins identified, B_n is just the *disjoint* union of the lines ℓ. In B_n, we have replaced the one origin by a whole variety of origins, a different one for each line ℓ containing the original origin. This explains why B_n is called the result of blowing up $\mathbb{O}$ in $\mathbb{A}^n$. The locus of origins, $\{\mathbb{O}\} \times \mathbb{P}^{n-1}$, is called the *exceptional divisor* of B_n, and will be denoted by E.

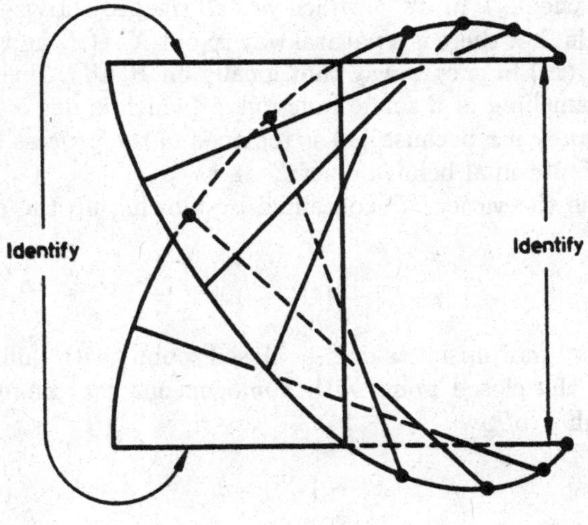

the case $n = 2$

To give a purely algebraic description of B_n, we must cover it by affine open pieces. Now B_n is embedded in $A^n \times \mathbb{P}^{n-1}$, and $\mathbb{P}^{n-1}$ is covered by the usual n pieces $U_1, \ldots, U_n$. Take $X_1, \ldots, X_n$ as affine coordinates in $\mathbb{A}^n$ and $Y_1, \ldots, Y_n$ as homogeneous coordinates in $\mathbb{P}^{n-1}$. Then $\mathbb{A}^n \times \mathbb{P}^{n-1}$ is the union of the open affines:

$$\mathbb{A}^n \times U_i = \operatorname{Spec} k\left[X_1, \ldots, X_n, \frac{Y_1}{Y_i}, \ldots, \frac{Y_n}{Y_i}\right].$$

The projection p is defined by setting the ratios $X_1 : \ldots : X_n$ and $Y_1 : \ldots : Y_n$ equal, i.e., we let $X_i\,(Y_1/Y_i) = X_1, \ldots, X_i\,(Y_n/Y_i) = X_n$. Therefore

$$B_n \cap (\mathbb{A}^n \times U_i) = V\left(\left(\ldots, X_i \cdot \frac{Y_j}{Y_i} - X_j, \ldots\right)\right),$$

and

$$B_n \cap (\mathbb{A}^n \times U_i) = \operatorname{Spec} k\left[X_1, \ldots, X_n, \frac{Y_1}{Y_i}, \ldots, \frac{Y_n}{Y_i}\right] \Big/ \left(\ldots, X_i \cdot \frac{Y_j}{Y_i} - X_j, \ldots\right).$$

This means that Y_j/Y_i, as an element of the function field $k(B_n) \cong k(\mathbb{A}^n) = k(X_1, \ldots, X_n)$, equals X_j/X_i, and that if we identify the affine ring $\Gamma\left(B_n \cap (\mathbb{A}^n \times U_i), \underline{o}_{B_n}\right)$ with its isomorphic image to $k(X_1, \ldots, X_n)$, we obtain:

$$B_n \cap (\mathbb{A}^n \times U_i) = \operatorname{Spec} k\left[X_i, \frac{X_1}{X_i}, \ldots, \frac{X_n}{X_i}\right].$$

Call this piece of B_n, $B_n^{(i)}$. We see from this description that each $B_n^{(i)}$, as a scheme in its own right, is isomorphic to $\mathbb{A}^n$. On the other hand, inside B_n, $B_n^{(i)}$ and $B_n^{(j)}$ are patched together along the common open subset:

$$\begin{aligned}\left[B_n^{(i)}\right]_{X_j/X_i} &= \operatorname{Spec} k\left[X_i, \frac{X_1}{X_i}, \ldots, \frac{X_n}{X_i}, \frac{X_i}{X_j}\right] \\ &= \operatorname{Spec} k\left[X_j, \frac{X_1}{X_j}, \ldots, \frac{X_n}{X_j}, \frac{X_j}{X_i}\right] = \left[B_n^{(j)}\right]_{X_i/X_j}.\end{aligned}$$

The birational morphism q corresponds, on the ring level, to the inclusions:

$$k[X_1, \ldots, X_n] \subset k\left[X_i, \frac{X_1}{X_i}, \ldots, \frac{X_n}{X_i}\right].$$

Moreover, $E \cap B_n^{(i)}$ is the subset $V((X_1, \ldots, X_n))$ in $B_n^{(i)}$. But all the X's are multiples of X_i in $\Gamma\left(B_n^{(i)}, \underline{o}_{B_n}\right)$, therefore

$$E \cap B_n^{(i)} = V((X_i)) \cong \operatorname{Spec} k\left[\frac{X_1}{X_i}, \ldots, \frac{X_n}{X_i}\right].$$

In this way, E itself is just $\mathbb{P}^{n-1}$.

The process of blowing up can be generalized to an arbitrary scheme X of finite type over k, and an arbitrary closed point $x \in X$. The quickest way to define this new scheme $B_x(X)$ is to choose an affine open neighbourhood U of x, and a closed immersion

$$i : U \longrightarrow \mathbb{A}^n$$

such that $i(x) = \mathbb{O}$. Let $i(U) = \text{Spec } (k[X_1, \ldots, X_n]/A)$. Set-theoretically, we want to define $B_x(X)$ as the union of $X - \{x\}$, and the closure of $q^{-1}[i(U) - \{\mathbb{O}\}]$ in B_n, suitably identified. More precisely, for all $i = 1, \ldots, n$, define an ideal A_i^* in $k[X_i, X_1/X_i, \ldots, X_n/X_i]$ by

$$A_i^* = \{f \in k[X_i, X_1/X_i, \ldots, X_n/X_i] \mid f \cdot X_i^N \in A, \text{ if } N \gg 0\} .$$

It is clear that A_i^* and A_j^* induce the same ideal in $k[X_i, X_1/X_i, \ldots, X_n/X_i, X_i/X_j]$, which is the affine ring of $B_n^{(i)} \cap B_n^{(j)}$. Therefore, the ideals $\{A_i^*\}$ define a coherent $\underline{o}_{B_n}$-ideal $\underline{\mathcal{Q}}^* \subset \underline{o}_{B_n}$. Let $U^* \subset B_n$ be the corresponding closed subscheme. Note that via q, we get a morphism q'

$$\begin{array}{ccc} U^* & \subset & B_n \\ \Big\downarrow q' & & \Big\downarrow q \\ U & \xrightarrow{\;\;i\;\;} & \mathbb{A}^n \end{array}$$

corresponding to the ring homomorphisms

$$\begin{array}{ccc} k[X_i, X_1/X_i, \ldots X_n/X_i]/A_i^* & \twoheadleftarrow & k[X_i, X_1/X_i, \ldots X_n/X_i] \\ \Big\uparrow & & \Big\uparrow \\ k[X_1, \ldots, X_n]/A & \twoheadleftarrow & k[X_1, \ldots, X_n] \end{array} .$$

Note finally that just as q restricts to an isomorphism from $B_n - E$ to $\mathbb{A}^n - \{0\}$, so q' restricts to an isomorphism of $U^* - U^* \cap E$ to $U - \{x\}$.

Definition 3. $B_x(X)$ is the union of U^* and $X - \{x\}$, patched along their isomorphic open subsets $U^* - U^* \cap E$ and $U - \{x\}$.

Then q' extends to a morphism $Q : B_x(X) \to X$. Let

$$\mathcal{E} = Q^{-1}(\{x\}) \underset{\text{as a set}}{=} U^* \cap E :$$

this is called the *exceptional subscheme* of $B_x(X)$. This set-up has the following properties:

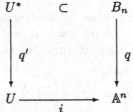

(I.) Q induces an isomorphism of $B_x(X) - \mathcal{E}$ with $X - \{x\}$.
(II.) $B_x(X)$ is a scheme (not just a prescheme) and Q is proper.

Proof. Let $f, g : K \rightrightarrows B_x(X)$ be a pair of test morphisms. Let $Z_1 = \{s \in K \mid f(s) \equiv g(s)\}$ and let $Z_2 = \{s \in K \mid Q(f(s)) \equiv Q(g(s))\}$. Z_2 is closed since X is a scheme, and Z_2 is covered by its intersection with the 2 open pieces $U_1 = \{s \in K \mid f(s) \text{ and } g(s) \notin \mathcal{E}\}$, and $U_2 = \{s \in K \mid f(s) \text{ and } g(s) \in U^*\}$. It suffices to show that Z_1 is closed in $Z_2 \cap U_1$ and in $Z_2 \cap U_2$. But $Z_1 \cap U_1 = Z_2 \cap U_1$ since Q is an isomorphism over $X - \{x\}$. And on U_2, instead of f and g, we can consider the compositions

$$\begin{array}{ccc} f', g' : K \overset{f}{\underset{g}{\rightrightarrows}} & U^* & \longrightarrow B_n \subset \mathbb{A}^n \times \mathbb{P}^{n-1} . \\ & \cap & \\ & B_x(X) & \end{array}$$

Then $Z_1 \cap U_2 = \{s \in K \mid f'(s) \equiv g'(s)\}$, and this is closed since $\mathbb{A}^n \times \mathbb{P}^{n-1}$ is a scheme. This shows that $B_x(X)$ is a scheme. To show that Q is proper, it suffices to show that the restrictions of Q to $Q^{-1}(X - \{x\})$ and to $Q^{-1}(U)$ are proper. But the first is an isomorphism and the second is the restriction of the proper morphism q to a closed subscheme. □

(III.) For all $y \in \mathcal{E}$, there is an affine open neighbourhood $V \subset B_x(X)$ of y such that the ideal of $\mathcal{E} \cap V$ in $\Gamma\left(V, \underline{o}_{B_x(X)}\right)$ is equal to (f), for some non-zero divisor f.

Proof. Suppose that the image of y in B_n is in $E \cap B_n^{(i)}$. Then let V equal $U^* \cap B_n^{(i)}$. Let $x_i \in \Gamma(U, \underline{o}_X)$ be the restriction of the function X_i to U. Since the ideal of $\{x\}$ in $\Gamma(U, \underline{o}_X)$ is $(x_1, \ldots, x_n)$, the ideal of $\mathcal{E}$ in $\Gamma(V, \underline{o}_X)$ is also $(x_1, \ldots, x_n)$. But all the X's are multiples of X_i in $\Gamma\left(B_n^{(i)}, \underline{o}_{B_n}\right)$, so this ideal is just (x_i). Referring to the definition of A_i^*, it is obvious that x_i is not a 0-divisor in $\Gamma\left(V, \underline{o}_{B_x(X)}\right)$. □

(IV.) If X is a variety of dimension r, and $r \geq 1$, then $B_x(X)$ is a variety of dimension r, Q is birational, and $\mathcal{E}$ is non-empty and pure $(r-1)$-dimensional. In general, if
$r = \sup\{\dim Z \mid Z \text{ a component of } X \text{ through } x\} \geq 1$, then
a) $r = \sup\{\dim Z \mid Z \text{ a component of } B_x(X) \text{ meeting } \mathcal{E}\}$
b) $r - 1 = \sup\{\dim Z \mid Z \text{ a component of } \mathcal{E}\}$.

Proof. First of all, by (III.), no component of $B_x(X)$ is contained in $\mathcal{E}$. Therefore, Q induces an isomorphism between non-empty open subsets of all components of $B_x(X)$ and of X. To prove assertion a), we must check that if Z is a component of X containing x (with dimension ≥ 1), then the component

$$Z' = \overline{Q^{-1}(Z - \{x\})}$$

of $B_x(X)$ meets $\mathcal{E}$. But $Q(Z')$ is closed since Q is proper, and it contains $Z - \{x\}$, therefore $Q(Z') = Z$, so Z' meets $Q^{-1}(x)$. This proves (a) and then (b) follows from (III.) and the results of Ch. I, §7. Now assume X is a variety. Then $B_x(X)$ is irreducible. Also A is a prime ideal, and it follows immediately from the definition of A_i^* that it is a prime ideal too. Therefore $B_x(X)$ is reduced too. The fact that $\mathcal{E}$ is *pure* $(r-1)$-dimensional follows again using Ch. I, §7. □

(V.) *(the whole point)*: $\mathcal{E} \cong$ the projectivized tangent cone to X at x.

Proof. By construction, $\mathcal{E}$ is a closed subscheme of the projective space E. Its ideal in $B_n^{(i)}$ is $A_i^* + (X_i)$, and its ideal in the ith affine piece $E \cap B_n^{(i)}$ of E is:

$$\overline{A_i} = A_i^* + (X_i) \big/ (X_i) \subset k\left[\frac{X_1}{X_i}, \ldots, \frac{X_n}{X_i}\right] .$$

But A_i^* is just the ideal of quotients f/X_i^r, where $f \in A$ and the leading term f^* of f has degree $\geq r$. Therefore

$$A_i^* + (X_i) = \left\{ \frac{f^*}{X_i^r} \;\middle|\; \begin{array}{l} f^* = \text{ leading term of} \\ \text{some } f \in A, \\ \text{degree } f^* = r \end{array} \right\} + (X_i) ,$$

so if B is the ideal in $k[X_1, \ldots, X_n]$ generated by the leading terms of elements $f \in A$,

$$\overline{A_i} = \left\{ \frac{g}{X_i^r} \middle| g \in B, \text{ homogeneous of degree } r \right\} .$$

The ideal on the right is nothing but the ideal defining the ith affine piece of the projectivized tangent cone. □

Corollary. *If X is an r-dimensional variety, then the tangent cone to X at x is pure r-dimensional. In general*

$$\sup \left\{ \dim Z \;\middle|\; \begin{array}{l} Z \textit{ a component} \\ \textit{of } X \textit{ at } x \end{array} \right\} = \sup \left\{ \dim Z \;\middle|\; \begin{array}{l} Z \textit{ a component of the} \\ \textit{tangent cone to} \\ X \textit{ at } x \end{array} \right\} .$$

Example E bis. If we blow up $\mathbb{A}^2 = \operatorname{Spec} k[x, y]$, we get a B_2 covered by:

$$\begin{aligned} U &= \operatorname{Spec} k[x, y/x] \\ V &= \operatorname{Spec} k[y, x/y] . \end{aligned}$$

E is a projective line, and $E \cap U$ and $E \cap V$ are defined by setting $x = 0$ and $y = 0$ respectively. Now if $f = x^2 - y^2 + x^3$, and $X = V((f))$, then $B_x(X)$ will be a curve in B_2 entirely contained in U and defined by the equation

$$\frac{f}{x^2} = 1 - (x/y)^2 + x \qquad \text{in } U\,.$$

The picture is, roughly:

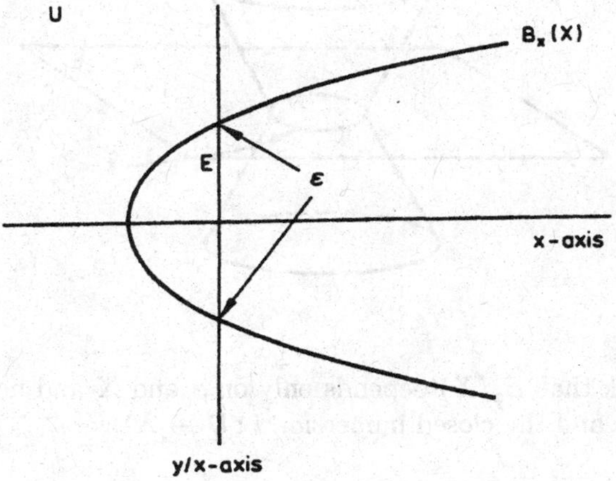

Now if $f = Y^3 - x^3$, then $B_x(X)$ is entirely contained in U and is the curve:

$$\left(\frac{y}{x}\right)^2 = x\,.$$

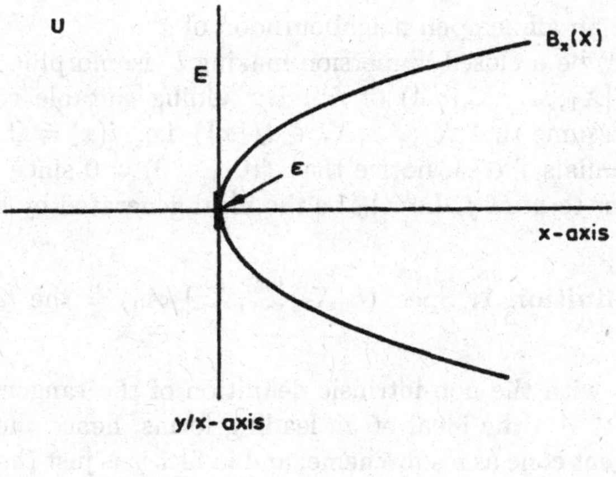

Note that the subscheme of $B_x(X)$ defined by $x = 0$ is the point P *doubled* since $B_x(X)$ is tangent to the exceptional curve E at P. And, indeed, the tangent cone in this case is a *double* line, as (V) requires.

Example F bis. Let $X = \text{Spec } k[x,y,z]/\left(x^2 - y^2 - z^2\right) \subset \mathbb{A}^3$. Then B_3 contains a copy of the projective plane as its exceptional divisor E, and $B_x(X)$ is a surface in B_3 that meets E in the circle: $(y/x)^2 + (z/x)^2 = 1$. Roughly, it looks like this:

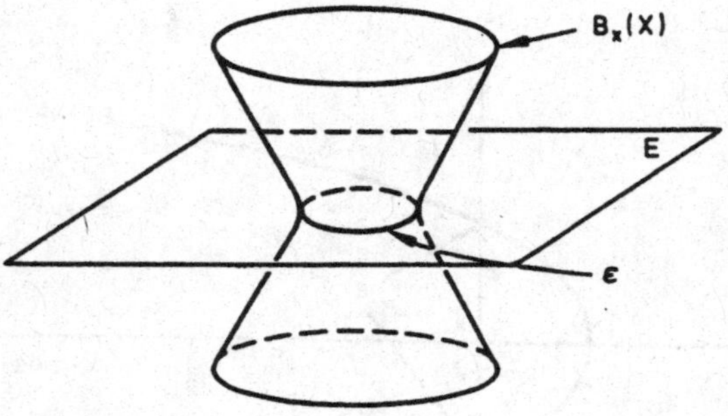

Problem. Check that $B_x(X)$ depends only on x and X and not on the neighbourhood U of x and the closed immersion $i : U \to \mathbb{A}^n$.

§4. Non-singularity and differentials

From a technical point of view, the big drawback about the tangent cone is that it is non-linear. It is always easier to handle essentially linear objects. The most natural way around this is to study the "linear hull" of the tangent cone, which we will call the tangent *space*. Let X be a scheme of finite type over k, and let $x \in X$ be a closed point.

1. Let $U \subset X$ be an affine open neighbourhood of x.
2. Let $i : U \to \mathbb{A}^n$ be a closed immersion making U isomorphic with the closed subscheme $(k[X_1, \ldots, X_n]/A)$ of $\mathbb{A}^n$. By adding suitable constants to the X_i's, we can assume that $X_1, \ldots, X_n \in I(\{x\})$, i.e., $i(x) = \mathbb{O}$.
3. For all polynomials $f \in A$, notice that $f(0, \ldots, 0) = 0$ since $\mathbb{O} \in V(A)$. Let f^ℓ be the *linear* term of f. Let A_0 be the ideal generated by the linear forms f^ℓ, all $f \in A$.

Provisional Definition 1. Spec $(k[X_1, \ldots, X_n]/A_0)$ = the *tangent space* to X at x.

Compare this with the non-intrinsic definition of the tangent cone in §3. It is clear that $A_0 \subset A^*$, the ideal of *all* leading forms, hence the tangent space contains the tangent cone as a subscheme; and in fact it is just the smallest *linear* subscheme of $\mathbb{A}^n$ containing the tangent cone as a subscheme. For this reason, we can say that the tangent space is just the "linear hull" of the tangent cone.

When dealing with linear subspaces of $\mathbb{A}^n$, and, more generally, with any "affine spaces", i.e., schemes isomorphic to $\mathbb{A}^n$, it is possible to get confused between the underlying vector space, and the whole scheme. It is important to realize that one can *canonically* attach to every k-vector space V a scheme whose set of closed points is just V, and which is isomorphic to $\mathbb{A}^n$; and that under this canonical correspondence, the set of sub-vector spaces of k^n is "equal" to the set of linear subschemes of $\mathbb{A}^n$. This is simply a matter of making the inclusion of the vector space k^n in the scheme $\mathbb{A}^n$ *coordinate-invariant.* It is also the special case when the base scheme is Spec (k) of the functorial bundle, locally free sheaf correspondence outlined in §2. In particular, in dealing with the tangent space to a scheme X one sometimes wants to deal with it as a scheme, and sometimes as a vector space. This correspondence goes as follows: suppose V is a k-vector space. Let R_V be the ring of polynomial functions from V to k (equivalently, the symmetric algebra on the dual space $\hom_k(V,k)$). Then let

$$V^{\text{sch}} = \text{Spec } (R_V) \quad .$$

Notice that V "equals" the set of closed points of V^{sch}. In fact, by the Nullstellensatz, every maximal ideal of R_V is the kernel of a homomorphism

$$f \longmapsto f(a_1, \ldots, a_n) \in k$$

for some $(a_1, \ldots, a_n) \in k^n$. If we call this ideal M_a, the correspondence $a \to [M_a]$ maps k^n isomorphically onto the set of closed points of V^{sch}.

Bearing in mind, then, that linear subspaces of $\mathbb{A}^n$ are essentially the same as subvector spaces of k^n, I want to show how easy it is to compute the tangent space to an affine scheme Spec $(k[X_1, \ldots, X_n]/A)$ in $\mathbb{A}^n$ at any closed point. First look at the origin. Note the lemma:

(*) *If* $A = (f_1, \ldots, f_n)$, *then* $A_0 = (f_1^\ell, \ldots, f_n^\ell)$.

Proof. In fact, if f is any element of A, then $f = \sum_{i=1}^{n} g_i f_i$, some $g_i \in k[X_1, \ldots, X_n]$. Then the linear term f^ℓ of f is just $\sum_{i=1}^{n} g_i(0, \ldots, 0) f_i^\ell$, so $f^\ell \in (f_1^\ell, \ldots, f_n^\ell)$. □

(Whereas it is *not* always true that the ideal A^* of *all* leading forms is generated by the leading forms of generators of A.)

Proposition 1. *Let* $X =$ Spec $(k[X_1, \ldots, X_n]/(f_1, \ldots, f_m))$ *be a closed subscheme of* $\mathbb{A}^n$. *Let* $x \in X$ *be a closed point with coordinates* $(a_1, \ldots, a_n)$. *Then the tangent space to* X *at* x *is naturally isomorphic to the linear subspace of* $\mathbb{A}^n$ *defined by:*

$$\sum_{i=1}^{n} \frac{\partial f_1}{\partial X_i}(a_1, \ldots, a_n) \cdot X_i = \ldots = \sum_{i=1}^{n} \frac{\partial f_m}{\partial X_i}(a_1, \ldots, a_n) \cdot X_i = 0 \ .$$

(Note that the linear equations here define *both* a linear subscheme of $\mathbb{A}^n$, and the corresponding subvector space of k^n.)

Proof. If we translate X so that $(a_1, \ldots, a_n)$ is shifted to the origin, it is then defined by the equations

$$f_1(X_1 + a_1, \ldots, X_n + a_n) = \ldots = f_m(X_1 + a_1, \ldots, X_n + a_n) = 0 .$$

The tangent space of the original X at (a) is isomorphic to the tangent space of the shifted X at $\mathbb{O}$, and this, by $(*)$, is the locus of zeroes of the linear terms of $f_i(X_1 + a_1, \ldots, X_n + a_n)$. But the linear term of this polynomial is:

$$\sum_{j=1}^{n} \frac{\partial f_i}{\partial X_j}(a_1, \ldots, a_n) \cdot X_j$$

so the Prop. follows. □

The intrinsic definition of the tangent space is this:

Final Definition 1 (due to Zariski). Let x be a closed point of the scheme X of finite type over k. Let

$$T_{x,X} = \hom_k(m_x/m_x^2, k) .$$

Then the vector space $T_{x,X}$ and the corresponding scheme $T_{x,X}^{\mathrm{sch}}$ will both be called the *tangent space* to X at x. The dual vector space m_x/m_x^2, and its scheme $(m_x/m_x^2)^{\mathrm{sch}}$ are both the *cotangent space* to X at x.

Here is the canonical isomorphism of this tangent space with the previous one:

As before, let $U \subset X$ be isomorphic to Spec $(k[X_1, \ldots, X_n]/A)$. Let

$$R = k[X_1, \ldots, X_n]/A = \Gamma(U, \underline{o}_X) .$$

Assume that

$$M = (X_1, \ldots, X_n)/A = I(\{x\}) ,$$

so

$$R_M = \underline{o}_x .$$

Then

$$\begin{aligned} m_x/m_x^2 &\cong M/M^2 \\ &\cong (X_1, \ldots, X_n)/(X_1, \ldots, X_n)^2 + A \\ &\cong (X_1, \ldots, X_n)/(X_1, \ldots, X_n)^2 + A_0 . \end{aligned}$$

Therefore, since A_0 is generated by linear forms:

$$\left\{ \begin{array}{l} \text{Symmetric algebra} \\ \text{on } m_x/m_x^2 \end{array} \right\} \cong k[X_1, \ldots, X_n]/A_0 .$$

Taking Spec of both sides,

$$\begin{aligned} T^{\text{sch}}_{x,X} &\underset{\text{def.}}{=} \text{Spec}\left\{\begin{array}{l}\text{Symmetric algebra}\\ \text{on } m_x/m_x^2\end{array}\right\} \\ &\cong \text{Spec}\,(k[X_1,\ldots,X_n]/A_0) \underset{\text{def.}}{=} \left\{\begin{array}{l}\text{Non-intrinsic tangent}\\ \text{space}\end{array}\right\}.\end{aligned}$$

We can even embed the tangent cone inside the tangent space in a completely intrinsic way. There is a canonical surjection:

$$\left\{\begin{array}{l}\text{Symmetric algebra}\\ \text{on } m_x/m_x^2\end{array}\right\} \longrightarrow \text{gr}\,(\underline{o}_x) \longrightarrow 0\,.$$

This defines a closed immersion:

$$\left\{\begin{array}{l}\text{Tangent cone to}\\ X \text{ at } x\end{array}\right\} = \text{Spec}\,(\text{gr}\,(\underline{o}_x)) \subset T^{\text{sch}}_{x,X}\,.$$

Once again, we see that $F^{\text{sch}}_{x,X}$ is exactly a "linear hull" of the tangent cone.

Definition 2. The (closed) point x is a *non-singular point of* X, or X is *non-singular at* x, if the tangent space to X at x equals the tangent cone to X at x (i.e., the tangent cone is itself linear).

Look at this condition algebraically: it means that $\text{gr}\,(\underline{o}_x)$ is isomorphic to the symmetric algebra on m_x/m_x^2. In other words:

A. $\quad m_x^k/m_x^{k+1} \cong \left\{\begin{array}{l}k^{\text{th}} \text{ symmetric power}\\ \text{of } m_x/m_x^2\end{array}\right\}$

or

B. if $f(X_1,\ldots,X_n) \in \underline{o}_x[X_1,\ldots,X_n]$ is homogeneous of degree k and if $x_1,\ldots,x_n \in m_x$ are independent mod m_x^2, then $f(x_1,\ldots,x_n) \in m_x^{k+1}$ only if all coefficients of f are in m_x.

Note that this implies that $\underline{o}_x$ is an integral domain: for if $a,b \in \underline{o}_x$, $a \neq 0$, $b \neq 0$, then for some integers k and ℓ we would get a $a \in m_x^k - M - x^{k+1}$, $b \in m_x^\ell - m_x^{\ell+1}$. Then a and b would have non-zero images $\bar{a}, \bar{b}$ in m_x^k/m_x^{k+1} and $m_x^\ell/m_x^{\ell+1}$. Therefore $\bar{a}\,\bar{b} \in m_x^{k+\ell}/m_x^{k+\ell+1}$ would be non-zero, i.e., $a\,b \notin m_x^{k+\ell+1}$, hence *a fortiori* $a \cdot b \neq 0$.[10] On the scheme X, this means that some neighbourhood U of x is a variety [to be precise: x is in the closure of only one associated point z of $\underline{o}_X$, namely the one corresponding to the ideal $(0) \subset \underline{o}_x$; and if W is the union of the closures of the other associated points of $\underline{o}_X$, then $X - W$ is an open neighbourhood of x which is a variety.] Therefore, in discussing non-singularity, we may usually restrict our attention to *varieties*.

[10] But conversely, $\underline{o}_x$ may be a domain even if $\text{gr}\,(\underline{o}_x)$ is not: see Ex. E, §3.

Proposition 2. *Let X be an n-dimensional variety. Then for all closed points $x \in X$,*

$$\dim T_{x,X} \geq n \ ,$$

equality holding if and only if X is non-singular at x.

Proof. In fact, we saw in §3 that the tangent cone to X at x is pure n-dimensional. Therefore the dimension of the tangent space is at least n, and if it equals n, the tangent space must equal the tangent cone.

Corollary 1. *Let $X = \text{Spec }(k[X_1,\ldots,X_N]/(f_1,\ldots,f_m))$. Assume that X is an n-dimensional variety. Then for all closed points $x = (a_1,\ldots,a_N) \subset X$,*

$$\text{rank}\left[\frac{\partial f_i}{\partial X_j}(a_1,\ldots,a_N)\right] \leq N - n \ ,$$

equality holding if and only if X is non-singular at x.

Proof. This is just Prop. 1 + Prop. 2.

Note that this criterion was exactly what we used to define non-singular points on hypersurfaces in Ex. E,F of §3.

Corollary 2. *If $k = \mathbb{C}$, X is a variety over $\mathbb{C}$, and $x \in X$ is a closed point, then X is non-singular at x if and only if the analytic space $\mathcal{X}$ corresponding to X (as in Ch. I, §9) is an analytic manifold at x.*

Proof. Use Cor. 1 and the fact that an n-dimensional analytic subspace of $\mathbb{C}^n$ defined by $f_1 = \ldots = f_m = 0$ is a *manifold* at (a) if and only if

$$\text{rank}\,[\partial f_i/\partial X_k\,(a_1,\ldots,a_N)] = N - n \ .$$

(Cf. Gunning-Rossi, Ch. V, A 13 and A 14 for example.) □

Theorem 3. *Let $x \in X$ be a closed point of a scheme X of finite type over k. The following k-vector spaces are canonically isomorphic:*

1) the tangent space $T = \hom_k\left(m_x/m_x^2, k\right)$ to X at x,
2) the space of point derivations $D : \underline{o}_x \to k$ over k, i.e., k-linear maps such that $D(fg) = f(x)\cdot Dg + g(x)\cdot Df$.
3) $\hom_{\underline{o}_x}\left(\left(\Omega_{X/k}\right)_x, \Bbbk(x)\right)$.

As a set, these are also isomorphic to:

4) the set of morphisms $f : I \to X$ with image x, where $I = \text{Spec }\left(k[\epsilon]/(\epsilon^2)\right)$.

Proof. (1) and (2) are isomorphic, since every point derivation D kills m_x^2, and therefore induces a linear functional $\ell : m_x/m_x^2 \to k$. Conversely, given any such ℓ, define D by

$$D(f) = \ell[f - f(x)]$$

and we get a point derivation. To compute (3), let $U = \mathrm{Spec}\,(R)$ be an affine open neighbourhood of x, and assume $x = [M]$. Then

$$\mathrm{hom}_{\underline{o}_x}\left(\Omega_{X/k}, \mathbb{k}(x)\right) \cong \mathrm{hom}_R\left(\Omega_{R/k}, R/M\right) .$$

On the other hand, we know from §1 that for all R-modules A

$$\mathrm{hom}_R\left(\Omega_{R/k}, A\right) \cong \left\{ \begin{array}{c} \text{module of } k\text{-derivations} \\ D : R \to A \end{array} \right\} .$$

Applying this with $A = R/M$, we find that (2) and (3) are isomorphic. Finally, according to Prop. 3. §6, Ch. II, morphisms $f : I \to X$ with image X correspond to local homomorphisms

$$f^* : \underline{o}_x \longrightarrow k[\epsilon]/(\epsilon^2) .$$

Such f^* always kill m_x^2, hence define and are defined by linear functionals $\ell : m_x/m_x^2 \to (\epsilon) \cong k$. Therefore, the sets (1) and (4) are isomorphic. □

Corollary. *For all closed points $x \in X$, m_x/m_x^2 is canonically isomorphic to $\left(\Omega_{X/k}\right)_x \otimes_{\underline{o}_x} \mathbb{k}(x)$. In this map, df (modulo $m_x \cdot \left(\Omega_{X/k}\right)_x$) corresponds to $f - f(x)$ (modulo m_x^2).*

Proof. Dualize the isomorphism of vector spaces (2) and (3) in the theorem. □

This corollary shows a very significant and far-reaching thing: that the vector spaces m_x/m_x^2, which *a priori* are a collection of unrelated vector spaces, one for each closed point x, can all be derived from one coherent sheaf $\Omega_{X/k}$ on X. This enables us a) to use the machinery of Nakayama's lemma (§2), and b) to handle non-closed points.

Definition 3. Let X be a scheme of finite type over k. Then for all (not necessarily closed) points $x \in X$, the *cotangent space* to X at x is:

$$\left(\Omega_{X/k}\right)_x \otimes_{\underline{o}_x} \mathbb{k}(x) .$$

We abbreviate this to $\Omega_{X/k}(x)$, or $\Omega(x)$. Moreover, let

$$d(x) = \dim_{\mathbb{k}(x)}[\Omega(x)] ,$$

sometimes called the "embedding dimension" of X at x.

Note that d is upper-semi-continuous by version II of Nakayama's lemma. Now assume that X is a variety and $n = \dim X$. Then

$$\begin{aligned} d\left(\begin{array}{l} \text{generic point} \\ \text{of } X \end{array} \right) &= \dim_{k(X)}\left[\Omega_{k(X)/k}\right] \\ &= \dim_{k(X)}\left[\begin{array}{l} \text{vector space of } k\text{-derivations} \\ \text{from } k(X) \text{ to } k(X) \end{array} \right] \\ &= n . \end{aligned}$$

(Cf. Ex. C, §1). It follows that $\{x \in X \mid d(x) = n\}$ is an open dense subset U of X. Again by version II, U is the maximal open subset of X on which $\Omega_{X/k}$ is a locally free $\underline{o}_X$-module. Since for *closed* points $x \in X$,

$$\begin{aligned} d(x) = n \quad &\Longleftrightarrow \quad \dim_k (m_x/m_x^2) = n \\ &\Longleftrightarrow \quad X \text{ is non-singular at } x, \end{aligned}$$

it is natural to extend the definition of non-singularity to *all* points x of X to mean that $d(x) = n$, or that $\Omega_{X/k}$ is a free $\underline{o}_X$-module near x. The conclusion is:

Proposition 3. *Let X be a variety over k of dimension n. Then the set of non-singular points $x \in X$ is an open dense subset of X on which $\Omega_{X/k}$ is a locally free $\underline{o}_X$-module of rank n.*

In particular, if X is a non-singular variety (i.e., non-singular everywhere), then $\Omega_{X/k}$ is a locally free $\underline{o}_X$-module. In the correspondence between locally free modules and vector bundles discussed in §2, the bundle version of $\Omega_{X/k}$ is then exactly the usual *cotangent bundle* of X over k, and conversely the sheaf of differentials $\Omega_{X/k}$ is then the sheaf of sections of the cotangent bundle.

The operation of taking the tangent space is a *covariant functor*, while the cotangent space is a *contravariant functor*:

(A)

Given $f : X \to Y$, X and Y schemes of finite type$/k$.
For all closed points $x \in X$, let $y = f(x)$.
We get this

$$\underline{o}_x \overset{f^*}{\longleftarrow} \underline{o}_y$$

and this:

$$m_x/m_x^2 \longleftarrow m_y/m_y^2 ,$$

and this:

$$df_x : T_{x,X} \longrightarrow T_{y,Y} .$$

(B)

Given $f : X \to Y$, X and Y schemes of finite type over k.
For *all* $x \in X$, let $y = f(x)$.
We get this

$$\underline{o}_x \overset{f^*}{\longleftarrow} \underline{o}_y ,$$

hence this

$$(\Omega_{X/k})_x \overset{df^*}{\longleftarrow} (\Omega_{Y/k})_y$$

and this

$$\Omega_X(x) \longleftarrow \Omega_Y(y) : df_x^* .$$

Of course, at closed points, $\Omega_X(x)$ and $\Omega_Y(y)$ are the dual spaces of $T_{x,X}$ and $T_{y,Y}$ respectively; and df_x^* is the transpose of df_x. For example, suppose Y is a closed subscheme of a scheme X. Then for all $x \in Y$, we get a *surjection* $f^* : \underline{o}_{x,X} \to \underline{o}_{x,Y}$, hence $\Omega_X(x)$ is a quotient of $\Omega_Y(x)$. More precisely, one checks that:

$$\Omega_Y(x) \cong \Omega_X(x) / \sum_{f \in I} \mathbb{k}(x) \cdot df$$

where

$$I = \ker \left(\underline{o}_{x,X} \longrightarrow \underline{o}_{x,Y} \right) .$$

On the other hand, if x is a closed point, we get a dual *injection*:

$$T_{x,Y} \hookrightarrow T_{x,X}$$

such that the image of $T_{x,Y}$ is the subspace of $T_{x,X}$ perpendicular to the differentials df, $(f \in I)$.

Example G. Let's compute the function d for a hypersurface $H \subset \mathbb{P}^n$. Since $\mathbb{P}^n$ itself is non-singular, for all $x \in \mathbb{P}^n$, $\dim_{\mathbb{P}^n}(x) = n$. Since $\Omega_H(x)$ is always a quotient of $\Omega_{\mathbb{P}^n}(x)$, its dimension is at most n. But a point $x \in H$ is non-singular if this dimension is $n-1$, so we get:

$$d(x) = \begin{cases} n-1, & \text{if } x \text{ is a non-singular point} \\ n, & \text{if } x \text{ is a singular point .} \end{cases}$$

Now let $I = \operatorname{Spec} k[\epsilon]/(\epsilon^2)$. I itself has a one-dimensional tangent space with a canonical generator $\mathcal{L}$, given by the linear functional

$$\alpha \cdot \epsilon \longmapsto \alpha$$

from (ϵ) to k. We can give a suggestive interpretation to definition 4 of the tangent space, in Theorem 3: it means that for all closed points $x \in X$ and all tangent vectors $t \in T_{x,X}$, there is one and only one morphism $f : I \to X$ with image x such that $df(\mathcal{L}) = t$.

[Proof: Such f's correspond to possible local homomorphisms $f^* : \underline{o}_x \to k[\epsilon]/(\epsilon^2)$, hence to possible k-linear maps $m_x/m_x^2 \to (\epsilon)$. But $df(\mathcal{L})$ is the composite linear functional

$$m_x/m_x^2 \xrightarrow{f^*} (\epsilon) \longrightarrow k ,$$

and since $\mathcal{L}$ is an isomorphism, the set of f^*'s correspond $1-1$ with the set of linear functionals $m_x/m_x^2 \to k$.]

In other words, I is a sort of disembodied tangent vector which can be embedded in any scheme X so as to lie along any given tangent vector to X. The set of all morphisms from I to X is a sort of set-theoretic tangent bundle to X, being isomorphic to $\bigcup_{\substack{\text{closed}\\ \text{points } x}} T_{x,X}$. Using this approach to the tangent space, we can give a more geometric explanation of the realization of $T_{x,X}$ as an actual linear subspace of $\mathbb{A}^m$, when $X \subset \mathbb{A}^m$.

Suppose $X = \text{Spec } (k[X_1, \ldots, X_m] / (g_1, \ldots, g_k))$ and $x = \mathbb{O}$ is a zero of the g_i's. Then $T_{\mathbb{O},\mathbb{A}^m}$ is isomorphic to the set of morphisms

$$f : I \longrightarrow \mathbb{A}^m, \qquad \text{Im } f = \mathbb{O}$$

and these are determined by the n-tuple $(\alpha_1, \ldots, \alpha_m)$ such that $f^*(X_i) = \alpha_i \cdot \epsilon$. $T_{\mathbb{O},X}$ will be the subvector space of $T_{\mathbb{O},\mathbb{A}^m}$ corresponding to the morphisms f that factor through X. But for f to factor through X, it is necessary and sufficient that $f^*(g_1) = \ldots = f^*(g_k) = 0$. In general, though,

$$\begin{aligned} f^*(h) &= h(f^*(X_1), \ldots, f^*(X_m)) \\ &= h(\alpha_1\epsilon, \ldots, \alpha_m\epsilon) \\ &= h(0, \ldots, 0) + \left(\sum_{f=1}^{m} \alpha_j \cdot \frac{\partial h}{\partial X_j}(0, \ldots, 0) \right) \cdot \epsilon \end{aligned}$$

since $\epsilon^2 = 0$. Therefore

$$f^*(g_i) = \left(\sum_{j=1}^{m} \alpha_j \cdot \frac{\partial g_i}{\partial X_j}(0, \ldots, 0) \cdot \epsilon \right),$$

and the vector space of $(\alpha_1, \ldots, \alpha_m)$'s making $f^*(g_i) = 0$, $1 \leq i \leq k$, is exactly the space of solutions of the linear equations:

$$\sum_{j=1}^{m} \frac{\partial g_i}{\partial X_j}(0, \ldots, 0) \cdot X_j = 0, \qquad 1 \leq i \leq k.$$

If we identify the m-tuples $(\alpha_1, \ldots, \alpha_m)$ with the closed points of $\mathbb{A}^m$ itself, we have again the tangent space to X at x as described in Proposition 1.

The property defining non-singularity requires that the given scheme X be locally the set of zeroes of functions in $\mathbb{A}^N$ with enough independent differentials. It is even true that if X is non-singular of dimension n and $f_1, \ldots, f_{N-n}$ are functions of $\mathbb{A}^N$ which vanish on X with independent differentials, then the locus of zeroes of the f's is *exactly* X, plus perhaps some other components disjoint from X. The general result is this:

Theorem 4. *Let X be a non-singular variety, $Y \subset X$ a closed subscheme, and $x \in Y$ any point. Then Y is non-singular at x if and only if there exists an affine open neighbourhood $U \subset X$ of x and elements $f_1, \ldots, f_k \in R = \Gamma(U, \underline{o}_X)$ such that*

1) $Y \cap U = \text{Spec } [R/(f_1, \ldots, f_k)]$
2) $df_1, \ldots, df_k$ *define independent elements of* $\Omega_{X/k}(x)$.

In this case, $\dim Y = \dim X - k$.

Proof. We begin with a special case:

> **Lemma.** *Let* $X = \operatorname{Spec}(R)$ *be a non-singular n-dimensional affine variety. Let* $f \in R$ *be an element such that the image of df in* $\Omega_X(x)$ *is not zero, for all* $x \in X$. *Then the subscheme* $Y = \operatorname{Spec}(R/(f))$ *is a disjoint union of non-singular subvarieties of dimension* $n-1$.
>
> *Proof.* Let $x \in Y$ be a closed point. We saw above that $T_{x,Y}$ is isomorphic to the subspace of $T_{x,X}$ perpendicular to df. Since $df \neq 0$ in $\Omega_X(x)$, $\dim T_{x,X} = \dim T_{x,X}^{-1} = n-1$. But since $Y = V((f))$, Y is pure $n-1$ dimensional, hence its tangent cone at x has a component of dimension $n-1$. But the only subscheme of an $(n-1)$-dimensional affine space with an $(n-1)$-dimensional component is the whole space. Therefore the tangent space and tangent cones are equal and Y is non-singular at x. □

Now assume $Y \cap U = \operatorname{Spec}\,[R/(f_1, \ldots, f_k)]$ and that $df_1, \ldots, df_k$ are independent in $\Omega_{X/k}(y)$ for all $y \in U$ (replace U by a smaller neighbourhood of x if necessary). Let $Z_i = \operatorname{Spec}\,[R/(f_1, \ldots, f_i)]$. We can use induction to show that each z_i is a disjoint union of non-singular varieties. In fact, suppose Z_i is such a union. Note that for all $z \in Z_i$,

$$\Omega_{Z_i/k}(z) \cong \Omega_{X/k}(z) / \sum_1^i \Bbbk(z) \cdot \overline{df}_i \,.$$

Therefore the image of df_{i+1} in $\Omega_{Z_i/k}(z)$ is not zero. Therefore by the lemma, Z_{i+1} is also a disjoint union of non-singular varieties. Also, since the dimension goes down by 1 each time, $Y \cap U = Z_k$ is pure $(\dim X - k)$-dimensional.

Conversely, assume Y is non-singular at x. Let the ideal of Y as a subscheme of X be generated at x by $f_1, \ldots f_N$. As before

$$\Omega_{Y/k}(x) \cong \Omega_{X/k}(x) / \sum_{i=1}^N \Bbbk(x) \cdot \overline{df}_i \,.$$

If $k = \operatorname{codim}_X(Y)$, then these $\overline{df}_i$ must span a subspace of $\Omega_{X/k}(x)$ of dimension exactly k. Choose $f_1, \ldots, f_k$ such that $\overline{df}_1, \ldots, \overline{df}_k$ in $\Omega_{X/k}(x)$ are independent. Extend these f_i's to functions in some neighbourhood $U = \operatorname{Spec}(R)$ of x in X. Let $Y^* = \operatorname{Spec}\,(R/(f_1, \ldots, f_k))$. If U is small enough, these f_i's will still vanish on $Y \cap U$, so $Y \cap U$ will be a closed subvariety of Y^*. But by the first half of the proof, if U is small enough, Y^* will be a non-singular subvariety of U of codimension k. In particular, it is irreducible and reduced. Then since $\dim Y^* = \dim Y \cap U$, $Y \cap U = Y^*$. □

Problems

1. If X is non-singular at x and Y is non-singular at y, then show that $X \times Y$ is non-singular at $x \times y$.
2. Describe $\Omega_{X/k}(x)$ for a non-closed point $x \in X$ as follows: let Ω_0 be the subspace generated by the elements df, $f \in m_x$. Show
   a) $\Omega_0 \cong m_x/m_x^2$, with df corresponding to $f \pmod{m_x^2}$, all $f \in m_x$.
   b) $\Omega_1/\Omega_0 \cong \Omega_{\mathbb{k}(x)/k}$, with $df \pmod{\Omega_o}$ corresponding to $d\overline{f}$, all $f \in \underline{o}_x$, ($\overline{f}$ being the image of f in $\mathbb{k}(x)$).

   Hence show X is non-singular at x if and only if $\dim\left(m_x/m_x^2\right) = \operatorname{codim} \overline{\{x\}}$.
3. Let x be *any* point of a variety X (not necessarily closed), show that X is non-singular at x if and only if $\underline{o}_x$ is a regular local ring.

   [This result, although pretty, has historically been rather a red herring. The concepts of non-singularity and regularity diverge over imperfect ground fields.]

§5. Étale morphisms

In the last section, we have seen that many familiar concepts involving differentials can be transferred from differentials and analytic geometry to algebraic geometry. But one very important theorem in the differential and analytic situations is *false* in the algebraic case – the implicit function theorem. This asserts that if we are given k differentiable (resp. analytic) functions $f_1, \ldots, f_k$ near a point x in $\mathbb{R}^{n+k}$ (resp. $\mathbb{C}^{n+k}$) such that

$$\det_{1 \le i,j \le k} \left(\partial f_i / \partial X_j\right)(x) \neq 0\,,$$

then the restriction of the coordinate projection

$$\begin{array}{ccc} \{\text{Locus } f_1 = \ldots = f_k = 0\} & \longrightarrow & \mathbb{R}^n \quad (\text{resp. } \mathbb{C}^n) \\ (x_1, \ldots, x_{n+k}) & \longmapsto & (x_{k+1}, \ldots, x_{k+n}) \end{array}$$

is locally an isomorphism near x. However, to take a typical algebraic situation, look at the projection:

$$\begin{array}{ccc} V\left(x_1^2 - x_2\right) & \longrightarrow & \mathbb{A}^1 \\ (x_1, x_2) & \longmapsto & x_2\,. \end{array}$$

At $x = (1,1)$, $\frac{\partial}{\partial x_1}\left(x_1^2 - x_2\right)$ is not zero, but the projection is not even $1-1$ in *any* Zariski-open subset U of $V\left(x_1^2 - x_2\right)$ since for all but a finite set of values $x_2 = a$, U will contain both points $(+\sqrt{a}, a)$ and $(-\sqrt{a}, a)$. This indicates that there exists in algebraic geometry a *non-trivial* class of morphisms that are nonetheless "local isomorphisms" in both a differential-geometric and analytic sense. These are known as étale[11] morphisms and are defined by mimicking the implicit function theorem as follows (we are now dealing with arbitrary schemes):

Definition 1. 1st the particular morphisms:

$$\begin{array}{rcl} X & = & \operatorname{Spec} R[X_1,\ldots,X_n]/(f_1,\ldots,f_n) \\ & & \downarrow \\ Y & = & \operatorname{Spec}(R) \end{array}$$

are *étale* at a point $x \in X$ if

$$(*) \qquad \det(\partial f_i/\partial X_j)(x) \neq 0\,.$$

2nd an arbitrary morphism $f : X \to Y$ of finite type is *étale*, if for all $x \in X$, there are open neighbourhoods $U \subset X$ of x and $V \subset Y$ of $f(x)$ such that $F(U) \subset V$ and such that f, restricted to U, looks like a morphism of the above type:

$$\begin{array}{ccc} U & \xrightarrow[\text{open immersion}]{} & \operatorname{Spec} R[X_1,\ldots,X_n]/(f_1,\ldots,f_n) \\ {\scriptstyle \operatorname{res}(f)}\downarrow & & \downarrow \\ V & \hookrightarrow & \operatorname{Spec}(R) \end{array}$$

where $\det(\partial f_i/\partial X_j)(x) \neq 0$.

This intuitively reasonable definition, like the provisional ones we made for the tangent cone and tangent space, is not really intrinsic.

One has a right to ask for an equivalent form involving only the local rings of X and Y and not dragging in affine space. There is such a reformulation, but it involves the concept of flatness so we have to put it off until §10. This clumsy form is adequate for the present.

The condition on the partials of the f_i's means exactly that $\Omega_{X/Y} = (0)$. In fact, if $S = R[X_1,\ldots,X_n]/(f_1,\ldots,f_n)$, then

$$\Omega_{X/Y} = \widetilde{\Omega}_{S/R}$$

$$\Omega_{S/R} = \left\{ \begin{array}{l} S\text{-module generated by } dX_1,\ldots,dX_n \\ \text{modulo the relations} \\ \sum_{j=1}^n (\partial f_i/\partial X_j)\cdot dX_j = 0, \quad 1 \le i \le n \end{array} \right\}$$

and this module is (0) exactly when $\det(\partial f_i/\partial X_j)$ is a unit in S.

Notice that if $f : X \to Y$ is étale, and we take any fibre product:

[11] The word apparently refers to the appearance of the sea at high tide under a full moon in certain types of weather.

$$\begin{array}{ccc} X' = X \times_Y Y' & \longrightarrow & X \\ {\scriptstyle f'}\downarrow & & \downarrow{\scriptstyle f} \\ Y' & \longrightarrow & Y \end{array}$$

then f' is still étale. This follows immediately from the definition. Thus, for example, the fibre of an étale morphism $f : X \to Y$ over a point $y \in Y$ must be a scheme étale over Spec $\mathbb{k}(y)$. We can easily work out the definition in this case:

Proposition 1. *Let X be a scheme of finite type over a field k. Then X is étale over k if and only if X is the union of a finite set of points $x_i =$ Spec (k_i), where each k_i is a separable finite algebraic extension of k.*

Proof. First of all, assume X is étale over k. Then if $U =$ Spec (R) is an affine piece of X, $\Omega_{R/k} = (0)$. By the Problem in §1, this shows that $R = \oplus k_i$, k_i being finite separable/k. Conversely, if K/k is a finite separable extension, then by the theorem of the primitive element,

$$K \cong k[X]/(f(X))$$

where $\partial f/\partial X \not\equiv 0$. But then $\partial f/\partial X \notin (f(X))$ either, so if Spec $K = \{x\}$, $\partial f/\partial X(x) \neq 0$. This shows that Spec (K) is étale over Spec (k), (taking $n = 1$ in the definition). □

Corollary. *If $f : X \to Y$ is étale, then*

1) *for all $y \in Y$, the fibre $f^{-1}(y)$, (as a scheme over* Spec $(\mathbb{k}(y))$*) is the union of a finite set of points* Spec (k_i), *k_i finite separable over $\mathbb{k}(y)$.*
2) *for all $x \in X$, the maximal ideal $m_x \subset \underline{o}_x$ is generated by $f^*(m_{f(x)})$ and the residue field $\mathbb{k}(x)$ is finite separable over $\mathbb{k}(f(x))$.*

Proof. The Prop. implies (1) since $f^{-1}(y)$ is étale over Spec $(\mathbb{k}(y))$. If $x \in X$, and $y = f(x)$, then (2) is seen to be a restatement of (1) since the point x defines a point x' in the fibre $f^{-1}(y)$ with local ring:

$$\underline{o}_{x',f^{-1}(y)} = \underline{o}_{x,X}/f^*(m_y) \cdot \underline{o}_{x,X} \, .$$

□

This condition on f looks nicer if stated in terms of *geometric* fibres instead of ordinary fibres – i.e., fibre products

$$\begin{array}{ccc} F & \longrightarrow & X \\ \downarrow & & \downarrow{\scriptstyle f} \\ \mathrm{Spec}\,(\Omega) & \xrightarrow[i]{} & Y \end{array}$$

where Ω is an algebraically closed field. In fact, if y = Image of i, then

$$F = f^{-1}(y) \times_{\mathbf{k}(y)} \Omega \, ,$$

so (1) is the same as

1') all geometric fibres of f are finite sets of reduced points.

Corollary 2. *Let X and Y be varieties over an algebraically closed field k. Let $f : X \to Y$ be an étale morphism. Then f is dominating, and $k(X)$ is a finite separable extension of $k(Y)$.*

Proof. Locally, X is the locus of roots of n-equations in $Y \times \mathbb{A}^n$, for some n. Therefore $\dim X \geq \dim Y$. Now if $\overline{f(X)}$ were a proper subvariety of Y, then the fibres of f would have positive dimension and would be infinite. Therefore f must be dominating. But then Spec $(k(X))$ is the fibre of f over the generic point of Y, so Cor. 1 says that $k(X)$ is finite separable over $k(Y)$. □

One of the key facts about étale morphisms is Hensel's lemma, of which here is a variant:

Theorem 2. *Let $\mathcal{O}$ be a complete (noetherian) local ring, with maximal ideal m, residue field k. Let f be a morphism of finite type:*

$$\begin{aligned} X \xrightarrow{f} Y &= \text{Spec } (\mathcal{O}) \\ x \longmapsto y &= [m] \, . \end{aligned}$$

Assume $k = \mathbf{k}(y) \xrightarrow{\sim} \mathbf{k}(x)$. Then f étale near $x \Rightarrow f$ a local isomorphism near x.

Proof. We may assume that

$$\begin{aligned} X &= \text{Spec } (\mathcal{O}\,[X_1, \ldots, X_n] \,/\, (f_1, \ldots, f_n)) \\ x &= [m + (X_1 - a_1, \ldots, X_n - a_n)] \\ & \det\,(\partial f_i / \partial X_j)\,(x) \neq 0 \, , \end{aligned}$$

where $a_1, \ldots, a_n \in \mathcal{O}$. In fact, if $x = [P]$, then modulo P, each X_i is equal to an element of k, hence for some $a_i \in \mathcal{O}$, $X_i \equiv a_i \pmod P$. Therefore if $P_0 = m + (X_1 - a_1, \ldots, X_n - a_n)$, $P \supseteq P_0$. But since $\mathcal{O}\,[X_1, \ldots, X_n]\,/P_0 \cong \mathcal{O}/m = k$, P_0 is maximal so $P = P_0$. The main part of the proof consists in constructing a *section* $s : Y \to X$ of the morphism f such that $s(y) = x$. This is equivalent to constructing a subscheme $Z \subset X$ isomorphic to Y under the restriction of f. We do this by a classical form of Hensel's lemma:

Hensel's Lemma. *Let $\mathcal{O}$ be a complete local ring, and let $f_1, \ldots, f_n \in \mathcal{O}\,[X_1, \ldots, X_n]$. Assume $a_1, \ldots, a_n$ satisfy*

$$f_1\,(a_1, \ldots, a_n) \equiv \ldots \equiv f_n\,(a_1, \ldots, a_n) \equiv 0 \pmod m$$

$$(\det \partial f_i / \partial X_j)\,(a_1, \ldots, a_n) \notin m \, .$$

Then there exist $\alpha_1, \ldots, \alpha_n \in \mathcal{O}$ *such that*

$$\alpha_i \equiv a_i \pmod{m}$$
$$f_1(\alpha_1, \ldots, \alpha_n) = \ldots = f_n(\alpha_1, \ldots, \alpha_n) = 0 .$$

Proof. We "refine" the approximation root $(a_1, \ldots, a_n)$ by induction. Suppose at the rth stage that we have got an n-tuple $\left(a_1^{(r)}, \ldots, a_n^{(r)}\right)$ such that

$$a_i^{(r)} \equiv a_i \pmod{m}$$
$$f_1\left(a_1^{(r)}, \ldots, a_n^{(r)}\right) \equiv \ldots \equiv f_n\left(a_1^{(r)}, \ldots, a_n^{(r)}\right) \equiv 0 \pmod{m^r} .$$

Vary the $a_i^{(r)}$ a little to $a_i^{(r)} + \epsilon_i$, where $\epsilon_i \in m^r$. Then

$$f_i\left(a_1^{(r)} + \epsilon_1, \ldots, a_n^{(r)} + \epsilon_n\right)$$
$$\equiv f_i\left(a_1^{(r)}, \ldots, a_n^{(r)}\right) + \sum_{j=1}^{n} \frac{\partial f_i}{\partial X_j}(a_1, \ldots, a_n) \cdot \epsilon_j \pmod{m^{r+1}} .$$

But since $\det(\partial f_i/\partial X_j)(a_1, \ldots, a_n)$ is a unit in $\mathcal{O}$, we can find a matrix $B = \{B_{ij}\}$, with $B_{ij} \in \mathcal{O}$ such that

$$B \cdot [\partial f_i/\partial X_j(a_1, \ldots, a_n)] = I_n$$

(I_n = the identity $n \times n$ matrix). If we set

$$\epsilon_i = -\sum_{j=1}^{n} B_{ij} \cdot f_j\left(a_1^{(r)}, \ldots, a_n^{(r)}\right) ,$$

it follows that $f_i\left(a_1^{(r)} + \epsilon_1, \ldots, a_n^{(r)} + \epsilon_n\right) \equiv 0 \pmod{m^{r+1}}$ for all i. Then $a_i^{(r+1)} = a_i^{(r)} + \epsilon_i$ is closer approximate to a solution, and if we set $\alpha_i = \lim_{r\to\infty} a_i^{(r)}$, the lemma follows. □

Applying this to the theorem, we find elements $\alpha_1, \ldots, \alpha_n \in \mathcal{O}$, such that

$$Z = \operatorname{Spec} \mathcal{O}[X_1, \ldots, X_n] / (X_1 - \alpha_1, \ldots, X_n - \alpha_n)$$

is a subscheme of X through the point x. Since

$$\mathcal{O}[X_1, \ldots, X_n] / (X_1 - \alpha_1, \ldots, X_n - \alpha_n) \cong \mathcal{O} ,$$

Z is isomorphic to Y under the restriction of f. What remains is to prove that *near* x, $X = Z$.

Let $\mathcal{Q}$ be the $\underline{o}_X$-ideal defining Z, and notice that the stalks at x fit into a diagram:

$$\begin{array}{ccc}
\underline{o}_{x,X} & \longrightarrow & \underline{o}_{x,X}/\mathcal{Q}_x = \underline{o}_{x,Z} \\
\uparrow f^* & \nearrow \approx & \\
\underline{o}_{y,Y} & & \\
\| & & \\
\mathcal{Q} & &
\end{array}$$

Thus $\underline{o}_{x,X} \cong f^*(\mathcal{Q}) \oplus \mathcal{Q}_x$. In particular, its maximal ideal $m_{x,X}$ is $f^*(m_y) \oplus \mathcal{Q}_x$. On the other hand, since f is étale, $m_{x,X} \subset f^*(m_y) \cdot \underline{o}_{x,X}$ (Cor. 1 just above). Therefore

$$\mathcal{Q}_x \subset f^*(m_y) \cdot \mathcal{Q}_x \subset m_x \cdot \mathcal{Q}_x \ .$$

Therefore, by Nakayama's lemma, $\mathcal{Q}_x = (0)$, hence $\mathcal{Q} \equiv (0)$ near x, hence $Z = X$ near x. □

Using this, we prove the following connecting the concept of étale with power series:

Theorem 3. *Let $f : X \to Y$ be a morphism of fnite type of noetherian schemes. Let $x \in X$, $y = f(x)$, and assume that the induced map*

$$f_x^* : \mathbb{k}(y) \longrightarrow \mathbb{k}(x)$$

is an isomorphism. Then f is étale in some neighbourhood of x if and only if the induced map:

$$\widehat{f_x^*} : \widehat{\underline{o}}_y \longrightarrow \widehat{\underline{o}}_x$$

of complete local rings is an isomorphism.

Proof. First we assume that f is étale near x, and prove that $\widehat{f_x^*}$ is an isomorphism. This part can be reduced to Theorem 2 by making a fibre product:

$$\begin{array}{ccc}
X' = X \times_Y \mathrm{Spec}\,(\widehat{\underline{o}}_y) & \longrightarrow & X \\
\downarrow f' & & \downarrow f \\
Y' = \mathrm{Spec}\,(\widehat{\underline{o}}_y) & \xrightarrow{\ i\ } & Y
\end{array}$$

where, by definition, i takes the closed point $y' = [\hat{m}_y]$ of y' to y, and i^* is the canonical inclusion $\underline{o}_y \hookrightarrow \hat{\underline{o}}_y$. Let $x' \in X'$ lie over $x \in X$ and $y' \in Y'$. Then f' is étale near x', so by Theorem 2, we get a diagram:

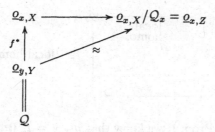

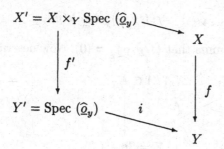

$$\begin{array}{ccc} \underline{o}_{x',X'} & \xleftarrow[\text{homo.}]{\text{local}} & \underline{o}_{x,X} \\ \approx \uparrow & & \uparrow f_x^* \text{ (local homo.)} \\ \hat{\underline{o}}_y & \hookleftarrow & \underline{o}_y \end{array} .$$

Also, from Cor. 1 of Prop. 1, we know that $m_{x,X} = f_x^*(m_y) \cdot \underline{o}_{x,X}$. Therefore the theorem follows from the elementary fact:

> Given 2 local rings $\mathcal{O}_1$, $\mathcal{O}_2$ and local homomorphisms
> $$\mathcal{O}_1 \xrightarrow{f} \mathcal{O}_2 \longrightarrow \hat{\mathcal{O}}_1$$
> if $m_2 = f(m_1) \cdot \mathcal{O}_2$, then
> $$\hat{\mathcal{O}}_1 \xrightarrow{\hat{f}} \hat{\mathcal{O}}_2$$
> is an isomorphism .

Now conversely, assume $\hat{\underline{o}}_y \xrightarrow{\sim} \hat{\underline{o}}_x$. Since for any ideal $A \subset \underline{o}_x$, $(A \cdot \hat{\underline{o}}_x) \cap \underline{o}_x = A$ (Zariski-Samuel, vol. 2, p.   ), we find that

$$\begin{aligned} f^*(m_y) \cdot \underline{o}_x &= [f^*(m_y) \cdot \hat{\underline{o}}_x] \cap \underline{o}_x \\ &= \left[\hat{f}^*(\hat{m}_y) \cdot \hat{\underline{o}}_x\right] \cap \underline{o}_x \\ &= \hat{m}_x \cap \underline{o}_x \\ &= m_x . \end{aligned}$$

In other words, the fibre of f over y, near x, is just a copy of Spec $(\mathbb{k}(y))$. Therefore $\Omega_{f^{-1}(y)/\text{Spec }\mathbb{k}(y)}$ is (0) at x, and since

$$\left(\Omega_{f^{-1}(y)/\text{Spec }\mathbb{k}(y)}\right)_x = \left(\Omega_{X/Y}\right)_x \otimes_{\underline{o}_y} \mathbb{k}(y) ,$$

it follows from Nakayama's lemma that $\left(\Omega_{X/Y}\right)_x = (0)$. Now describe

$$\begin{array}{ccc} X = \text{Spec } R[X_1, \ldots, X_n]/A & \subset & \mathbb{A}_R^n \\ & \searrow^{f} & \downarrow \\ & & Y = \text{Spec }(R) \end{array} .$$

Then

$$(0) = \left(\Omega_{X/Y}\right)_x \cong \left[\begin{array}{l} \text{free } \underline{o}_x\text{-module generated} \\ \text{by } dX_1, \ldots, dX_n \end{array}\right] \Big/ \begin{array}{l} \text{relations} \\ \sum_i \frac{\partial f}{\partial X_i} \cdot dX_i = 0 \\ \text{all } f \in A \end{array} .$$

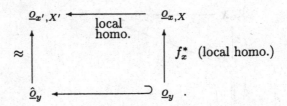

Therefore, there must be elements $f_1, \ldots, f_n \in A$ such that $[\partial f_i / \partial X_j]$ forms an invertible matrix over $\underline{o}_x$, i.e., $\det(\partial f_i / \partial X_j)(x) \neq 0$. Define a subscheme $\widetilde{X}$:

$$X \subset \widetilde{X} \subset \mathbb{A}^n_R$$

by Spec $R[X_1, \ldots, X_n] / (f_1, \ldots, f_n)$. By definition, $\widetilde{X}$ is étale over Y near x. Now compare all the local rings at x. Applying the 1st half of the theorem to $\widetilde{X}/Y$, we get the diagram:

$$\begin{array}{ccccc} \underline{o}_{x,X} & & \longleftarrow & & \underline{o}_{x,\tilde{X}} \\ \cap & & & & \cap \\ \hat{\underline{o}}_{x,X} & & & & \hat{\underline{o}}_{x,\tilde{X}} \\ & \nwarrow \approx & & \approx \nearrow & \\ \text{by assumption} & & \hat{\underline{o}}_y & & \text{since } \tilde{f} \text{ étale} \end{array}$$

Therefore, $\underline{o}_{x,\widetilde{X}}$ is a subring of $\underline{o}_{x,X}$.

But since X is a closed subscheme of $\widetilde{X}$, $\underline{o}_{x,X}$ is isomorphic to a quotient of $\underline{o}_{x,\widetilde{X}}$. Thus $\underline{o}_{x,\widetilde{X}}$. In other words, A and $(f_1, \ldots, f_n)$ must generate the same ideal in the local ring of x on $Y \times \mathbb{A}^n$. Therefore X and X are locally, near x, the same subscheme of $Y \times \mathbb{A}^n$, and f must also be étale near x. □

For the rest of this paragraph, we return to the geometric case: we fix an *algebraically closed* ground field k.

Corollary 1. *Let X and Y be schemes of finite type over k. Let $f : X \to Y$ be a morphism. Then f is étale if and only if for all closed points $x \in X$, the induced map:*

$$f_x^* : \hat{\underline{o}}_{f(x)} \longrightarrow \hat{\underline{o}}_x$$

is an isomorphism. Moreover, if f is étale, then for all closed points $x \in X$, there is a natural isomorphism of the tangent spaces to X and Y at x, $f(x)$ taking the tangent cone into each other:

$$\begin{array}{ccc} \textit{Tangent cone to } X \textit{ at } x & \subset & \textit{Tangent space to } X \textit{ at } x \\ \| \big\downarrow & & \| \big\downarrow \\ \textit{Tangent cone to } Y \textit{ at } f(x) & \subset & \textit{Tangent space to } Y \textit{ at } x. \end{array}$$

In particular, X is non-singular at x if and only if Y is non-singular at $f(x)$.

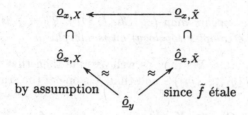

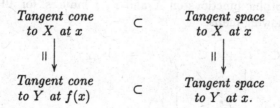

Proof. The first statement follows from the theorem since $\mathbb{k}(x) = \mathbb{k}(f(x)) = k$. The second statement follows from the first since $\operatorname{gr}(\underline{o}_x) \cong \operatorname{gr}\left(\hat{\underline{o}}_{f(x)}\right) \cong \operatorname{gr}\left(\underline{o}_{f(x)}\right)$ as graded rings, and the tangent space and cone set-up is deduced formally from the graded rings $\operatorname{gr}(\underline{o})$. □

Corollary 2. *Let X and Y be varieties of $\mathbb{C}$, and let $f: X \to Y$ be a morphism. Let $\mathcal{X}$ and $\mathcal{Y}$ be the corresponding analytic spaces and let*

$$\mathcal{F}: \mathcal{X} \longrightarrow \mathcal{Y}$$

be the corresponding holomorphic map (cf. Ch. I, §10). Then f is étale if and only if $\mathcal{F}$ is locally (in the complex topology) an isomorphism.

Proof. First assume f is étale. We may as well assume then that Y is affine – say $(\operatorname{Spec} \mathbb{C}[Y_1,\ldots,Y_k]/(g_1,\ldots,g_\ell))$ – and that f is one of the standard maps:

$$\begin{array}{c}\operatorname{Spec} \mathbb{C}[Y_1,\ldots,Y_k,X_1,\ldots,X_n]/(g_1,\ldots,g_\ell,f_1,\ldots,f_n)\\ \downarrow \\ \operatorname{Spec} \mathbb{C}[Y_1,\ldots,Y_k]/(g_1,\ldots,g_\ell)\ ,\end{array}$$

where $\det(\partial f_i/\partial X_j)$ is nowhere 0 on X. Then $\mathcal{F}$ has the form:

$$\begin{array}{ccccccc}\mathcal{X} = & V(f_1,\ldots,f_n)\cap V(g_1,\ldots,g_\ell) & \subset & V(f_1,\ldots,f_n) & \subset & \mathbb{C}^{n+k} \\ \mathcal{F}\downarrow & & & & & \downarrow \mathcal{F}' \\ \mathcal{Y} = & V(g_1,\ldots,g_\ell) & & \subset & & \mathbb{C}^k\ .\end{array}$$

For all $x \in \mathcal{X}$, since $\det(\partial f_i/\partial X_j)(x) \neq 0$, $\mathcal{F}'$ is a local isomorphism near x by the implicit function theorem. Since $\mathcal{F}$ is the restriction of $\mathcal{F}'$ to the inverse image of $V(g_1,\ldots,g_\ell)$, it is a local isomorphism too.

Conversely, assume $\mathcal{F}$ is a local isomorphism. Let Ω_X and Ω_Y denote the sheaves of holomorphic functions on X and Y. f induces, for all closed points $x \in X$, all the following maps:

$$\begin{array}{ccccccc}\underline{o}_{Y,y} & \subset & \Omega_{y,Y} & \subset & \hat{\underline{o}}_{y,Y} & = & \hat{\Omega}_{y,Y} \\ \downarrow f_x^* & & \downarrow \mathcal{F}_x^* & & \downarrow f_x^* & & \\ \underline{o}_{x,X} & \subset & \Omega_{x,X} & \subset & \hat{\underline{o}}_{x,X} & = & \hat{\Omega}_{x,X}\end{array}$$

where $y = f(x)$. But $\mathcal{F}_x^*$ is an isomorphism. Therefore f_x^* is an isomorphism, and f is étale by Cor. 1. □

In case X and Y are non-singular varieties, the concept of étale simplifies remarkably:

Theorem 4. *Let $f : X \to Y$ be a morphism of non-singular n-dimensional varieties over k. If for all closed points $y \in Y$, the fibre $f^{-1}(y)$ is a finite set of reduced points (or, equivalently, if for all closed points $x \in X$, $m_x = f^*\left(m_{f(x)}\right) \cdot \underline{o}_x$), then f is étale.*

Proof. Everything being local, we may assume that X and Y are affine. Choose a closed immersion $i : X \hookrightarrow \mathbb{A}^N$. This defines a closed immersion

$$(i, f) : X \hookrightarrow \mathbb{A}^N \times Y\,.$$

Then here we have an n-dimensional non-singular variety embedded in an $N+n$-dimensional non-singular variety. By Theorem 4, §4, x is locally the subscheme of zeroes of exactly N functions on $\mathbb{A}^N \times Y$. Therefore, if $R = \Gamma(Y, \underline{o}_Y)$ for all closed points $x \in X$, there is an open neighbourhood $U \subset X$ of x and elements $f_1, \ldots, f_N \in R[X_1, \ldots, X_n]$ such that

$$\begin{array}{ccc} U & \xrightarrow[\text{restriction}]{\text{open}} & \operatorname{Spec} R[X_1, \ldots, X_N] / (f_1, \ldots, f_N) \\ {\scriptstyle \text{restriction of } f}\downarrow & & \downarrow \\ Y & = & \operatorname{Spec} R\;. \end{array}$$

Let $y = f(x)$, and let $\overline{f}_1, \ldots, \overline{f}_n \in k[X_1, \ldots, X_n]$ denote the images of the f_i's when their coefficients are evaluated at y. Then the fibre $f^{-1}(y)$ looks locally near x like Spec $k[X_1, \ldots, X_n] / (\overline{f}_1, \ldots, \overline{f}_N)$. By assumption, this fibre has no tangent space at all at x. Therefore the differentials $d\overline{f}_i$ must be independent at x, i.e.,

$$\det\left(\partial f_i / \partial X_j\right)(x) = \det\left(\partial \overline{f}_i / \partial X_j\right)(x) \neq 0\,.$$

Therefore f is étale. □

§6. Uniformizing parameters

In the differential and analytic geometry, n-dimensional manifolds M are distinguished from singular spaces by the existence of coverings $\{U_i\}$ such that each U_i is isomorphic, in the appropriate sense, to an open ball in $\mathbb{R}^n$ or in $\mathbb{C}^n$. In our case, we cannot hope in general to find any Zariski open sets U in a variety V

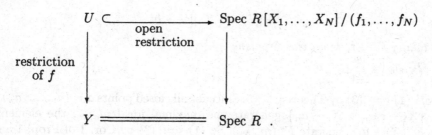

which are isomorphic to Zariski open sets in $\mathbb{A}^n$ because this would imply that the function fields are isomorphic:

$$k(V) \cong k(\mathbb{A}^n) = k(X_1, \ldots, X_n) .$$

What we can do, if V is non-singular, is to find sets of n functions $f_1, \ldots, f_n$ defined in Zariski open sets $U \subset V$, such that the morphism

$$F = f_1 \times \ldots \times f_n : U \longrightarrow \mathbb{A}^n$$

induced by the f_i's is étale. Because the implicit function theorem is false, this is the closest we can come to local coordinates, and such f_i's are known as "uniformizing parameters".

Theorem 1. *Let X be an n-dimensional non-singular variety over k (algebraically closed). Let $f_1, \ldots, f_n \in \Gamma(U, \underline{o}_X)$ for some open set $U \subset X$. The following are equivalent:*

(1) Let $F = f_1 \times \ldots \times f_n : U \to \mathbb{A}^n$ be the induced map. Then f is étale.
(2) For all closed points $x \in U$, let $t_1 = f_1 - f_1(x), \ldots, t_n = f_n - f_n(x)$. Then $t_1, \ldots, t_n$ generate m_x/m_x^2.
(3) For all closed points $x \in U$, the natural map

$$k[[T_1, \ldots, T_n]] \longrightarrow \hat{\underline{o}}_x$$

taking $T_i \to t_i$ is an isomorphism.
(4) $\Omega_{X/k|U} \cong \bigoplus_{i=1}^{n} \underline{o}_X \cdot df_i$.

Proof. (1) $\Longleftrightarrow$ (3) by Theorem 2, §5. Since for all closed points $a = (a_1, \ldots, a_n) \in \mathbb{A}^n$, $\{X_1 - a_1, \ldots, X_n - a_n\}$ generate the maximal ideal $m_{a,\mathbb{A}^n}$, the elements $t_i = F^*(X_i - a_i)$ generate $F^*(m_{a,\mathbb{A}^n})$. So (1) $\Longleftrightarrow$ (3) by Cor. 1 of Prop. 1 and Theorem 3, §5. Finally, for all closed points $x \in U$,

$$\begin{aligned} \{t_1, \ldots, t_n\} \text{ generate } m_x/m_x^2 &\Longleftrightarrow \{df_1, \ldots, df_n\} \text{ generate } \Omega_{X/k}(x) \\ &\Longleftrightarrow \{df_1, \ldots, df_n\} \text{ generate } \Omega_{X/k} \text{ in} \\ &\qquad \text{some neighbourhood of } x . \end{aligned}$$

Therefore (2) is equivalent to $\Omega_{X/k}$ being a quotient of $\bigoplus_{i=1}^{n} \underline{o}_X \cdot df_i$. But since it is locally free of rank n, it can only be a quotient of $\bigoplus_{i=1}^{n} \underline{o}_X \cdot df_i$ if it actually equals $\bigoplus_{i=1}^{n} \underline{o}_X \cdot df_i$. □

Definition 1. Elements $f_1, \ldots, f_n \in \Gamma(U, \underline{o}_X)$ with these properties are called *uniformizing parameters* in U.

Since $\Omega_{X/k}$ is locally free of rank n and is spanned by the df_i's, any non-singular variety of X can be covered by open sets for which a set of uniformizing parameters exists. Also, n elements $f_1, \ldots, f_n \in k(X)$ are uniformizing parameters in *some* non-empty open set U if and only if $df_1, \ldots, df_n$ generate $\Omega_{k(X)/k}$, i.e., $k(X)$ is separable and algebraic over $k(f_1, \ldots, f_n)$.

Example H. Let $X \subset \mathbb{A}^2$ be the curve $y^2(1-x^2) = 1$ and assume $\mathrm{char}(k) \neq 2$. Then $\Omega_{X/k}$ is generated by dx and dy with the relation:

$$2y\left(1-x^2\right)dy - 2y^2 x dx = 0\ .$$

Multiplying by $y/2$, this becomes $dy = xy^3 dx$, so dx alone generates $\Omega_{X/k}$. Therefore x is a uniformizing parameter everywhere on X and the projection p_1 onto the x-axis is étale:

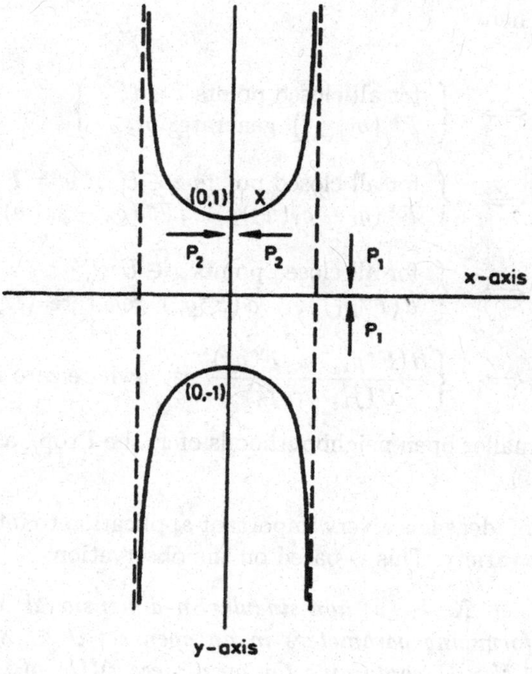

On the other hand, dy only generates Ω when $x \neq 0$ (y is nowhere 0). So if $U = X - \{(0,1),(0,-1)\}$, the projection $p_2 : X \to \{y\text{-axis}\}$ is étale and y is a uniformizing parameter only on U.

All the standard machinery of Calculus can be carried over to our algebraic setting. For example, one can generalize the Jacobian criterion for a map to be a local isomorphism as follows: First of all, if X is a non-singular variety, $f_1, \ldots, f_n$

are uniformizing parameters in $U \subset X$, and $g_1, \ldots, g_n$ are any functions in $\Gamma(U, \underline{o}_X)$, define the Jacobian as follows:

$$\frac{\partial(g_1, \ldots, g_n)}{\partial(f_1, \ldots, f_n)} = \det(a_{ij}), \quad \text{if} \quad df_i = \sum_{j=1}^{n} a_{ij} df_j$$
$$a_{ij} \in \Gamma(U, \underline{o}_X) .$$

Proposition 2. *Let $F : X \to Y$ be a morphism of non-singular n-dimensional varieties. Let $x \in X$ and $y = F(x)$. Choose uniformizing parameters $f_1, \ldots, f_n$ and $g_1, \ldots, g_n$ in some neighbourhoods of x and y respectively. Then f is étale in some neighbourhood of x if and only if*

$$\frac{\partial(F^* g_1, \ldots, F^* g_n)}{\partial(f_1, \ldots, f_n)}(x) \neq 0 .$$

Proof. Suppose the f_i's and g_i's are parameters in U and V respectively, and that $F(U) \subset V$. Then

$$f \text{ is étale on } U \iff \left\{ \begin{array}{l} \text{for all closed points } z \in U , \\ F^*\left(m_{f(Z)}\right) \text{ generates } m_z \end{array} \right\}$$

$$\iff \left\{ \begin{array}{l} \text{for all closed points } z \in U, \text{ if } w = F(z) , \\ F^*(g_1 - g_1(w)), \ldots, F^*(g_n - g_n(w)) \text{ generate } m_z \end{array} \right\}$$

$$\iff \left\{ \begin{array}{l} \text{for all closed points } z \in U , \\ d(F^* g_1), \ldots, d(F^* g_n) \text{ generate } \left(\Omega_{X/k}\right)_z \end{array} \right\}$$

$$\iff \left\{ \frac{\partial(F^* g_1, \ldots, F^* g_n)}{\partial(f_1, \ldots, f_n)} \text{ is nowhere zero in } U \right\} .$$

Replacing U by smaller open neighbourhoods of x, the Prop., as stated, follows. □

This complex of ideas has a very important application to intersection theory on a non-singular variety. This is based on the observation:

Proposition 3. *Let X be a non-singular n-dimensional variety, and let $f_1, \ldots, f_n$ be uniformizing parameters in an open set $U \subset X$. Then there is an open set $U^* \subset U \times U$ containing the open piece $\Delta(U)$ of the diagonal such that the elements $f_i \otimes 1 - 1 \otimes f_i \in \Gamma\left(U \times U, \underline{o}_{X \times X}\right)$, for $1 \leq i \leq n$, generate the ideal $\mathcal{Q} \subset \underline{o}_{X \times X}$ of the diagonal in U^*.*

Proof. By definition, the $\underline{o}_X$-module $\Omega_{X/k}$ is just the $\underline{o}_{X \times X}$-module $\mathcal{Q}/\mathcal{Q}^2$, regarded as a module over $\underline{o}_{X \times X}/\mathcal{Q}$ and carried over to X. In the process, the differential df comes from $f \otimes 1 - 1 \otimes f$. Since we assumed that $df_1, \ldots, df_n$ generate $\Omega_{X/k}$ in U, it follows that $f_1 \otimes 1 - 1 \otimes f_1, \ldots, f_n \otimes 1 - 1 \otimes f_n$ generate $\mathcal{Q}/\mathcal{Q}^2$ in $\Delta(U)$, i.e., for all $x \in \Delta(U)$, they generate the stalk $\left(\mathcal{Q}/\mathcal{Q}^2\right)_x$ over $\underline{o}_{x, X \times X}$. But $\left(\mathcal{Q}/\mathcal{Q}^2\right)_x = \mathcal{Q}_x/\mathcal{Q}_x^2$, so *a fortiori* they generate $\mathcal{Q}_x/m_x \cdot \mathcal{Q}_x$.

Therefore, by Nakayama's lemma, the elements $f_i \otimes 1 - 1 \otimes f_i$ generate $\mathcal{Q}$ itself in some neighbourhood of each of these x's. □

The weaker fact that $\Delta(X)$ is locally on $X \times X$ the scheme of zeroes of *some* n functions follows also from Th. 4, §4.

Proposition 4. *Let X be a non-singular variety of dimension n. Let Y and Z be irreducible closed subsets of X and let W be a component of $Y \cap Z$. Then*

$$\dim W \geq \dim Y + \dim Z - n$$

(or, $\operatorname{codim} W \leq \operatorname{codim} Y + \operatorname{codim} Z$*).*

Proof. $Y \cap Z$ is isomorphic to the intersection $Y \times Z \cap \Delta(X)$ taken in $X \times X$. Therefore $\Delta(W)$ is a component of $Y \times Z \cap \Delta(X)$. But for all $x \in \Delta(W)$, $\Delta(X)$ is defined in some neighbourhood U^* of x by the vanishing of n functions $g_1, \ldots, g_n$ (Prop. 3). Therefore

$$\begin{aligned} \dim W &= \dim \Delta(W) \\ &\geq \dim(Y \times Z) - n \\ &= \dim Y + \dim Z - n \end{aligned}$$

by Cor. 3 of Th. 2, Ch. I, §6. □

$\dim W$ can, of course, be bigger than $\dim Y + \dim Z - n$: e.g., if $Y = Z$. When equality holds, Y and Z are said to *intersect properly* on X.

Example I. The Proposition really requires the non-singularity of X. To see this, let $X \subset \mathbb{A}^4$ be the cone by $xy = zw$. X is a 3-dimensional variety. Let Y be the plane $x = z = 0$ and let Z be the plane $y = w = 0$. Then $Y \cap Z$ consists of the origin alone, so

$$0 = \dim Y \cap Z < \dim Y + \dim Z - 3 = 1 .$$

§7. Non-singularity and the UFD property

We want to prove in this section the important:

Theorem 1. *Let X be a non-singular variety over k. Then for all $x \in X$, $\underline{o}_{x,X}$ is a UFD.*

Varieties X such that all $\underline{o}_{x,X}$'s are UFD's are sometimes called *factorial.* The geometric meaning of this property is this:

Theorem 1'. *Let X be a non-singular variety over k. For all irreducible closed subsets $Z \subset X$ of codimension 1, there is an open affine covering $\{U_i\}$ of X and elements $f_i \in \Gamma(U_i, \underline{o}_X)$ such that $(f_i) = I(Z \cap U_i)$.*

In fact, $\underline{o}_x$ is a UFD if and only if all minimal prime ideals $P \subset \underline{o}_x$ are principal; and these minimal prime ideals are the ideals $I(Z)$, for irreducible closed $Z \subset X$ of codimension 1, containing x. Notice that to prove the geometric property, it suffices to prove that $\underline{o}_x$ is a UFD for all *closed* points x. Therefore Theorem 1 for all x follows if it is proven for closed points.

One case of the Theorem is very elementary: when Z is non-singular too, it follows from Th. 4, §4.

To prove Theorem 1 all 3 of the methods mentioned in the introduction can be used. There is a projective method, based on Severi's idea of projecting cones. Or using complete local rings, we can reduce the UFD property for $\underline{o}_x$ (x closed) to the UFD property for $\hat{\underline{o}}_x$ which is isomorphic by Th. 10, §6, to $k[[t_1, \ldots, t_n]]$. The most far-reaching method is the cohomological one, due to Auslander and Buchsbaum, by which it can be proven that all regular local rings are UFD's. We shall present the first two methods. For the last, cf. Zariski-Samuel, appendix, vol. 2.

Proof No. 1, via completions.

As we saw at the beginning of this section, it suffices to prove that $\underline{o}_x$ is a UFD, for closed points x. We shall assume then that $k[[t_1, \ldots, t_n]]$ and hence $\hat{\underline{o}}_x$ is known to be a UFD (cf. Zariski-Samuel, vol. 2, p.).

Lemma 1. *Let R be a noetherian integral domain. Then R is a UFD if and only if, for all $f, g \in R$, $f \neq 0$, $g \neq 0$, the ideal*

$$(f) : (g) \underset{\text{def.}}{=} \{h \in R \mid hg \in (f)\}$$

is principal.

Proof. Assume R is a UFD. Let

$$f = \epsilon f_1^{r_1} \ldots f_n^{r_n}$$
$$g = \eta f_1^{s_1} \ldots f_n^{s_n}$$

where ϵ, η are units, $f_1, \ldots, f_n$ are prime, and $r_i, s_i \geq 0$. Then $hg \in (f)$ if and only if h is divisible by

$$e = \prod_{i=1}^{n} f_i^{\max(r_i - s_i, o)},$$

so $(f) : (g) = (e)$. Conversely, assaume this property of principal ideals. We need to show:

$$f \mid gh \implies f = g'h' \text{ where } g' \mid g, h' \mid h$$
$$\text{all } f, g, h \in R, \text{ non-zero}.$$

Assume that $bf = gh$. We know that $(f) : (g) = (h')$ for some $h' \in R$. But f and h are in the ideal $(f) : (g)$. Therefore $f = g'h'$ and $h = a_1 h'$, for some g', $a_1 \in R$. Also $h' \in (f) : (g)$ means $h'g = a_2 f$, some $a_2 \in R$. Therefore

$$\begin{aligned} h'g &= a_2 f = a_2 g' h' \\ g &= a_2 g' . \end{aligned}$$

Thus $h' \mid h$ and $g' \mid g$. □

Lemma 2. *If $\mathcal{O}$ is a noetherian local ring such that $\hat{\mathcal{O}}$ is a UFD, then $\mathcal{O}$ is a UFD.*

Proof. We use the basic fact that $\hat{\mathcal{O}}$ is *flat* over $\mathcal{O}$, i.e., if

$$M \longrightarrow N \longrightarrow P$$

is an exact sequence of $\mathcal{O}$-modules, then

$$M \otimes_{\mathcal{O}} \hat{\mathcal{O}} \longrightarrow N \otimes_{\mathcal{O}} \hat{\mathcal{O}} \longrightarrow P \otimes_{\mathcal{O}} \hat{\mathcal{O}}$$

is an exact sequence of $\hat{\mathcal{O}}$-modules (cf. §9 below, for a summary of the basic facts about flatness). Let $f, g \in \mathcal{O}$ be non-zero elements. Let $A = f \cdot \mathcal{O} : g \cdot \mathcal{O}$. (We write $f \cdot \mathcal{O}$ instead of (f) to distinguish between the ideal generated by f in $\mathcal{O}$ and in $\hat{\mathcal{O}}$). First, I claim

$$A \cdot \hat{\mathcal{O}} = f \cdot \hat{\mathcal{O}} : g \cdot \hat{\mathcal{O}} .$$

But by definition A is the kernel of multiplication by g as a map from $\mathcal{O}$ to $\mathcal{O}/f \cdot \mathcal{O}$, i.e.,

$$0 \longrightarrow A \longrightarrow \mathcal{O} \xrightarrow{g} \mathcal{O}/f \cdot \mathcal{O}$$

is exact. Therefore,

$$0 \longrightarrow A \otimes_{\mathcal{O}} \hat{\mathcal{O}} \longrightarrow \hat{\mathcal{O}} \xrightarrow{g} \hat{\mathcal{O}}/f \cdot \hat{\mathcal{O}}$$

is exact. Therefore, $A \otimes_{\mathcal{O}} \hat{\mathcal{O}}$ is isomorphic to its image $A \cdot \hat{\mathcal{O}}$ in $\hat{\mathcal{O}}$, and this is the kernel of multiplication by g as a map from $\hat{\mathcal{O}}$ to $\hat{\mathcal{O}}/f \cdot \hat{\mathcal{O}}$, i.e., it equals $f\hat{\mathcal{O}} : g\hat{\mathcal{O}}$.

Now since $\hat{\mathcal{O}}$ is a UFD, $f \cdot \hat{\mathcal{O}} : g \cdot \hat{\mathcal{O}}$ is generalized by some element $H \in \hat{\mathcal{O}}$. We must deduce that A is generated by an element of $\mathcal{O}$. But $A \cdot \hat{\mathcal{O}}$ principal implies that $A \cdot \hat{\mathcal{O}}/A \cdot m \cdot \hat{\mathcal{O}}$ is one-dimensional over the residue field k of $\mathcal{O}$ ($m =$ maximal ideal of $\mathcal{O}$). Also,

$$A/mA \cong A \cdot \hat{\mathcal{O}}/A \cdot m \cdot \hat{\mathcal{O}} .$$

If $h \in A$ generates $A/m \cdot A$ over k, then by Nakayama's lemma, h generates A. Therefore A is a principal ideal. By Lemma 1, this shows that $\mathcal{O}$ is a UFD. □

Proof No. 2, via projections.

We base this proof on the useful fact:

Theorem 2. *Let X be a variety of dimension n and let $x \in X$ be a non-singular point. Then there exists an open neighbourhood $U \subset X$ of x, an irreducible hypersurface $H \subset \mathbb{P}^{n+1}$, and an isomorphism of U with an open subset $U' \subset H$:*

$$X \overset{\text{open}}{\supset} U \cong U' \overset{\text{open}}{\subset} H .$$

Proof. First we can replace X by an affine neighbourhood of x. Then embedding X in $\mathbb{A}^N$ we can replace this affine X by its closure in $\mathbb{P}^N$. Therefore assume that X is a closed subvariety of $\mathbb{P}^N$. Also we can assume that x is a closed point. Now suppose $L \subset \mathbb{P}^N$ is a linear space of dimension $N-n-2$ disjoint from X. Projecting from L gives a morphism $p_0 : \mathbb{P}^N - L \to \mathbb{P}^{n+1}$. We saw in §7, Ch. 2, that p_0 restricts to a finite morphism from X to $\mathbb{P}^{n+1}$. Let $H = p_0(X)$: H is a closed n-dimensional subvariety of $\mathbb{P}^{n+1}$, i.e., an irreducible hypersurface. Let p denote the restriction of p_0:

$$x \xrightarrow{p} H .$$

The real point of the proof is to show that if L is chosen suitably, there will be an open neighbourhood $U \subset H$ of $p(x)$ such that p is an *isomorphism* from $p^{-1}(U)$ to U. We use the following criterion for this:

Lemma 1. *Let $f : X \to Y$ be a finite morphism of noetherian schemes. For some $y \in Y$, assume that the fibre $f^{-1}(y)$ consists of one reduced point, with sheaf $\Bbbk(y)$ on it. Then there is an open neighbourhood $U \subset Y$ of y such that*

$$\operatorname{res}(p) : p^{-1}(U) \longrightarrow U$$

is a closed immersion.

Proof of lemma. Let $U = \operatorname{Spec}(R)$ be an affine open neighbourhood of y. Then $f^{-1}(U)$ is affine and if $S = \Gamma\left(f^{-1}(U), \underline{o}_X\right)$, S is an R-algebra which is a finite R-module. Let $y = [P]$. Let $L = \Bbbk(y)$, the quotient field of $R(P)$. Then, since $f^{-1}(y) = \operatorname{Spec}\ (S \otimes_R L)$, our assumption is that the natural homomorphism $\phi : R \to S$ induces an isomorphism

$$L = R \otimes_R L \longrightarrow S \otimes_R L .$$

Let $M = S/\phi(R)$. Then M is a finite R-module such that $M \otimes_R L = (0)$. By Nakayama's lemma, this means that $f \cdot M = (0)$ for some $f \in R - P$. Then $M \otimes_R R_f = (0)$, so the natural homomorphism

$$R_f \longrightarrow S \otimes_R R_f = S_f$$

is surjective. This means that if p is a closed immersion over U_f,

QED for lemma.

Therefore, the theorem will follow if we can construct L such that p has the 2 properties:

$$\begin{aligned} 1)\quad \{x\} &= p^{-1}(p(x)), \\ 2)\quad m_{x,X} &= p^*\left(m_{p(x),H}\right)\underline{o}_x . \end{aligned}$$

The next step is to translate these into synthetic-geometric conditions on L. 1st of all, for any closed point $y \in \mathbb{P}^N - L$, $p_0^{-1}(p_0(y))$ is the join of L and $\{y\}$, i.e., the smallest linear space containing L and y. Write this $J(L, \{y\})$. Therefore, (1) is equivalent to

$$1^*) \quad J(L, \{x\}) \cap X = \{x\}\,.$$

As for (2), it is equivalent to:

$$m_{x,X}/m_{x,X}^2 \xleftarrow{p^*} m_{p(x),H}/m_{p(x),H}^2$$

being surjective. Dualizing, this is the same as

$$T_{x,X} \xrightarrow{dp} T_{p(x),H}$$

being injective (T's being the tangent spaces). Since $T_{p(x),H} \subset T_{p(x),\mathbb{P}^{n+1}}$, this is the same as

$$T_{x,X} \xrightarrow{\text{restriction of } dp_0} T_{p(x),\mathbb{P}^{n+1}}$$

being injective. But the set of linear subspaces of P^N has the following property:

* | for all closed points $y \in \mathbb{P}^N$, and all subvector spaces $V \subset T_{y,\mathbb{P}^N}$, there is a unique linear subspace $M \subset \mathbb{P}^N$ containing y whose tangent space at y is V.

For this reason, if $Z \subset \mathbb{P}^N$ is any variety and $y \in Z$ is any non-singular closed point of Z, one can talk of *the linear subspace $M \subset \mathbb{P}^N$ tangent to Z at y*: i.e., that linear space through y with the same tangent space at y as Z. This gives us a global formulation of the last statement. Let M be the subspace of $\mathbb{P}^N$ tangent to X at x. Then either

a) $M \cap L = \emptyset$, in which case p_0 restricts to a linear isomorphism of M with a subspace of $\mathbb{P}^{n+1}$, hence $dp_0 : T_{x,M} \to T_{p(x),\mathbb{P}^{n+1}}$ is injective.

or

b) $M \cap L \neq \emptyset$, in which case p_0 restricts to a projection of M onto a lower dimensional subspace of $\mathbb{P}^{n+1}$, hence it maps the whole join $J = J(M \cap L, \{x\})$ to one point. Then dp_0 is 0 on $T_{x,J}$, and is not injective on $T_{x,M}$.

Therefore (2) is equivalent to:

2*) if $M \subset \mathbb{P}^N$ is the linear subspace tangent to X at x, then $M \cap L = \emptyset$.

It remains to show that there is a linear space L satisfying 1*), 2*), and disjoint from X too. Let

$$q : \mathbb{P}^N - \{x\} \longrightarrow \mathbb{P}^{N-1}$$

be the projection with center x. Then $q(X - \{x\})$ is a constructable set of $\mathbb{P}^{N-1}$ whose closure is an n-dimensional variety W. And $q^{-1}(W)$ is an $(n+1)$-dimensional subvariety of $\mathbb{P}^n - \{x\}$. (In fact, if W_0 is W minus any hyperplane, $q^{-1}(W_0) \cong W_0 \times \mathbb{A}^1$). Let

$$S = X \cup M \cup q^{-1}(W)\,.$$

This is a closed subset of $\mathbb{P}^N$, all of whose components have dimension $\leq n+1$. Now use:

Lemma 2. *Let $S \subset \mathbb{P}^N$ be a closed subset, whose components have dimension $\leq k$. Then there exists a linear subspace $L \subset \mathbb{P}^N$ of dimension $N-k-1$ disjoint from S.*

(Proof left to the reader.)

Now choose $L \subset \mathbb{P}^N$ disjoint from S, of dimension $N-n-2$. (2*) is obvious. If (1*) were false, then there would be a closed point y in $X \cap J(L, \{x\})$. Then line ℓ joining x and y would be in S. But the line ℓ is in $J(L, \{x\})$, so it also meets L which has codimension 1 in $J(L, \{x\})$. Therefore L meets S, a contradiction. So (1*) is true and we are finished. □

Now to prove the UFD property for the rings $\underline{o}_x$, we can assume that x is a non-singular closed point of a hypersurface $H \subset \mathbb{P}^{n+1}$. Let $Z \subset H$ be an irreducible closed subset of dimension $n-1$, containing x. The first part of our proof is an extension of the argument just given, involving choosing a projection:

$$\mathbb{P}^{n+1} - \{y\} \xrightarrow{p_0} \mathbb{P}^n .$$

To be precise, choose y subject to 3 conditions:

a) $y \notin H$
b) if $\ell =$ line joining x, y, $\ell \cap Z = \{x\}$
c) if $H_0 \subset \mathbb{P}^{n+1}$ is the hyperplane tangent to H at x, then $y \notin H_0$.
[It is obvious that we can realize conditions a) and c). As for b), if $q : \mathbb{P}^{n+1} - \{x\} \to \mathbb{P}^n$ is the projection, it amounts to $y \notin q^{-1}[q(Z - \{x\})]$, and since

$$\dim \left\{\text{closure of } \ q^{-1}[q(Z - \{x\})]\right\} \leq n ,$$

this is possible too.]

Then by (a) p_0 restricts to a finite morphism

$$p : H \longrightarrow \mathbb{P}^n .$$

By (c), interpreted exactly as above, $p^*\left(m_{p(x)}\right)$ generates m_x, so by Th. 3, §5, p is étale at x. And by (b), $p^{-1}(p(x)) \cap Z = \{x\}$. Now $p(Z)$ is an irreducible hypersurface in $\mathbb{P}^n$: let $F = 0$ be the irreducible homogeneous polynomial defining it. Upstairs in $\mathbb{P}^{n+1}$, $F = 0$ defines the hypersurface $\{y\} \cup p_0^{-1}(p(Z))$, which is the cone over $p(Z)$. Restricted to H, $F = 0$ defines a subscheme $Z^* \subset H$, whose underlying set is the locus $F = 0$, i.e., $p^{-1}(p(Z))$, and whose ideal is generated everywhere by one of the functions F/X_i^d, $d =$ degree (F). Then Z is a closed subscheme of Z^*. Our claim is that:

$$(*) \qquad Z = Z^* \ \text{ near } \ x .$$

Since $I\left(Z^*\right)_x$ is principal, then $I\left(Z\right)_x$ will have to be principal too, and the Theorem follows.

Intuitively, Z^* should equal Z because of the fact that p is étale at x and Z stays away from all points of $p^{-1}(p(x))$ except x:

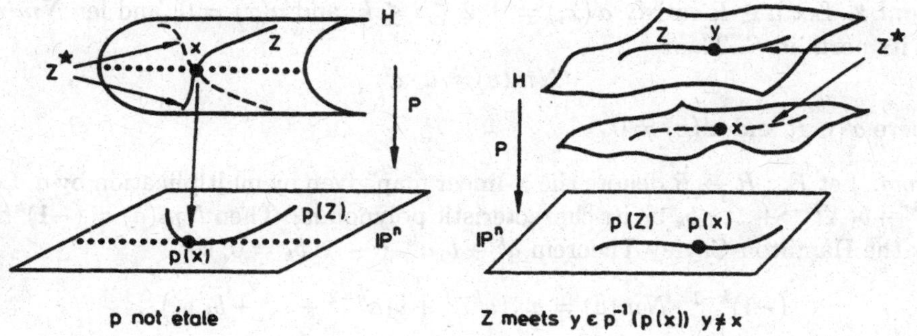

We can give a rigorous proof using norms. Introduce coordinates:

$$\begin{aligned} y &= (0,0,\ldots,1) \\ p &: (X_0,\ldots,X_{n+1}) \longrightarrow (X_0,\ldots,X_n), \quad x_i = X_i/X_0 \\ x &= (1,0,\ldots,0) \\ U &= \text{Spec } (k[x_1,\ldots,x_{n+1}]/(g)) = \left\{ \begin{array}{l} \text{open affine of} \\ \text{where } X_0 \neq 0 \end{array} \right\} \\ \mathbb{A}_0 &= \text{Spec } k[x_1,\ldots,x_n] = \left\{ \begin{array}{l} \text{open affine of } \mathbb{P}^n \\ \text{where } X_0 \neq 0 \end{array} \right\}. \end{aligned}$$

p is dual to the inclusion of rings:

$$\begin{aligned} R &= k[x_1,\ldots,x_{n+1}]/(g) \\ &\cup \\ S &= k[x_1,\ldots,x_n]\,. \end{aligned}$$

Since $y \notin H$, the equation g of H has the form

$$g(x_1,\ldots,x_{n+1}) = x_{n+1}^r + a_1(x_1,\ldots,x_n)\,x_{n+1}^{e-1} + \ldots + a_e(x_1,\ldots,x_n)\,.$$

In particular, R is a free S-module, with basis $1, x_{n+1}, \ldots, x_{n+1}^{e-1}$. Now let $f = F/X_0^d \in S$ be the affine equation of $p(Z)$. Then in U, $I(Z^*) = f \cdot R$ by definition. Our contention is that $I(Z)_x = f \cdot \underline{o}_x$.

Point 1. Let $p^{-1}(p(x)) = (x, x_2, \ldots, x_k)$. Let $P = I(Z)$ be the ideal of Z in R. Then P is generated by elements $a \in R$ such that $a(x_i) \neq 0$, $2 \leq i \leq k$.

Proof. Let $M_i = I(x_i)$. Since $x_i \notin Z, P \not\subset M_i$. Let $P_0 \subseteq P$ be the ideal generated by elements $a \in P - \bigcup_{i=2}^{k} M_i$. Then every element of $P - P_0$ must be in some M_i, so

$$P \subset P_0 \cup M_2 \cup \ldots \cup M_k\,.$$

But then $P \subset P_0$ (cf. Zariski-Samuel, vol. 1, p.), hence $P = P_0$. □

Point 2. Let $a \in R$ satisfy $a(x_i) \neq 0$, $2 \leq i \leq k$, and $a(x) = 0$, and let $Nm(a)$ be its norm in S. Then

$$Nm(a) = a \cdot a' ,$$

where $a' \in R$ and $a'(x) \neq 0$.

Proof. Let $T_a : R \to R$ denote the S-linear map given by multiplication by a. Let $X^{k^a} + b_1 X^{k-1} + \ldots + b_k$ be its characteristic polynomial. Then $Nm(a) = (-1)^k b_k$. By the Hamilton-Cayley Theorem $a^k - b_1 a^{k-1} + \ldots b_k = 0$, so

$$(-1)^{k-1} \cdot Nm(a) = a \cdot \left(a^{k-1} + b_1 a^{k-2} + \ldots + b_{k-1}\right) .$$

Therefore $a' = (-1) \cdot \left[a^{k-1} + \ldots + b_{k-1}\right]$, and since $a(x) = 0$, $a'(x) = (-1) \cdot b_{k-1}(p(x))$. Let M be the ideal of $p(x)$ in S and take tensor products:

$$\begin{array}{ccc} R & \twoheadrightarrow & R/M \cdot R \\ \cup & & \cup \\ S & \twoheadrightarrow & S/M \end{array} .$$

To compute $b_{k-1} \bmod M$, we may as well use the characteristic polynomial $\overline{T_a}$ of multiplication by a in $R/M \cdot R$. But Spec $(R/M \cdot R) = p^{-1}(p(x))$, and this is $\{x\}$ with reduced structure and some other stuff. Thus

$$R/M \cdot R = \mathbb{k}(x) \oplus A$$

where Spec $(A) = \{x_2, \ldots, x_k\}$. Thus the component of a in $\mathbb{k}(a)$ is 0, and in A is a unit. Thus $\overline{T_a}$ breaks up into a 0 map and an invertible map $\overline{T_a}\big|_A$. Thus

$$\begin{aligned} X^k + b_1(p(x))X^{k-1} + \ldots + b_k(p(x)) &= \left\{ \begin{array}{l} \text{Char.polyn.} \\ \text{of } \overline{T_a} \end{array} \right\} \\ &= X \cdot \left\{ \begin{array}{l} \text{Char.polyn.} \\ \text{of } \overline{T_a}\big|_A \end{array} \right\} \\ &= X \cdot \left(X^{k-1} + \ldots + c_{k-1}\right) \end{aligned}$$

where $c_{k-1} \neq 0$. Thus $b_{k-1}(p(x)) \neq 0$. □

Let's put these together. We know that $Z \subset Z^*$, hence $f \cdot \underline{o}_x \subset I(Z)_x$. To show the converse, it suffices to take an element $a \in I(Z)_x$ such that $a(x_i) \neq 0$, $2 \leq i \leq k$, and show that $a \in f \cdot \underline{o}_x$ (by Pt. 1). But $Nm(a)$ is zero on $p(Z)$, so $Nm(a) \in f \cdot S$. Thus by Pt. 2, in $\underline{o}_x$, $a = \frac{Nm(a)}{a'} \in f \cdot \underline{o}_x$.

QED and end of proof no. 2

What happens to the UFD property at singular points? In one case, there is a converse to our Theorem:

Proposition 2. *Let X be a variety, and let $x \in X$. Then if there is one closed subvariety Z of codimension 1 through x which is non-singular at x and whose ideal $I(Z)_x$ in $\underline{o}_{x,X}$ is principal, X is non-singular at x.*

Proof. Let $I(Z)_x = (f)$. Then

$$\Omega_{Z/k}(x) \cong \Omega_{X/k}(x)/\mathbb{k}(x) \cdot df .$$

If Z is non-singular at x, then $\dim \Omega_{Z/k}(x) = \dim Z$, so

$$\dim \Omega_{X/k}(x) \leq \dim \Omega_{Z/k}(x) + 1 = \dim X$$

so X is non-singular at x. □

Example J. Assume that char $(k) \neq 2$ and let X be the subscheme of $\mathbb{A}^n$ defined by:

$$\sum_{i=1}^{n} X_i^2 = 0 .$$

Since char $\neq 2$, $\sum X_i^2$ is a non-degenerate quadratic form, and therefore, if $n \geq 3$, it is irreducible and X is a variety. Since

$$2X_i = \frac{\partial}{\partial X_i}\left(\sum_{j=1}^{n} X_j^2\right) ,$$

the origin $\mathbb{O}$ is the only singular point. Let $\mathcal{O}$ be the local ring of the origin on X. Then $\mathcal{O}$ is not a UFD if and only if there are subvarieties $W \subset X$ of codimension 1 through the origin, whose ideal in $\mathcal{O}$ is not principal. Note first of all, that every linear form $\sum a_i X_i$ is an *irreducible* element in $\mathcal{O}$ (in fact, if $m =$ maximal ideal of $\mathcal{O}$, then $\sum a_i X_i \notin m^2$, so if $\sum a_i X_i = f \cdot g$, then either f or g is $\notin m$ and is a unit).

Case 1: $n = 3$. Then X is an ordinary cone in $\mathbb{A}^3$, with apex at the origin. Let W be one of the lines on X, e.g., $X_1 = iX_2$, $X_3 = 0$. Then X is singular, but W is non-singular at $\mathcal{O}$. Thus $\mathcal{O}$ is not a UFD by Prop. 2. Put another way, we get 2 factorizations into irreducible elements

$$X_1^2 + X_2^2 = (X_1 + iX_2)(X_1 - iX_2) = -X_3 \cdot X_3$$

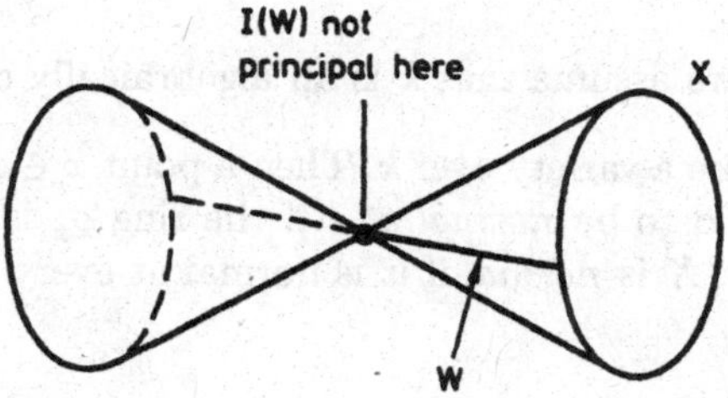

Case 2: $n = 4$. Again we have problems because if $a = X_1^2 + X_2^2$, then

$$a = (X_1 + iX_2)(X_1 - iX_2) = -(X_3 + iX_4)(X_3 - iX_4) \ .$$

Similarly, if W is the plane $X_1 = iX_2$, $X_3 = iX_4$, then this plane has a non-principal ideal at $\mathbb{O}$ on our 3-dimensional X, since it is non-singular at $\mathbb{O}$, while X itself is singular at $\mathbb{O}$.

Case 3: $n \geq 5$. Now we're OK. We can use:

Lemma (Nagata). *Let R be a noetherian domain, and let $x \in R$ be an element such that $x \cdot R$ is prime. If $R[1/x]$ is a UFD, then R is a UFD also.*

Proof. Let $P \subset R$ be a minimal prime ideal. If $P = x \cdot R$, it is principal. If $P \neq x \cdot R$, then $P = P \cdot R\left[\frac{1}{x}\right] \cap R$. Moreover $P \cdot R\left[\frac{1}{x}\right]$ is principal, since $R\left[\frac{1}{x}\right]$ is a UFD, so for some $y \in P$, $P \cdot R\left[\frac{1}{x}\right] = y \cdot R\left[\frac{1}{x}\right]$. Write $y = y_0 \cdot x^r$ where $y_0 \notin x \cdot R$. Then one checks easily that $\left(y \cdot R\left[\frac{1}{x}\right]\right) \cap R = y_0 \cdot R$, hence $P = y_0 \cdot R$. Thus all minimal prime ideals are principal and R is a UFD. □

Apply this to

$$\begin{aligned} R &= k[X_1, \ldots, X_n] \Big/ \sum_{i=1}^{n} X_i^2 \\ x &= X_1 + iX_2 \ . \end{aligned}$$

Let $x' = iX_2 - X_1$, and note that $\sum_{i=1} X_i^2 = \sum_{i=3}^n X_i^2 - x \cdot x'$. Thus $R/(x) \cong k[x', X_3, \ldots, X_4] / \sum_{i=3}^n X_i^2$, and this is an integral domain when $n > 5$, so (x) is a prime ideal. But

$$R\left[\frac{1}{x}\right] \cong \frac{k\left[x, \frac{1}{x}, x', X_3, \ldots, X_n\right]}{\left(\sum_3^n X_i^2 - x \cdot x'\right)} \cong k\left[x, \frac{1}{x}, X_3, \ldots, X_n\right]$$

since the equation asserts that $x' = \sum_3^n X_i^2 / x$. Therefore by Nagata's lemma, R is a UFD and *a fortiori* $\mathcal{O}$ is a UFD.

§8. Normal varieties and normalization

In this entire section, we assume that k is an algebraically closed field.

Definition 1. Let X be a variety over k. Then a point $x \in X$ is called a *normal point of* X, or X is said to be *normal at* z if the ring $\underline{o}_x$ is integrally closed in its quotient field $k(X)$. X is *normal* if it is normal at every point.

Note, for example, that a factorial variety X (i.e., $\underline{o}_x$ always a UFD) is normal, since all UFD's are integrally closed in their quotient fields. In particular, non-singular varieties are normal. If a variety X is normal, its affine coordinate rings are integrally closed in $k(X)$ too. In fact, an intersection of integrally closed rings is integrally closed and for any domain R,

$$R = \bigcap_{\substack{\text{all prime}\\ \text{ideals } p}} R_p{}^{12} .$$

Recall the important:

Structure theorem of integrally closed noetherian rings. *If R is a noetherian integral domain, then*

$$R \text{ is integrally closed} \iff \left\{ \begin{array}{l} (i)\ R = \bigcap_{\substack{\text{minimal prime}\\ \text{ideals } P}} R_P , \\ (ii)\ \text{for all minimal prime ideals } p, \\ \quad R_P \text{ is a principal valuation ring} \end{array} \right\} .$$

This shows that every integrally closed noetherian ring is an intersection of valuation rings, and reduces the study of such rings to the theory of fields with a distinguished family of valuation rings in them. This result is the essential step in the classical ideal theory of Dedekind rings. (For a proof cf. Zariski-Samuel, vol. I, Ch. V, §6; or Bourbaki, Ch. 7). In our context, it shows us that if X is a normal variety, and Z is a subvariety of codimension 1, with generic point x, then $\underline{o}_x$ is a principal valuation ring. If ord_Z is the associated valuation, then this means that for all functions $f \in k(X)$, we can define

$$\text{ord}_Z(f) ,$$

known as *the order of vanishing of the function f along Z*. It is negative if $f \notin \underline{o}_x$, in which case we say that f has a pole along Z; it is 0 if $f \in \underline{o}_x - m_x$ so that f is defined on an open set U meeting Z but does *not* vanish on $U \cap Z$; and it is positive if $f \in m_x$, i.e., if f vanishes on Z. We also find:

Proposition 1. *Let X be a normal variety and let $S \subset X$ be its singular locus (i.e., the set of singular points). Then*

$$\text{codim}_X(S) \geq 2 .$$

[12] In fact, assume that $x, y \in R$, $y \notin 0$, and $x/y \in R_p$ for all prime ideals P. Let $A = (y) : (x) = \{z \in R \mid x/y = w/z, \text{ some } w \in R\}$. Since $x/y \in R_p$, x/y can be written w/z, with $w \in R$, $Z \in R-P$. Therefore $A \not\subset P$. Therefore A is not contained in any prime ideal, so $A = R$. This means that $1 \in A$, i.e., $x/y \in R$.

Proof. If x is the generic point of a subvariety $Z \subset X$ of codimension 1, then we must show that X is non-singular at x. But the ideal of Z at x is the maximal ideal of $\underline{o}_x$, which is principal since $\underline{o}_x$ is a principal valuation ring. And Z itself is non-singular at its generic point. So by Prop. 2, §7, X is non-singular at x. □

Corollary. *Let X be a 1-dimensional variety. Then*

$$X \text{ is non-singular} \iff X \text{ is factorial}$$
$$\iff X \text{ is normal}.$$

In this proposition, we have used only the property that $\underline{o}_x$ is a principal valuation ring, if $\overline{\{x\}}$ has codimension 1. Conversely, if $\operatorname{codim}_X(S) \geq 2$, then X is non-singular at such points x, hence normal at such x, hence $\underline{o}_x$ is a principal valuation ring. Let's give this property a name:

Definition 2. A variety X is *non-singular in codimension* 1 if it is non-singular at all points x such that $\overline{\{x\}}$ has codimension 1.

In some cases, such a variety is normal:

Proposition 2. *Let $X \subset \mathbb{A}^n$ be an irreducible affine hypersurface. If X is non-singular in codimension 1, then X is normal.*

Proof. Let $R = \Gamma(X, \underline{o}_X)$. We must show that R is integrally closed. By the structure theorem, we need only show that

$$R = \bigcap_{\substack{\text{minimal prime} \\ \text{ideals } P}} R_P .$$

Let $f, g \in R$ and assume $f/g \in R_P$ for all minimal prime ideals P. Let

$$(g) = Q_1 \cap \ldots \cap Q_\ell$$

be a decomposition of (g) into primary ideals. Let $P_i = \sqrt{Q_i}$. What we must show is that all these P_i are minimal. Because if P_i is minimal, then $f/g \in R_{P_i}$, hence $f \in g \cdot R_{P_i}$, hence $f \in (Q_i \cdot R_{P_i}) \cap R$. But since Q_i is primary, $Q_i = (Q_i \cdot R_{P_i}) \cap R$. Therefore, if *all* P_i are minimal, $f \in \bigcap_{i=1}^{\ell} Q_i$, hence $f \in (g)$, hence $f/g \in R$.

Now suppose $R = k[X_1, \ldots, X_n]/(h)$ and let $\pi : k[X_1, \ldots, X_n] \to R$ be the canonical map. Let g' be a polynomial such that $\pi(g') = g$. Then in $k[X_1, \ldots, X_n]$, the ideal (h, g') decomposes as follows:

$$\left| \begin{array}{l} (h, g') = \bigcap_{i=1}^{\ell} \pi^{-1}(Q_i) \\ \\ \pi^{-1}(Q_i) \text{ primary, with } \sqrt{\pi^{-1}(Q_i)} = \pi^{-1}(P_i) . \end{array} \right.$$

But now since h and g' are relatively prime (in fact, h is prime and $g' \notin (h)$ since $g \neq 0$), *Macauley's Unmixedness Theorem* asserts that all the associated prime ideals of (h, g') have codimension 2. (Cf. Zariski-Samuel, vol. 2, p. 203.) Therefore P_i has codimension 1 in R, i.e., is minimal. □

On the other hand, it is easy to construct varieties of any dimension which are normal at all but a finite set of points as follows:

Example K. Start with any normal affine variety $Y = \text{Spec}\,(S)$ – for example $\mathbb{A}^n$. Let $R \subset S$ be a finitely generated subring such that S/R, as a *vector space* over k, is finite dimensional. Then S will be integrally dependent on R and with the same quotient field. Therefore $X = \text{Spec}\,(R)$ is definitely not normal. Let the inclusion of R in S define the finite morphism

$$f : Y \longrightarrow X\,.$$

Then since S and R are so nearly equal, it turns out that there is a finite set of closed points $x_1, \ldots, x_n \in X$ such that

$$\text{res}\, f : Y - f^{-1}\left(\{x_1, \ldots, x_n\}\right) \longrightarrow X - \{x_1, \ldots, x_n\}$$

is an isomorphism. In particular, X is normal at all points except $x_1, \ldots, x_n$.

Proof. Consider S/R as R-module. Since it is finite-dimensional over k, it is annihilated by an ideal $A \subset R$ of finite codimension. Then $V(A)$ is a finite set of closed points. And if $f \in A$, then $R_f \cong S_f$, so the restriction of f means Y_f isomorphically onto X_f. □

To be more specific, take

(A)
$$\begin{aligned} S &= k[x], \qquad Y = \mathbb{A}^1 \\ R &= k\left[x^2, x^3\right] = k + x^2 \cdot S\,. \end{aligned}$$

Then if we let $y = x^2$, $z = x^3$,

$$R \cong k[y,z]/\left(y^3 - z^2\right),$$

so $X = \text{Spec}\,(R)$ is the affine cubic curve with a cusp at the origin – cf. Ex. E in this Chapter and Ex. O, Ch. I. X has one singular point, which is also its one non-normal point.

(B)
$$\begin{aligned} S &= k[x,y], \qquad Y = \mathbb{A}^2 \\ R &= k\left[x, xy, y^2, y^3\right]\,. \end{aligned}$$

Then $X = \text{Spec}\,(R)$ is an affine surface in $\mathbb{A}^4$, with one non-normal point – the origin.

We can summarize the discussion by drawing a picture illustrating the hierarchy of good and bad varieties:

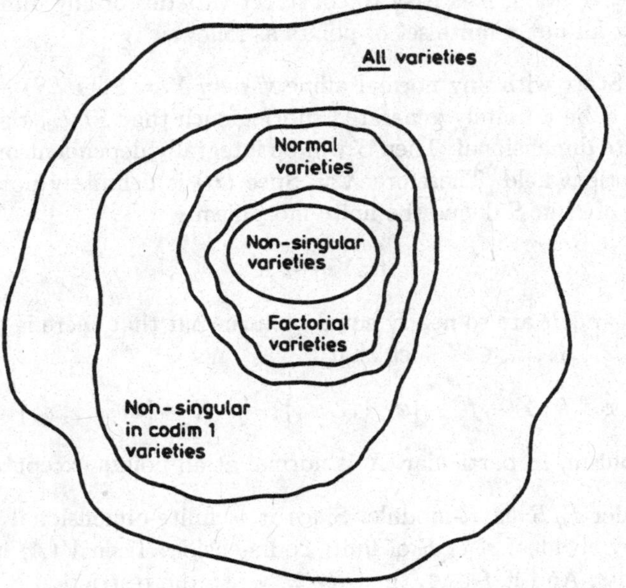

Let's list examples showing how each successive class does include more varieties than the one before:

1. A factorial variety which is singular
   – take $\sum_{i=1}^{5} X_i^2 = 0$ in $\mathbb{A}^5$ (Ex. J).
2. A normal variety which is not factorial
   – take the cone $xy = z^2$ in $\mathbb{A}^3$. It is not factorial (cf. Ex. J), but it is normal by Prop. 2.
3. A variety non-singular in codimension 1, but not normal.
   – Cf. Ex. K.
4. A variety not non-singular in codimension 1
   – take $y^2 = x^3$ in $\mathbb{A}^2$.

Just as one can associate to every integral domain its integral closure in its quotient field, so to every variety, there is another variety called its "normalization". This normalization is at the same time normal, yet birationally equivalent to the original variety. The existence of such a simple way of "making" every variety normal is one of the reasons why normal varieties are an important class. Life would be much simpler if there were an analogous way of canonically constructing a non-singular variety birationally equivalent to any given variety.

Definition 3. Let X be a variety and let L be a finite algebraic extension of $k(X)$. A *normalization of X in L* is a normal variety Y with function field $k(Y) = L$, plus a finite surjective morphism:

$$\begin{array}{c} Y \\ \pi \big\downarrow \\ X \end{array}$$

such that the induced map $\pi^* : k(X) \to k(Y) = L$ is the given inclusion of $k(X)$ in L. If $L = k(X)$, so that π is birational, Y and π are simply called a *normalization of* X.

Theorem 3. *For every variety X and every finite algebraic extension $L \supset k(X)$, there is one and (essentially) only one normalization of X in L: i.e., if $\pi_i : Y_i \to X$ were 2 normalizations, there is a unique isomorphism t:*

$$\begin{array}{ccc} Y_1 & \xrightarrow[\sim]{\;t\;} & Y_2 \\ & {\scriptstyle\pi_1}\searrow \quad \swarrow{\scriptstyle\pi_2} & \\ & X & \end{array}$$

such that $\pi_1 = \pi_2 \cdot t$ and such that t^* is the identity map from L to L.

Proof. To show uniqueness of (Y, π), cover X by affines $U_i = \text{Spec}\,(R_i)$. If $S_{i,1} = \Gamma\left(\pi_1^{-1}(U_i), \underline{o}_{Y_1}\right)$, $S_{i,2} = \Gamma\left(\pi_2^{-1}(U_i), \underline{o}_{Y_2}\right)$, then both $S_{i,1}$ and $S_{i,2}$ are integrally dependent on R, are subrings of $L = k(Y_1) = k(Y_2)$, and are integrally closed. Therefore $S_{i,1} = S_{i,2} =$ integral closure of R in L. Let $t_i : \pi_1^{-1}(U_i) \xrightarrow{\sim} \pi_2^{-1}(U_i)$ be defined by the identity map from $S_{i,1}$ to $S_{i,2}$. Since t_i is the only isomorphism inducing the identity map $L \to L$ of function fields, the t_i's patch together into the required t.

To show existence, first assume X is affine. If $X = \text{Spec}\,(R)$, let S be the integral closure of R in the field L. By a classical (but not easy) theorem, S is a finite R-module (cf. Zariski-Samuel, vol. 1, p. 267). Let $Y = \text{Spec}\,(S)$ and let $\pi : Y \to X$ be dual to the inclusion of R in S. Then π is finite, Y is normal, and $k(Y) =$ quotient field of $S = L$. To show existence in general, cover X by affines U_i. Let $\pi_i : V_i \to U_i$ be a normalization of U_i in L. By the uniqueness part of the result, we can patch the V_i's together into a Y and the π_i's to a π so that (Y, π) is a normalization of X in L. □

Corollary. *Let X be a variety. Then $\{x \in X \mid X$* is normal at *$x\}$ is open.*

Proof. Let $\pi : Y \to X$ be a normalization of X. Then $\pi_*(\underline{o}_Y)$ is a coherent $\underline{o}_X$-module. The map f^* from sections of $\underline{o}_X$ to sections of $\underline{o}_Y$ defines a homomorphism

$$f^* : \underline{o}_X \longrightarrow \pi_*(\underline{o}_Y) .$$

Let $\mathcal{K} =$ cokernel (f^*), and let $S =$ Support $(\mathcal{K}) = \{x \in X \mid \mathcal{K}_x \neq (0)\}$. Then S is a closed subset of X and

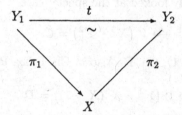

$$x \notin S \Longleftrightarrow \left\{ \underline{o}_{x,X} \xrightarrow{f_x^*} \pi_* \left(\underline{o}_Y\right)_x \quad \text{is surjective} \right\} .$$

But $\pi_* \left(\underline{o}_Y\right)_x$ is just the integral closure of $\underline{o}_{x,X}$ in $k(X)$ so f_x^* is surjective if and only if $\underline{o}_x$ is integral closed. □

Example L. Let's look back at various non-normal varieties we have seen and work out their normalizations.

In Ex. K, just above, we started with a normal $Y = \text{Spec}\,(S)$ and constructed a morphism

$$f : Y \longrightarrow X = \text{Spec}\,(R)$$

via a ring R of finite codimension. (Y, f) is just the normalization of X. In particular, we have repeatedly looked at the special case

$$\mathbb{A}^1 \xrightarrow{f} V\left(X^3 - Y^2\right) = C$$

where $f(t) = \left(t^2, t^3\right)$ (cf. Ex. O, Ch. I). Also in Ch. I, Ex. P, we looked at

$$\mathbb{A}^1 \xrightarrow{f} V\left(Y^2 - X^2(X+1)\right) = D .$$

Again $\left(\mathbb{A}^1, f\right)$ is the normalization of D. This is a case of Ex. K, corresponding to the ring $k[t]$ and the subring $k\left[t^2 - 1, t\left(t^2 - 1\right)\right]$ of finite codimension. We looked at these same 2 plane curves in Ex. E in §3 of this chapter, and if you look back at Ex. E bis you will see that we actually constructed the morphism f by blowing them up, i.e.,

$$\begin{array}{ccc} \mathbb{A}^1 & \xrightarrow{\quad f \quad} & C \\ \sim \Big\| & \nearrow \text{canonical map} & \\ B_{\mathbf{0}}(C) & & \end{array} \qquad \begin{array}{ccc} \mathbb{A}^1 & \xrightarrow{\qquad} & D \\ \sim \Big\| & \nearrow \text{canonical map} & \\ B_{\mathbf{0}}(D) & & \end{array} .$$

Another example. Take X to be the normal surface $xy = z^2$, but let $L = k\left(\sqrt{x}, z\right)$. Take coordinates u, v in $\mathbb{A}^2$ and define

$$f : \mathbb{A}^2 \longrightarrow X$$

by $f(u,v) = \left(u^2, v^2, uv\right)$. Via f^*, $\Gamma\left(X, \underline{o}_X\right)$ is identified with the subring $k\left[u^2, v^2, uv\right]$ in $k[u,v]$; since u and v depend integrally on this subring f is finite. $\mathbb{A}^2$ is non-singular, hence normal. And, via f^*, $u = \sqrt{x}$, and $v = z/\sqrt{x}$ so $k\left(\mathbb{A}^2\right) \cong k\left(\sqrt{x}, z\right)$. Thus $\left(\mathbb{A}^2, f\right)$ is the normalization of X in L. Incidentally, this gives a direct proof that X is normal, since

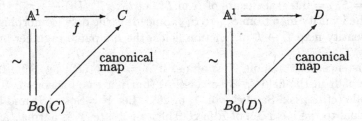

$$k[x,y,z]/\left(xy-z^2\right) \underset{f^*}{\cong} k[u,v]\cap k\left(u^2,u/v\right)$$

$$\cong \left\{\begin{array}{l}\text{subring of elements } f\in k[u,v]\\ \text{invariant under}\\ f(u,v)\longrightarrow f(-u,-v)\end{array}\right\}$$

and the intersection of an integrally closed ring with a subfield of its quotient field is always integrally closed.

Theorem 4. *If X is a projective variety, then its normalization in any finite algebraic extension $L\supset k(X)$ is a projective variety.*

Before proving this theorem, we must introduce the *Segre transformations* of a projective embedding. Start with $\mathbb{P}^n$ and let d be a positive integer. Let $X_0,\ldots,X_n$ be homogeneous coordinates in $\mathbb{P}^n$. Then take all the monomials in $X_0,\ldots,X_n$ of degree d and order them:

$$i^{\text{th}}\ \text{monomial} = \prod_{j=0}^{n} X_j^{r_{ij}}\,,\qquad \sum_{j=0}^{n} r_{ij} = d$$

for $0\le i\le N$. Here

$$N=\frac{(n+d)!}{n!d!}-1$$

is fact. Let $Y_0,\ldots,Y_n$ be homogeneous coordinates in $\mathbb{P}^N$. Define a morphism:

$$\Phi_d:\mathbb{P}^n\longrightarrow\mathbb{P}^N$$

so that if x is a closed point with homogeneous coordinates $(a_0,\ldots,a_n)$, then $\Phi(x)$ is a closed point with homogeneous coordinates

$$\left(\prod_j a_j^{r_{0j}},\ldots,\prod_j a_j^{r_{Nj}}\right)\,.$$

In fact, suppose we assume that the 1st $n+1$ monomials are the powers $X_0^d,X_1^d,\ldots,X_n^d$. Let $U_i\subset\mathbb{P}^n$ and $V_i\subset\mathbb{P}^N$ be the open sets $X_i\neq 0$ and $Y_i\neq 0$ respectively. Then we define Φ by requiring that $\Phi\left(U_i\right)\subset V_i$, for $0\le i\le n$, and that $\Phi_i^*:\Gamma\left(V_i,\underline{o}_{\mathbb{P}^N}\right)\to\Gamma\left(U_i,\underline{o}_{\mathbb{P}^n}\right)$ be:

$$\Phi_i^*\left(Y_k/Y_i\right)=\prod_{j=0}^{n}\left(X_j/X_i\right)^{r_{kj}}\,,\qquad \begin{array}{l}0\le k\le N\\ 0\le i\le n\end{array}\,.$$

In fact, $\Phi^{-1}\left(V_i\right)$ then turns out to be exactly U_i, and Φ_i^* is surjective since that Y_k/Y_i for which the kth monomial is $X_jX_i^{d-1}$ goes over to X_j/X_i. Therefore Φ is even a closed immersion.

Now if $X\subset\mathbb{P}^n$ is any projective variety and d is a positive integer, the image $\Phi_d(X)\subset\mathbb{P}^N$ is called its dth Segre transformation. This operation is very useful for simplifying a projective embedding. If the original X has a homogeneous coordinate ring:

$$R = k[X_0, \ldots, X_n]/P$$

and if R_k denotes the kth graded piece of R, then the new embedded variety $\Phi_d(X)$ has the homogeneous coordinate ring

$$\begin{aligned} R(d) &= \bigoplus_{k=0}^{\infty} R_{dk} \\ &= k\left[\prod_i X_i^{r_{0i}}, \ldots, \prod_i X_i^{r_{Ni}}\right]/P^* . \end{aligned}$$

We are now ready to begin:

Proof of Th. 4: Let $R = k[X_0, \ldots, X_n]/P$ be the homogeneous coordinate ring of a projective variety $X \subset \mathbb{P}^n$. Let α be a non-zero element of R_1, and let $\Sigma \subset R$ be the multiplicative system of non-zero homogeneous elements. Then the localization R_Σ is still graded and $k(X)$ is isomorphic to the subring $(R_\Sigma)_0$ of elements of degree 0. Moreover, R_Σ itself is just a simple polynomial ring $k(X)[\alpha]$. Let S be the integral closure of R in the ring $L[\alpha]$:

$$\begin{array}{ccc} S & \subset & L[\alpha] \\ \cup & & \cup \\ R & \subset & k(X)[\alpha] . \end{array}$$

Then S is a graded subring of $L[\alpha]$. Moreover, since R is finitely generated over k, so is S. We would like to construct a projective variety with the ring S, but there is one obstacle: S may not be generated by the vector space S_1 of elements of degree 1!

Here is where the Segre transformation comes in. I claim that if we replace X by a suitable Segre transformation $\Phi_d(X)$, the resulting S will be generated by S_1. First we need:

Lemma. *Let S be any finitely generated graded k-algebra, where $k = S_0$. Then there exists a positive integer d such that*

$$S(d) = \bigoplus_{k=1}^{\infty} S_{kd}$$

is generated by the elements $S(d)_1 = S_d$.

Proof of lemma. Let $x_1, \ldots, x_k$ be homogeneous generators of S and let $d_i =$ degree (x_i). Let d' be the least common multiple of the d_j's, let $d' = d_i e_i$, and let $d = k \cdot d'$. Now S_n is just the linear span of the monomials $\prod_{i=1}^{k} x_i^{f_i}$ such that $\sum f_i d_i = n$. Note that if $f_i < e_i$ for all i, then $n < d$. Therefore, if $n \geq d$, the monomial decomposes:

$$\prod_{i=1}^{n} x_i^{f_i} = x_{i_0}^{e_{i_0}} \cdot \prod_{i=1}^{n} x_i^{f'_i}$$

(where $f'_i = f_i$ if $i \neq i_0$, $f'_{i_0} = f_{i_0} - e_{i_0}$). Pursuing this inductively, it follows that if we start with a monomial of degree $n = \ell \cdot d'$, we can write it as the product of terms $x_i^{e_i}$ each of degree d', and a "remainder" monomial of degree d. In particular, if the monomial had degree $n = \ell \cdot d$, we could, by grouping these $x_i^{e_i}$'s, write it as a product of monomials *each* of degree d. Therefore $S_{\ell d}$ is spanned by the ℓth symmetric power of S_d. □

Now suppose that $S(d)$ is generated by $S(d)_1$. We replace X by the Segre transforms $\Phi_d(X)$. This means that its new homogeneous coordinate ring is just $R(d)$. Notice that $R_\Sigma(d)$ is the subring $k(X)[\alpha^d]$ of $k(X)[\alpha]$. Therefore $R(d) = R \cap k(X)[\alpha^d]$. I claim that $S(d)$ is the integral closure of $R(d)$ in $L[\alpha^d]$. Study the diagram:

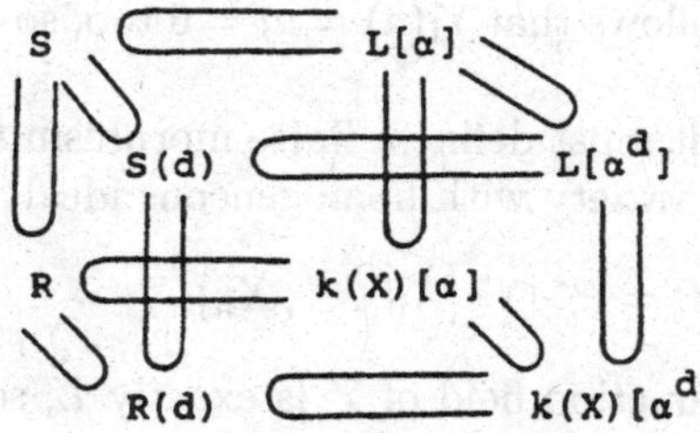

If $x \in L[\alpha^d]$ is integral over $R(d)$, then as an element of $L[\alpha]$, it is integral over R. Therefore $x \in S \cap L[\alpha^d]$, therefore $x \in S(d)$. Moreover, if $x \in S(d)$, then x is integral over R and writing out its equation of dependence, it follows that the coefficients have to lie in $R(d)$, so it is integral over $R(d)$.

The result of all of this is that we *can* assume S is generated by S_1. Therefore, choosing a basis of S_1, we can write

$$S = k[Y_0, \ldots, Y_m]/P^*$$

for some homogeneous prime ideal P^*, with the Y_i's all of degree 1. Let $Y \subset \mathbb{P}^m$ be the variety $V(P^*)$. We now have the following rings:

$$\begin{array}{ccc} k[Y_0, \ldots, Y_m] & \twoheadrightarrow & S \\ \cup & & \cup \\ k[X_0, \ldots, X_n] & \twoheadrightarrow & R \end{array}$$

Y is normal: In fact, S is integrally closed in $L[\alpha]$, hence in its quotient field. Therefore S_{Y_i} is also integrally closed. But if $U_i \subset Y$ is the open set $Y_i \neq 0$, then $\Gamma(U_i, \underline{o}_Y)$ equals

$$\{f/Y_i^m \mid f \in S_m\}$$

i.e., the subring of S_{Y_i} of elements of degree 0. Therefore it is the intersection $S_{Y_i} \cap L$ taken inside $L[\alpha]$, and this intersection is integrally closed in L.

Construction of $\pi : Y \to X$: Each X_i equals a linear form in the Y_i's. Let $M \subset \mathbb{P}^m$ be defined by $X_0 = \ldots = X_n = 0$. Then $(a_0, \ldots, a_m) \longmapsto (X_0(a), \ldots, X_n(a))$ defines a projection

$$p : \mathbb{P}^m - M \longrightarrow \mathbb{P}^n .$$

Note that $Y \cap M = \emptyset$: in fact, suppose $(a_0, \ldots, a_m) \in \mathbb{P}^m$ is a closed point on $Y \cap M$. Then $X_o(a) = \ldots = X_n(a) = 0$ and $f(a) = 0$ for all homogeneous $f \in P^*$. Since the image of Y_i in S is integrally dependent on R, it follows that for some polynomials $a_{ij} \in k\,[X_0, \ldots, X_n]$,

$$Y_i^N + a_{i1}\,(X_0, \ldots, X_n)\,Y_i^{N_1} + \ldots + a_{iN}\,(X_0, \ldots, X_N) \equiv 0 (\mathrm{mod} \quad P^*).$$

Substituting $a_0, \ldots, a_m$, it follows that $Y_i(a) = a_i = 0$ too, so $(a_0, \ldots, a_m)$ is not a real point.

Now restrict p to Y and it must define a finite morphism from Y to $\mathbb{P}^n$. The image will be the projective variety with homogeneous ideal

$$P = P^* \cap k\,[X_0, \ldots, X_n]$$

i.e., X. It is clear that the function field of Y is exactly L, so Y is the normalization of X in L. □

Theorem 5. *Let K be a finitely generated field extension of k such that $tr.d.K/k = 1$. Then there is a unique complete, non-singular "model" C of K, i.e., a variety C such that $k(C)$ is isomorphic/k to K. Moreover, C is a projective variety.*

Proof. Existence: Let $x \in K$ be a separating transcendence base. Then K is a finite separable extension of $k(x)$, so $K = k(x, y)$, for some $y \in K$. Therefore, K contains a subring $k[X, Y]/(f)$ where f is an irreducible polynomial. Homogenizing f, we obtain a plane curve $C_0 \subset \mathbb{P}^2$ defined by $f = 0$, such that $k(C_0) \cong K$. Let C be the normalization of C_0. Then C is normal and 1-dimensional, hence non-singular by the Cor. to Prop. 1.

Uniqueness: I claim that any complete non-singular model, regarded as a topological space plus a sheaf of rings, is nothing but the following:

i) take all principal valuation rings $R \subset K$ such that $k \subset R$: call this set $\mathcal{R}$.
ii) let C be the union of $\mathcal{R}$ and a generic point.
iii) topologize C by taking as open sets a) the empty set, and b) C minus a finite subset of $\mathbb{R}$.

iv) put a sheaf $\underline{o}_C$ on C via

$$\Gamma(C - \{R_1, \ldots, R_n\}, \underline{o}_C) = \bigcap_{R \in \mathcal{R} - \{R_1, \ldots, R_n\}} R$$

for all finite subsets $\{R_1, \ldots, R_n\} \subset \mathcal{R}$.

If C' is any complete non-singular model, then for all closed points $x \in C'$, $\underline{o}_x$ is a principal valuation ring in K, so we can map C' into C (taking the generic point of C' to the generic point of C' to the generic point of C). It is $1-1$ since the local rings of a variety are all distinct. It is onto since each ring $R \in \mathcal{R}$ must certainly dominate some $\underline{o}_x$ by the valuative criterion for completeness (Cor. of Prop. 3, §7, Ch. II.) And if one valuation ring dominates another, and they have the same quotient field, they are equal. It is clear that this is a homeomorphism and that $\underline{o}_C, \underline{o}_{C'}$ can be identified. Therefore $(C, \underline{o}_C) \cong (C', \underline{o}_{C'})$. □

This is completely false if tr.d.$K/k > 1$, essentially because a local ring $\underline{o}_x$ at a closed point of a variety X can be a valuation ring only when $\dim X = 1$.

Example M. Let $x \in \mathbb{P}^n$ be a closed point, and let $B = B_x(\mathbb{P}^n)$. Let $p : B \to \mathbb{P}^n$ be the canonical map. Then [(a)]both B and $\mathbb{P}^n$ are non-singular (in fact, B and $\mathbb{P}^n$ are both covered by open pieces isomorphic to $\mathbb{A}^n$), [(b)]p is proper and $\mathbb{P}^n$ is complete, so B is complete too, and [(c)]p is birational. In other words, we have 2 non-singular and complete varieties, and a birational map between them, which is not an isomorphism if $n \geq 2$.

Among higher dimensional varieties, however, we can put together the ideas of Th. 2, §7, and of this section in:

Proposition 6. *Let X be a normal projective n-dimensional variety. Then X is the normalization of a hypersurface $H \subset \mathbb{P}^{n+1}$.*

Proof. Assume X is a closed subvariety of $\mathbb{P}^N$. In the proof of Theorem 2 we constructed a projection

$$p_0 : \mathbb{P}^N - L \longrightarrow \mathbb{P}^{n+1}$$

such that $X \cap L = \emptyset$, and such that under p_0 an open set in X and an open set in $H = p_0(X)$ are isomorphic. Let $p : X \to H$ denote the restriction of p_0. Then p is finite since it is a projection, and p is birational since it is an isomorphism in an open set. Since X is normal, (X, p) is the normalization of H. □

§9. Zariski's Main Theorem

In §8 our discussion of normality has been largely a matter of carrying over into geometry the algebraic ideals and algebraic constructions involving integral closure. However, normality turns out to have a hidden geometric content as well, which is not so easy to discover. This involves the concept of the "branches" of a variety at a point. To understand this idea, let's first look at the case $k = \mathbb{C}$ and try to describe our naive topological notion of a branch:

Let X be a variety over $\mathbb{C}$.
Let $x \in X$ be a closed point.
Let $S \subset X$ be the singular locus.
Let $U \subset X$ be some sufficiently regular and sufficiently small neighbourhood of x in the *complex* (or strong) topology (consisting of closed points only).
Look at $U - U \cap S$, and decompose it into components in the complex topology:

$$U - U \cap S = V_1 \cup \ldots \cup V_n .$$

Then the closures $\overline{V_1}, \ldots, \overline{V_n}$ are the *branches* of X at x.

Example N. Let X be the plane curve

$$0 = x^2 - y^2 + x^3$$

in $\mathbb{C}^2$, and let U be the neighbourhood of the origin (= the one singular point)

$$\{(x, y) \mid |x| < \epsilon, \quad |y| < \epsilon\}$$

for some small ϵ. Then

$$U \cap X - \{(0,0)\}$$

always breaks up into 2 pieces. In fact, in $U \cap X$:

$$|x - y| \cdot |x + y| = |x|^3 < \epsilon \cdot |x|^2 .$$

Therefore, either $|x - y| < \sqrt{\epsilon} \cdot |x|$ or $|x - y| < \sqrt{\epsilon} \cdot |x|$. Obviously both cannot occur (if $\epsilon < 1$). Each piece separately turns out to be connected so we get 2 branches at $(0, 0)$, described by $|x - y| \ll |x|$ and $|x + y| \ll |x|$.

On the other hand, take X to be the cone $xy = z^2$. Let $U = \{(x, t, z) \mid |x| < \epsilon, \quad |y| < \epsilon, \quad |z| < \epsilon\}$. Then define a continuous surjective map:

$$\begin{array}{rcl} \{(s, t) \mid |s| < \sqrt{\epsilon}, \ |t| < \sqrt{\epsilon}\} & \longrightarrow & X \cap U - \{(0,0,0)\} \\ (s, t) & \longmapsto & (s^2, t^2, st) . \end{array}$$

Therefore $X \cap U - \{(0,0,0)\}$ is the continuous image of a connected set, hence connected. Thus X has only one branch.

Is there any purely algebraic way to get ahold of these branches? One way to detect the existence of several branches at a point $x \in X$ is to look for *covering spaces* of the following general type:

$$f : X' \longrightarrow X$$

such that

1) $f^{-1}(y)$ is a finite set, for all $y \in X$
2) f birational.

Then if $f^{-1}(x) = \{x_1, \ldots, x_n\}$, and $U \subset X$ is a small complex neighbourhood of x, we know that $f^{-1}(U)$ will have to break up:

$$f^{-1}(U) = U_1 \cup \ldots \cup U_n ,$$

where U_i is a small neighbourhood of x_i. Then in fact $\overline{f(U_i)}$ will be a union of some subset of the branches through x. In other words, in the set of branches of X at x, we will get n disjoint subsets, each one coming from the branches of X' at one of the x_i.

There is one canonical way to do this: let (X', f) be a normalization of X. In that case, if $S \subset X$ is the singular locus and $S' = f^{-1}(S)$, then f defines an isomorphism of $X' - S'$ and $X - S$. Therefore $U - U \cap S$ is homeomorphic to the disjoint union of the sets $U_i - U_i \cap S'$ and we get a canonical decomposition of the set of branches at x depending on the various points $x_i \in X'$ over x. The essential content of Zariski's Main Theorem, in all its forms, is once a variety is normal, there is *only one* branch at each of its points. Therefore, for any variety X and closed point $x \in X$, the set of points x_1 in its normalization X' over x is in $1-1$ correspondence with the set of branches of X at x. Now let k denote an algebraically closed field.

I. *Original form:* Let X be a normal variety over k and let $f : X' \to X$ be a birational morphism with finite fibres from a variety X' to X. Then f is an isomorphism of X' with an open subset $U \subset X$.

II. *Topological form:* Let X be a normal variety over $\mathbb{C}$, and let $x \in X$ be a closed point. Let S be the singular locus of X. Then there is a basis $\{U_i\}$ of complex neighbourhoods of x such that

$$U_i - U_i \cap S$$

is connected, for all i.

III. *Power series form:* Let X be a normal variety over k and let $x \in X$ be a normal point (not necessarily closed). Then the completion $\hat{\mathcal{o}}_x$ is an integral domain, integrally closed in its quotient field.

IV. *Grothendieck's form:* Let $f : X' \to X$ be a morphism of varieties over k with finite fibres. Then there exists a diagram:

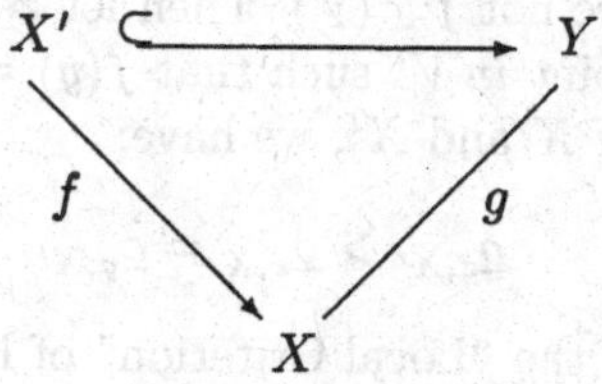

where Y is a variety, X' is an open set in Y and g is a *finite* morphism.

V. *Connectedness Theorem:* Let X be a variety over k normal at a closed point x. Let $f : X' \to X$ be a birational proper morphism. Then $f^{-1}(x)$ is a connected set (in the Zariski topology).

Let's consider first the original form (I). As a simple special case, it contains the important assertion that a bijective birational morphism between normal varieties is an isomorphism (compare Ch. I, §4, Ex. N,O, and P). Form (I) can be proven by a very direct attack involving factoring f through morphisms

$$\mathrm{Spec}\ R[X]/A \longrightarrow \mathrm{Spec}\ R$$

with only *one* new variable at a time. This is Zariski's original proof: cf. Lang, *Introduction to Algebraic Geometry*, p. 124. But these direct methods have never been generalized to other problems, so we will not present this proof. In the case where X is a *factorial* variety, however, a direct proof is so elementary that we do want to give this:

Proposition 1. *Let X be an n-dimensional factorial variety and let $f : X' \to X$ be a birational morphism. Then there is a non-empty open set $U \subset X$ such that*

1) $\mathrm{res}(f) : f^{-1}(U) \to U$ *is an isomorphism,*
2) if $E_1, \ldots, E_k$ are the components of $X' - f^{-1}(U)$, then $\dim E_i = n-1$, *for all i, while* $\dim \overline{f(E_i)} \leq n-2$.

In particular, if x is a closed point in $X - U$, all components of $f^{-1}(x)$ have dimension at least 1.

Proof. Let U be the set of points $x \in X$ such that x has a neighbourhood $V \subset X$ for which

$$\mathrm{res}(f) : f^{-1}(V) \longrightarrow V$$

is an isomorphism. U is clearly open and (1) holds. Now let $x' \in X'$ be any closed point and let $x = f(x')$.

(A) If $\underline{o}_{x,X} = \underline{o}_{x',X'}$, then $x \in U$. Let $V' \subset X'$ and $V \subset X$ be affine neighbourhoods of x' and x such that $f(V') \subset V$. Let $R = \Gamma(V, \underline{o}_X)$, $R' = \Gamma(V', \underline{o}_{X'})$, $x = [P]$, and $x' = [P']$. Then, via f^*, R is a subring of R', while the localizations R_P and $R'_{P'}$ are equal. Therefore, we can find a common denominator $g \in R$ for the generators of R' as R-algebra such that $g \notin P$. In particular, $R_g = R'_g$. Replacing V and V' by V_g and V'_g respectively, f becomes an isomorphism from V' to V. Suppose V' were not $f^{-1}(V)$. Then let $z \in f^{-1}(V) - V'$, let $x = f(z)$, and let $y \in X'$ be the point *in* V' such that $f(y) = x$. Then, as subrings of the common function field of X and X', we have:

$$\underline{o}_{z,X'} > \underline{o}_{x,X} = \underline{o}_{y,X'} \cdot$$

If $z \neq y$, this contradicts the "Local Criterion" of Hausdorffness (cf. Ch. II, §6). Thus $f^{-1}(V) = V'$ and $x \in U$.

(B) Suppose $\underline{o}_{x,X} \not\supseteq \underline{o}_{x',X'}$. Let $s \in \underline{o}_{x',X'} - \underline{o}_{x,X}$. Write $s = t_1/t_2$, where $z_1, t_2 \in \underline{o}_{x,X}$ are relatively prime (using the assumption: $\underline{o}_{x,X}$ a UFD). Let t_i extend to sections of $\underline{o}_X$ in a neighbourhood V of x, and let s extend to a section of $\mathcal{O}_{X'}$ in a neighbourhood V' of x', where $v' \subset f^{-1}(V)$. Then the subsets V_1 and V_2 of V defined by $t_1 = 0$ and $t_2 = 0$ are pure $n-1$-dimensional; and since t_1 and t_2 are relatively prime, they have no common component through x. Replacing V by a smaller neighbourhood of x, we can assume that V_1 and V_2 have no common components at all. Therefore $V_1 \cap V_2$ is pure $(n-2)$-dimensional. Upstairs in V', let V_2' be the locus $t_2 = 0$. Since in V', $t_1 = s \cdot t_2$, t_1 also vanishes on V_2'. Therefore $f(V_2') \subset V_1 \cap V_2$. But V_2' is pure $(n-1)$-dimensional, so every component of V_2' dominates under f a lower dimensional variety of X. Therefore all components $f^{-1}(f(x'))$ through x' have dimension ≥ 1, and

$$V_2' \subset X' - f^{-1}(U)\ .$$

This shows that every point of $X' - f^{-1}(U)$ is on an $(n-1)$-dimensional component of $X' - f^{-1}(U)$; hence $X' - f^{-1}(U)$ is pure $(n-1)$-dimensional. And that all its components are mapped onto subsets of X of codimension ≥ 2. □

The set-up of this Proposition should be familiar from the blowing-up morphisms of §3:

$$f : B_x(X) \longrightarrow X\ .$$

In this case, let $U = X - \{x\}$. Then $\operatorname{res}(f) : f^{-1}(U) \to U$ is an isomorphism, and we proved in §3 that all components E of $B_x(X) - f^{-1}(U)$ had dimension $n-1$. Since $\{x\} = f(E)$, $\dim \overline{f(X)} \leq n-2$ also, (if $n \geq 2$). The Proposition shows that these features are fairly typical of birational morphisms. Of course, if f has finite fibres, then no components E_i can occur, so $X' = f^{-1}(U)$ and Form (I) of Zariski's Main Theorem follows. If X is *not* factorial, the Proposition is false to the extent that it is stronger than the Main Theorem:

Example O. Let $X = V((X_1X_2 - X_3X_4)) \subset \mathbb{A}^4$. As we saw in §8, the origin is a normal, but not factorial point on X. Define

$$f : \mathbb{A}^3 \longrightarrow X$$

by

$$(y_1, y_2, y_3) \longmapsto (y_1, y_2y_3, y_2, y_1y_3)\ .$$

(What we are doing here is taking the closure of the graph of the rational function $y_3 = X_4/X_1 = X_2/X_3$.) If U is the locus $(X_1, X_3) \neq (0,0)$, then $f^{-1}(U)$ is the open set $(y_1, y_2) \neq (0,0)$ and is isomorphic to U. But if P is a closed point of $\mathbb{A}^3$ such that $y_1 = y_2 = 0$, then $f(P) = (0,0,0,0)$. Thus f shrinks a *line* to a point, and is $1-1$ elsewhere. But a line has codimension 2 on X, so this is a case where the conclusion of Prop. 1 is false.

The most powerful approach to the proof of (I.) is through the apparently innocuous form (III.). This method is also due to Zariski, who proved both (III.) and (III.) ⇒ (I.).

(III.) itself is actually quite a job to prove; however, it is a matter of pure local algebra, so we refer the reader to Zariski-Samuel, vol. 2, pp. 313–320. Here is how we get back to geometry:

Proof of (III.) ⇒ (I.): Let $f : X' \to X$ be a birational morphism with finite fibres, and assume X is normal. First, by argument A in the proof of Prop. 1, it will be enough to show that for all $x' \in X'$, if $x = f(x)$, then

$$f_{x'}^* : \underline{o}_{x,X} \longrightarrow \underline{o}_{x',X'}$$

is an isomorphism. If this is proven, we can proceed exactly as in Prop. 1 to construct a neighbourhood U of x such that $f^{-1}(U)$ and U are isomorphic, and (I.) follows. For simplicity we can even assume x and x' are closed, although this is not essential. The proof uses 2 lemmas:

Lemma 1. *Let $\mathcal{O}$ be a complete local ring, with maximal ideal m. Let M be an $\mathcal{O}$-module such that*

1) $\bigcap_n m^n \cdot M = (0)$

2) $M/m \cdot M$ *is finite-dimensional over* $\mathcal{O}/m$.

Then M is a finitely generated $\mathcal{O}$-module. (Cf. Zariski-Samuel, vol. 2, p. 259.)

Lemma 2. *Let $\mathcal{O}$ be a noetherian local ring with completion $\hat{\mathcal{O}}$. Then* $\dim(\mathcal{O}) = \dim(\hat{\mathcal{O}})$.

(For the definition of $\dim(\mathcal{O})$, cf. Ch. I, §7; for the lemma, cf. Zariski-Samuel, vol. 2, p. 288.)

Now look at the local homomorphism $f_{x'}^*$. It induces a homomorphism ϕ of the completed rings as follows:

$$\begin{array}{ccc} \underline{o}_{x,X} & \xrightarrow{f_{x'}^*} & \underline{o}_{x',X'} \\ \cap & & \cap \\ \hat{\underline{o}}_{x,X} & \xrightarrow[\phi]{} & \hat{\underline{o}}_{x',X'} \end{array} .$$

The basic fact we need is that $\hat{\underline{o}}_{x,X}$ is an integrally closed domain, by (III.). First of all, $\hat{\underline{o}}_{x',X'}$ is a finite module over $\hat{\underline{o}}_{x,X}$. This follows from Lemma 1 with $\mathcal{O} = \hat{\underline{o}}_{x,X}$, $M = \hat{\underline{o}}_{x',X'}$. In fact,

$$\begin{aligned} \bigcap_n m_x^n \cdot M &= \bigcap_n \phi(m_x)^n \cdot \hat{\underline{o}}_{x',X'} \\ &\subseteq \bigcap_n (m_{x'})^n \cdot \hat{\underline{o}}_{x',X'} \\ &= (0) \end{aligned}$$

by Krull's Theorem (Zariski-Samuel, vol. 1, p. 216). And

$$\begin{aligned} M/m_x \cdot M &= \hat{\underline{o}}_{x',X'}/\phi(m_x)\cdot \hat{\underline{o}}_{x',X'} \\ &\cong \widehat{\underline{o}_{x',X'}/f^*(m_x)\cdot \underline{o}_{x',X'}} \,. \end{aligned}$$

But $\underline{o}_{x',X'}/f^*(m_x)\cdot\underline{o}_{x',X'}$ is the local ring of x' in the fibre $f^{-1}(x)$. Since x' is an isolated point of $f^{-1}(x)$, this local ring is finite dimensional. Thus all hypotheses of Lemma 1 are satisfied.

Secondly, ϕ is injective. In fact, let $A = \hat{\underline{o}}_{x,X}/\ker(\phi)$. If ϕ is not injective, then $\dim A < \dim \hat{\underline{o}}_{x,X}$: in fact, for any chain of prime ideals $P_0 \subsetneq P_1 \subsetneq \ldots \subsetneq P_t \subset A$, we get a longer chain

$$(0) \subsetneq \phi^{-1}P_0 \subsetneq \ldots \subsetneq \phi^{-1}P_t \subset \hat{\underline{o}}_{x,X}$$

(using the fact that $\hat{\underline{o}}_{x,X}$ is a domain). Moreover, since $\hat{\underline{o}}_{x',X'}$ is integrally dependent on A, $\dim\left(\hat{\underline{o}}_{x',X'}\right) \leq \dim A$. Then, by Lemma 2:

$$\begin{aligned} \dim X &= \dim \underline{o}_{x,X} = \dim \hat{\underline{o}}_{x,X} > \dim A \geq \dim \hat{\underline{o}}_{x',X'} \\ &= \dim \underline{o}_{x',X'} = \dim X' \end{aligned}$$

and this is a contradiction.

We can now consider all our local rings as subrings of $\hat{\underline{o}}_{x',X'}$. Taking intersections in the total quotient ring of this big ring, we can deduce:

$$\underline{o}_{x',X'} \subset \left(\text{quotient field of } \hat{\underline{o}}_{x,X}\right) \cap \hat{\underline{o}}_{x',X'} \,.$$

Since $\hat{\underline{o}}_{x,X}$ is integrally closed in its quotient field, and $\hat{\underline{o}}_{x',X'}$ is integrally dependent on $\hat{\underline{o}}_{x,X}$, this means $\underline{o}_{x',X'} \subset \hat{\underline{o}}_{x,X}$. But then

$$\underline{o}_{x',X'} \subset \left(\text{quotient field of } \underline{o}_{x,X}\right) \cap \hat{\underline{o}}_{x,X} = \underline{o}_{x,X} \,.$$

□

As for form (II.), using results in Gunning and Rossi, that comes out of (III.) too:

Proof of (III.) ⇒ (II.): The basic fact is C. 16, Ch. III, p. 115 in Gunning-Rossi: they show that if $\mathcal{X}$ is the analytic space corresponding to the variety X, and Ω is the sheaf of holomorphic functions on $\mathcal{X}$, then (II.) is correct at a point $x \in \mathcal{X}$ if Ω_x is an integral domain. But all the rings:

$$\underline{o}_x \subset \Omega_x \subset \hat{\underline{o}}_x = \hat{\Omega}_x$$

are included in each other. If X is normal, (III.) tells us that $\hat{\underline{o}}_x$ is a domain; so Ω_x must be a domain too. □

Finally, forms (IV.) and (V.) of the Main Theorem are even deeper. (V.) is a much more global statement since the properness of f is involved. There is a cohomological proof, due to Grothendieck (cf. EGA, Ch. III, §4.3) and a proof using a combination of projective techniques and completions, due to Zariski. As for (IV.), the interesting point here is that it asserts the *existence* of plenty of finite morphisms, rather than asserting that normal varieties have some strong property. (The connection loosely speaking is that the more finite morphisms there are, the stronger restriction it is to be normal.) In fact, to see that (IV.) $\Rightarrow$ (I.) just apply (IV.) as it stands to a morphism $f : X' \to X$ in (I.). We may as well assume that the Y in (IV.) is a variety and that X' is dense in Y. Then g is birational and finite, hence its rings are integrally dependent on those of X. If X is normal, g must be an isomorphism, so X' is just an open subset of X as required. For a good proof of (IV.), cf. EGA, Ch. 4, (a proof relying very heavily on (III.)).

§10. Flat and smooth morphisms

The concept of flatness is a riddle that comes out of algebra, but which technically is the answer to many prayers. Let's recall the basic algebra first:

> Let R be a ring, M an R-module. Then M is *flat* over R if, for all elements $m_1, \ldots, m_n \in M$ and $a_1, \ldots, a_n \in R$ such that $\sum_{i=1}^{n} a_i \cdot m_i = 0$, there are equations
>
> $$m_i = \sum_{j=1}^{k} b_{ij} \cdot m'_j$$
>
> for some $m'_j \in M$, $b_{ij} \in R$ such that
>
> $$\sum_{i=1}^{n} b_{ij} a_i = 0, \qquad \text{all } j .$$

It is clear that a *free* R-module has this property (i.e., take the m'_j's to be part of a basis of M); and that any direct limit of flat R-modules again has this property, hence is flat. Conversely, it was recently proven that *any* flat R-module is a direct limit of free R-modules. Intuitively, one should consider flatness to be an abstraction embodying exactly that part of the concept of freeness which can be expressed in terms of linear equations. Or, one may say that flatness means that the linear structure of M preserves accurately that of R itself. On the other hand, to see an example of flat but not free modules, suppose R is a domain and let M be its quotient field. Since

$$M = \bigcup_{\substack{a \in R \\ a \neq 0}} \frac{1}{a} \cdot R ,$$

M is a direct limit of free R-modules, hence is flat. But M is certainly not a free R-module itself.

The defining property can be souped up without much difficulty to give the following:

Given an R-module M and an exact sequence of R-modules

$$0 \longrightarrow N_1 \longrightarrow N_2 \longrightarrow N_3 \longrightarrow 0$$

the sequence

$$0 \longrightarrow N_1 \otimes_R M \longrightarrow N_2 \otimes_R M \longrightarrow N_3 \otimes_R M \longrightarrow 0$$

is exact, if either M or N_3 is flat over R.

This is the form in which one usually uses the definition. In the special case where M is flat over R, $N_2 = R$ and N_1 is an ideal $I \subset R$, this implies that the natural map $I \otimes_R M \to I \cdot M$ is injective. Conversely, this special case implies that M is flat over R. Here are some of the basic facts:

A. If M is a flat R-module and S is an R-algebra, then $M \otimes_R S$ is a flat S-module.

B. If M is an R-module, then M is flat over R if and only if for all prime ideals $P \subset R$, the localization M_P is flat over R_P.

B'. (Stronger) If M is an S-module and S is an R-algebra, via the homomorphism $i : R \to S$, then M is flat over R if and only if for all prime ideals $P \subset S$, M_P is flat over $R_{i^{-1}(P)}$.

C. If R is a local ring and M is an R-module, *and either* R is artinian *or* M is a finitely presented R-module (i.e., of the type R^n/finitely generated submodule), then M flat over R implies M is a free R-module.

D. If R is a domain, and M is a flat R-module, then M is torsion-free. (More generally, every non-0-divisor $f \in R$ annihilates no non-0-elements in a flat R-module.) Conversely, if R is a valuation ring, then torsion-free R-modules are flat.

E. Let M be a B-module, and B an algebra over A. Let $f \in B$ have the property that for all maximal ideals $m \subset A$, multiplication by f is injective in $M/m \cdot M$. Then M flat over A implies $M/f \cdot M$ flat over A.

F. If $\mathcal{O}$ is a noetherian local ring, then its completion $\hat{\mathcal{O}}$ is a flat $\mathcal{O}$-module.

For a full discussion, cf. Bourbaki, *Alg. Comm.*, Ch. I. Putting A. and D. together reveals another important point about flatness: if M is flat over a domain R, then not only is M torsion-free, but for *all* homomorphisms $R \to S$, where S is a domain, $M \otimes_R S$ is still torsion-free as S-module, i.e., M is "universally torsion-free".

Definition 1. Let $f : X \to Y$ be a morphism of schemes and let $\mathcal{F}$ be a quasi-coherent $\underline{o}_X$-module. Then $\mathcal{F}$ is *flat* over $\underline{o}_Y$ for all $x \in X$, $\mathcal{F}_x$ is a flat $\underline{o}_{f(x)}$-module. The morphism f itself is *flat* if $\underline{o}_X$ is flat over $\underline{o}_Y$.

Notice that whether or not a morphism $f : X \to Y$ is flat or not involves only the $\underline{o}_Y$-module structure of $\underline{o}_X$ and not the ring structure of $\underline{o}_X$. Properties A. – D. above can be translated into geometric terms:

A*. Let

$$\begin{array}{ccc} X' & \xrightarrow{p} & X \\ {\scriptstyle f'}\downarrow & & \downarrow{\scriptstyle f} \\ Y' & \longrightarrow & Y \end{array}$$

be a fibre product, and let $\mathcal{F}$ be a quasi-coherent $\underline{o}_X$-module. If $\mathcal{F}$ is flat over $\underline{o}_Y$, the quasi-coherent $\underline{o}_{X'}$-module $p^*(\mathcal{F})$ is flat over $\underline{o}_{Y'}$.
[Here, $p^*(\mathcal{F})$ is an $\underline{o}_{X'}$-module defined locally on an affine

$$\begin{array}{ccc} \text{Spec}\,(R') & \subset & X' \\ \vdots & & \downarrow{\scriptstyle p} \\ \text{Spec}\,(R) & \subset & X \end{array}$$

by $\widetilde{M \otimes_R R'}$, if $M = \Gamma(\text{Spec}\,(R), \mathcal{F})$.]
In particular, if f is flat, then f' is flat.

B*. Let $f : \text{Spec}\,(R) \to \text{Spec}\,(S)$ be a morphism, and let $\mathcal{F} = \widetilde{M}$ be a quasi-coherent module on Spec (R). Then $\mathcal{F}$ is flat over $\underline{o}_{\text{Spec}\,(S)}$ if and only if M is flat over S.

C*. Let X be a noetherian scheme, and let $\mathcal{F}$ be a coherent $\underline{o}_X$-module. Then $\mathcal{F}$ is flat over $\underline{o}_X$ if and only if $\mathcal{F}$ is a locally free $\underline{o}_X$-module.

D*. Let $f : X \to Y$ be a morphism. Assume Y is irreducible and reduced with generic point y. Let $\mathcal{F}$ be a quasi-coherent $\underline{o}_X$-module flat over $\underline{o}_Y$. Then for all $x \in X$, $\mathcal{F}_x$ is a torsion-free $\underline{o}_{f(x),Y}$-module. If X is noetherian and $\mathcal{F}$ is a coherent $\underline{o}_X$-module, this means that all associated points of $\mathcal{F}$ lie over y. Conversely, this property implies that $\mathcal{F}$ is flat over $\underline{o}_Y$ if all stalks $\underline{o}_{y,Y}$ are valuation rings (e.g., Y a non-singular curve, or Spec $(\mathbb{Z})$).

The best intuitive description of when a morphism $f : X \to Y$ of finite type is flat is that this is the case when the fibres of f, looked at locally near any point $x \in X$, form a *continuously* varying family of schemes. Suppose, for example, X and Y are varieties. Then, by D, if f is flat, X dominates Y. We saw in Ch. I, §8, that there is an open set $U \subset Y$ such that all components of all fibres over points of U are n-dimensional, where $n = \dim X - \dim Y$. On the other hand, fibres over other points of Y may have dimension $> n$. This increase in dimension is clearly a big discontinuity of fibre type, and it can be shown that if f is flat, all components of *all* fibres of f have dimension n. In the other direction, for any morphism $f : X \to Y$ of varieties one would expect that almost all the fibres do form a continuous family, and indeed it can be shown that there always is some non-empty open $U \subset Y$ such that $\text{res}(f) : f^{-1}(U) \to U$ is flat.

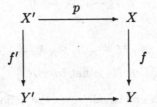

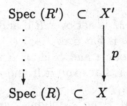

Another way of testing this intuitive description of flatness is via the fact that a continuous function has at most one extension from an open dense set to the whole space. The analogous fact about flatness is:

Proposition 1. *Suppose* $g : Z \to \text{Spec}\ (R)$ *is a morphism and* X_1, X_2 *are 2 closed subschemes of* Z. *Assume (1) that for some non-0-divisor* $f \in R$, X_1 *and* X_2 *are equal over* $\text{Spec}\ (R)_f$, *and (2) that the restrictions of* g *to* X_1 *and* X_2 *are flat morphisms from* X_1 *and* X_2 *to* $\text{Spec}\ (R)$. *Then* $X_1 = X_2$.

Proof. Let $U = \text{Spec}\ (S)$ be an open affine in Z, and let $X_i = \text{Spec}\ (A/A_i)$. By assumption (1), $A_1 \cdot S_f = A_2 \cdot S_f$, and by assumption (2), S/A_1 and S/A_2 are flat over R. I claim $A_i = S \cap (A_i \cdot S_f)$, hence

$$A_1 = S \cap (A - 1 \cdot S_f) = S \cap (A_2 \cdot S_f) = A_2\ .$$

But since f is not a 0-divisor in R, the sequences:

$$0 \longrightarrow R \xrightarrow{\text{mult.by } f} R$$

is exact. Therefore

$$0 \longrightarrow S/A_i \xrightarrow{\text{mult.by } f} S/A_i$$

is exact, i.e., if $f^n \cdot a \in A_i$, then $a \in A_i$. This means exactly that $A_i = S \cap (A_i \cdot S_f)$. □

Look back at Prop. 2 of Ch. II, §8. Here we had exactly the situation of the Prop. above with R a valuation ring, and we asserted that given a closed subscheme $\widetilde{X} \subset g^{-1}(\text{Spec}\ (R_f))$, there was one and only one closed subscheme $X \subset Z$ extending X and flat over R. In other words, when R is a valuation ring, we get existence as well as uniqueness.

Example P. Any "family" of affine hypersurfaces should be considered a "continuous" family, since it defines a flat morphism. To be precise, let

$$\sum_{0 \le i_1, \dots, i_n \le N} a_{i_1,\dots,i_n} X_1^{i_1} \cdot \cdots \cdot X_n^{i_n} = 0$$

be a hypersurface, with coefficients $a_{i_1,\dots,i_n}$ in an arbitrary ring R. Let

$$\mathcal{H} = \text{Spec}\ R[X_1, \dots, X_n] / \left(\sum a_{i_1,\dots,i_n} X_1^{i_1} \cdot \cdots \cdot X_n^{i_n}\right).$$

Then $\mathcal{H}$ is a scheme over $\text{Spec}\ (R)$ whose fibres over $\text{Spec}\ (R)$ are the hypersurfaces obtained by mapping the coefficients from R into a field by the various homomorphisms

$$R \longrightarrow R_P / P \cdot R_P$$

($P \subset R$ a prime ideal). Let's assume that none of these fibres is all of $\mathbb{A}^n$, i.e., that the equation doesn't vanish identically after applying any of these homomorphisms to its coefficients. Then $\mathcal{H}$ is flat over Spec (R) by Property E. above: in fact, $R[X_1, \ldots, X_n]$ is a free R-module, so it is flat over R, and for all maximal ideals $m \subset R$, $f = \sum a_{i_1,\ldots,i_n} X_1^{i_1} \cdot \cdots \cdot X_n^{i_n}$ is a non-0-divisor in $R/m[X_1, \ldots, X_n]$. Therefore, in E., take $B = M = R[X_1, \ldots, X_n]$ and $A = R$.

Now let's look at the case of a *finite* morphism $f : X \to Y$.

Proposition 2. *Let $f : X \to Y$ be finite, and assume Y noetherian. Then f is flat if and only if $f_*(\underline{o}_X)$ is a locally free $\underline{o}_Y$-module.*

Proof. To prove this, we may as well assume $Y = \text{Spec}(R)$. Then $X = \text{Spec}(S)$, where S is an R-algebra, finitely generated as R-module. Then $\widetilde{S} = f_*(\underline{o}_X)$ so

$$\begin{aligned} f \text{ is flat} &\iff S \text{ is flat over } R \ \ (\text{by B}^*) \\ &\iff \widetilde{S} \text{ is a flat } \underline{o}_Y\text{-module} \ \ (\text{by B}^*) \\ &\iff f_*(\underline{o}_X) \ \text{ is a locally free } \underline{o}_Y\text{-module} \ \ (\text{by C}^*)\,. \end{aligned}$$

□

Corollary. *Assume also that Y is reduced and irreducible. Then f is flat if and only if the integer*

$$\dim_{\mathbb{k}(y)} \left[f_*(\underline{o}_X) \otimes_{\underline{o}_y} \mathbb{k}(y)\right]$$

is independent of y.

Proof. Add version II of Nakayama's lemma to the Proposition.

Note that if $A_y = f_*(\underline{o}_X)_y \otimes_{\underline{o}_y} \mathbb{k}(y)$, then Spec (A_y) is the fibre of f over y. So the Corollary asserts that flatness is equivalent to the fibres of f being Spec's of Artin rings of constant length – again a natural continuity restriction.

Example Q. Let k be an algebraically closed field and define $f : \mathbb{A}^1 \to \mathbb{A}^1$ by $f(x) = x^2$. f is a finite morphism since it is dual to the inclusion of rings

$$\begin{gathered} k[X] \supset k[X] \\ X^2 \underset{f^*}{\longleftarrow} X\,. \end{gathered}$$

Let $a \in k$, and look at the fibre of f over the point $X = a$. It is Spec $(k[X]/(X^2 - a))$. So if $a \neq 0$, it is the disjoint union of 2 points; if $a = 0$, it is the disembodied tangent vector $I = \text{Spec}\ (k[\epsilon]/\epsilon^2)$ of §4. In all cases,

$$\dim_k k[X]/(X^2 - a) = 2$$

so f is flat.

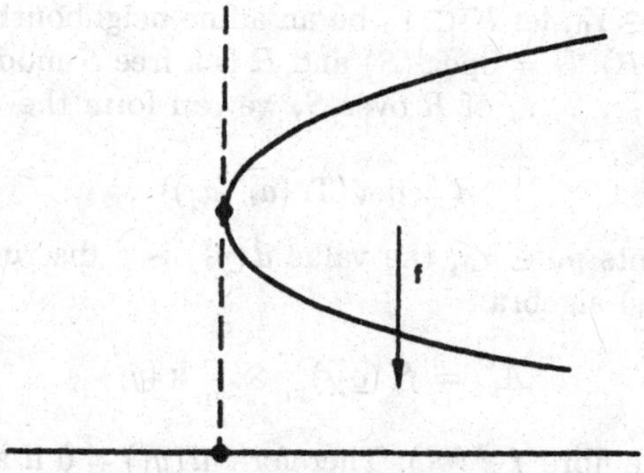

Example R. Look at the morphism $f : \mathbb{A}^2 \to V\left((X_1X_3 - X_2^2)\right) \subset \mathbb{A}^3$ defined by $f(x,y) = (x^2, xy, y^2)$ as in Ex. L. This is finite, and if $a, b, c \in k$ satisfy $ac = b^2$, the fibre of f over (a,b,c) is

$$\operatorname{Spec}\left(k[X,Y]/\left(X^2 - a, XY - b, Y^2 - c\right)\right) .$$

If a or c is not 0, this is the disjoint union of 2 points; but if $a = c = 0$, hence $b = 0$, it is

$$\operatorname{Spec} k[X,Y]/\left(x^2, XY, Y^2\right)$$

and this ring is 3-dimensional. Therefore f is *not* flat.

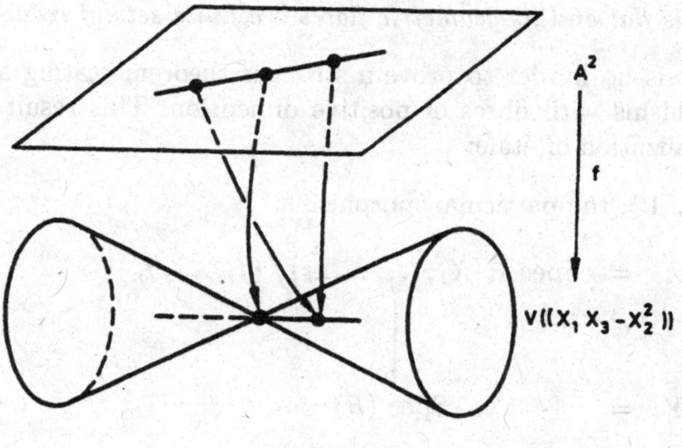

Given a finite morphism $f : X \to Y$, where Y is an irreducible and reduced noetherian scheme, the preceding discussion and examples suggest introducing several subsets of Y:

a) Let $Y_0 \subset Y$ be the open set over which $f_*(\underline{o}_X)$ is locally free, hence $\operatorname{res}(f) : f^{-1}(Y_0) \to Y_0$ is flat.

b) For any point $y \in Y_0$, let $U \subset Y_0$ be an affine neighbourhood of y such that $f^{-1}(U) = \text{Spec}(R)$, $U = \text{Spec}(S)$ and R is a free S-module.
Taking a basis $a_1, \ldots, a_n$ of R over S, we can form the *discriminant* in the usual way:

$$d = \det(\text{Tr}(a_i \cdot a_j)) \ .$$

Then for all points $y_1 \in Y_0$, the value $d(y_1)$ is a discriminant of the finite dimensional $\Bbbk(y_1)$-algebra

$$A_{y_1} = f_*(\underline{o}_X)_{y_1} \otimes_{\underline{o}_{y_1}} \Bbbk(y_1) \ ,$$

whose Spec is the fibre $f^{-1}(y_1)$. Therefore, $d(y_1) \neq 0$ if and only if the fibre $f^{-1}(y_1)$ is a union of the Spec's of separable extensions of $\Bbbk(y_1)$. Therefore, we have found an *open* subset

$$Y_1 \subset Y_0$$

of points $y_1 \in Y_0$ whose fibres are like this; equivalently whose geometric fibres are reduced.

Definition 2. $Y - Y_1$ is the *branch locus* of f, and points $y \in Y - Y_1$ are *ramification points* for f.

Moreover, I claim that Y_1 is the maximal open set such that $\text{res}(f) : f^{-1}(Y_1) \to Y_1$ is étale. To prove this, we need the main result of this section, which is the *intrinsic* characterization of étale morphisms referred to in §5:

Theorem 3. *Let $f : X \to Y$ be a morphism of finite type. Then f is étale if and only if f is flat and its geometric fibres are finite sets of reduced points.*

Actually it is no harder to prove a stronger theorem stating an equivalent fact for morphisms with fibres of positive dimension. This result involves the natural generalization of étale:

Definition 3. 1st, the particular morphisms:

$$\begin{array}{ccc} X & = & \text{Spec}\, R[X_1, \ldots, X_{n+k}] / (f_1, \ldots, f_n) \\ & & \big\downarrow \\ Y & = & \text{Spec}\,(R) \end{array}$$

are said to be *smooth* at a point $x \in X$ (of relative dimension k) if

$$(*) \qquad \text{rank}(\partial f_i / \partial X_k(x)) = n \ .$$

2nd, an arbitrary morphism $f : X \to Y$ of finite type is *smooth* (of relative dimension k), if for all $x \in X$, there are open neighbourhoods $U \subset X$ of x and $V \subset Y$ of $f(x)$ such that $f(U) \subset V$ and such that f restricted to U, looks like a morphism of the above type which is smooth at x:

$$\begin{array}{ccc} U & \xrightarrow[\text{open immersion}]{\subset} & \operatorname{Spec} R[X_1,\ldots,X_{n+k}]/(f_1,\ldots,f_n) \\ {\scriptstyle \operatorname{res}(f)}\downarrow & & \downarrow \\ V & \xrightarrow[\text{open immersion}]{\subset} & \operatorname{Spec}(R) \; . \end{array}$$

Theorem 3'. *Let $f : X \to Y$ be a morphism of finite type. Then f is smooth of relative dimension k if and only if f is flat and its geometric fibres are disjoint unions of k-dimensional non-singular varieties.*

(*N.b.* This makes sense because the geometric fibres of f are schemes of finite type over *algebraically closed* fields, and it is for such schemes that we have defined non-singularity.)

Proof. First assume f is smooth. It is clear that the definition of smoothness is such that whenever $f : X \to Y$ is smooth, all morphisms f' obtained by a fibre product:

$$\begin{array}{ccc} X' & \longrightarrow & X \\ {\scriptstyle f'}\downarrow & & \downarrow{\scriptstyle f} \\ Y' & \longrightarrow & Y \end{array}$$

are still smooth. In particular, if F is a geometric fibre of f over an algebraically closed field Ω, F is smooth over Ω. But by Theorem 4, §4, when $Y = \operatorname{Spec} \Omega$, Ω algebraically closed, a morphism of finite type $f : X \to Y$ is smooth if and only if X is a disjoint union of non-singular varieties.

To see that f is flat, it suffices to check that

$$S = R[X_1,\ldots,X_{n+k}]/(f_1,\ldots,f_n)$$

is a flat R-module when rank $(\partial f_i/\partial X_j(x)) = n$, all $x \in \operatorname{Spec}(S)$. Set

$$R_i = R[X_1,\ldots,X_{n+k}]/(f_1,\ldots,f_i) \; .$$

Then $R_{i+1} \cong R_i/f_{i+1}\cdot R_i$, and $R_0 = R[X_1,\ldots,X_{n+k}]$ is flat over R, so we are in a situation to use Property E. of flatness inductively. We need only check that for all maximal ideals $m \subset R$, f_{i+1} is a non-0-divisor in

$$R/m\,[X_1,\ldots,X_{n+k}]/(f_1,\ldots,f_i) \; .$$

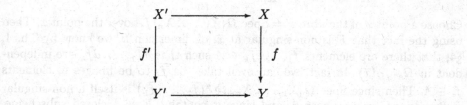

Let $\Omega \supset R/m$ be an algebraically closed field. Then we just saw that $\operatorname{Spec} \Omega[X_1,\ldots,X_{n+k}]/(f_1,\ldots,f_i)$ is a union of $n+k-i$-dimensional varieties V_{ij}. In particular, $\Omega[X_1,\ldots,X_{n+k}]/(f_1,\ldots,f_i)$ is a direct sum of integral domains. Therefore, f_{i+1} is a non-0-divisor here if its component in each factor is non-zero. Since all components of Spec $\Omega[X_1,\ldots,X_{n+k}]/(f_1,\ldots,f_{i+1})$ have lower dimension, f_{i+1} does not vanish on any of the V_{ij}. Thus Property E. is applicable and S is flat over R.

Conversely, assume f is flat with non-singular geometric fibres. Let $x \in X$, and express f locally near x by rings:

$$\begin{array}{ccccc} X & \supset & U & = & \operatorname{Spec} R[X_1,\ldots,X_{n+k}]/A \\ {\scriptstyle f}\big\downarrow & & & & \big\downarrow{\scriptstyle \operatorname{res}(f)} \\ Y & \supset & V & = & \operatorname{Spec}(R)\ . \end{array}$$

Let $f(x) = [P]$, and embed R/P in an algebraically closed field Ω. Look at the geometric fibre over Ω:

$$\begin{array}{ccccc} R[X_1,\ldots,X_{n+k}]/A & \longrightarrow & & & \Omega[X_1,\ldots,X_{n+k}]/\overline{A} \\ \big\uparrow & & & & \big\uparrow \\ R & \longrightarrow & R/P & \subset & \Omega\ . \end{array}$$

Choose a point $\bar{x}$ of the fibre $F = \operatorname{Spec} \Omega[X_1,\ldots,X_n]/\overline{A}$ over the point x. Then using the fact that F is non-singular at $\bar{x}$, of dimension k, we know by Th. 4, §4, that there are elements $\overline{f}_1,\ldots,\overline{f}_k \in \overline{A}$ such that $d\overline{f}_1,\ldots,d\overline{f}_k$ are independent in $\Omega_{\mathbf{A}^n/\Omega}(\bar{x})$. In fact, we can even take the $\overline{f}_i$ to be images of elements $f_i \in A$. Then since Spec $\Omega[X_1,\ldots,X_{n+k}]/(\overline{f}_1,\ldots,\overline{f}_k)$ is itself a non-singular k-dimensional variety near $\bar{x}$, and since it contains F as a closed subscheme, $F = \operatorname{Spec} \Omega[X_1,\ldots,X_{n+k}]/(\overline{f}_1,\ldots,\overline{f}_k)$ near $\bar{x}$. Now the fact that the $d\overline{f}_i$ are independent in $\Omega_{\mathbf{A}^n/\Omega}(\bar{x})$ means that

$$\operatorname{rank}\,(\partial\overline{f}_i/\partial X_j(\bar{x})) = k\ .$$

Hence

$$\operatorname{rank}\,(\partial f_i/\partial X_j(x)) = k\ ,$$

since $\bar{x}$ lies over x. Note that U is a closed subscheme of Spec $(R[X_1,\ldots,X_{n+k}]/(f_1,\ldots,f_k))$. It will suffice to show that these schemes are equal near x and then f will have been expressed in the standard form and hence is smooth. This last step follows from:

Lemma. *Given a diagram:*

$$\begin{array}{ccc} X_1 & \longrightarrow & X_2 \\ & {\scriptstyle f_1}\searrow \quad \swarrow{\scriptstyle f_2} & \\ & Y & \end{array}$$

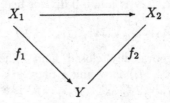

where X_1 is a closed subscheme of a noetherian scheme X_2 and f_1 is flat, then for all $x \in X_1$, if the geometric fibres of f_1 and f_2 over $f(x)$ are equal near some point $\bar{x}$ over x, then $X_1 = X_2$ near x.
(Note the analogy with Prop. 1).

Proof of lemma: Algebraically, we have the dual picture

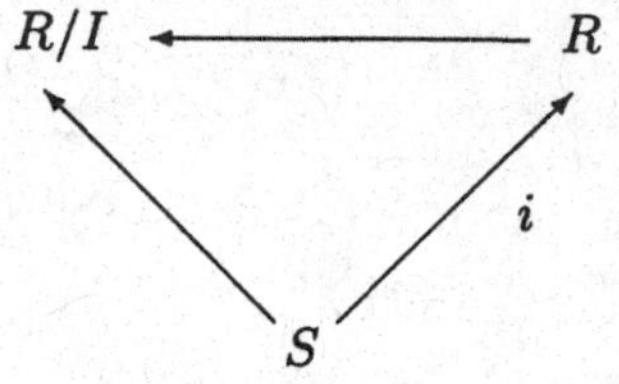

Let $x = [P]$, where $P \subset R$ is a prime ideal containing I. We want to prove that $I \cdot R_P = (0)$. By assumption, we can embed $S/I^{-1}(P)$ in an algebraically closed field Ω, and find a prime ideal $\overline{P} \subset R \otimes_S \Omega$ such that

1) $I \cdot (R \otimes_S \Omega)_{\overline{P}} = (0)$
2) if $j : R \to R \otimes_S \Omega$ is the canonical map,

$$j^{-1}(\overline{P}) = P \ .$$

Now tensor the exact sequence

$$0 \longrightarrow I \longrightarrow R \longrightarrow R/I \longrightarrow 0$$

over S with Ω. Since R/I is flat over Ω, it follows that $I \otimes_S \Omega \xrightarrow{\sim} I \cdot (R \otimes_S \Omega)$. Now let R_1 denote the local ring R_P and let R_2 denote the local ring $(R \otimes_S \Omega)_{\overline{P}}$. Let P_1 and P_2 denote their maximal ideals. j induces a local homomorphism $j' : R_1 \to R_2$, hence an injection $\overline{j}' : R_1/P_1 \subset R_2/P_2$. Since $I \cdot R_2 = (0)$, therefore

$$[I \cdot (R \otimes_S \Omega)] \underset{(R \otimes_S \Omega)}{\otimes} R_2/P_2 = (0) \ .$$

Using the fact that the 1st module is just $I \otimes_S \Omega$, this means that:

$$I \otimes_R R_2/P_2 = (0) \ .$$

But R_2/P_2 is just an extension field of R_1/P_1, so

$$I \otimes_R R_1/P_1 = (0)$$

also. But since $I \otimes_R (R_1/P_1)$ is the same as $I \cdot R_P/P \cdot (I \cdot R_P)$, Nakayama's lemma shows that $I \cdot R_P = (0)$ too.

QED for lemma and Th. 3'.

Problem. Here is a Main Theorem-type result for flat morphisms: Let $f : X \to Y$ be a birational flat morphism between varieties. Show that f is an open immersion.

Appendix

Curves and Their Jacobians

Preface[13]

This appendix is a slightly expanded version of the series of four Ziwet Lectures which I gave in November 1974 at The University of Michigan, Ann Arbor. The aim of the lectures and of this volume is to introduce people in the mathematical community at large – professors in other fields and graduate students beyond the basic courses – to what I find one of the most beautiful and what objectively speaking is at least one of the oldest topics in algebraic geometry: curves and their Jacobians. Because of time constraints, I had to avoid digressions on any foundational topics and to rely on the standard definitions and intuitions of mathematicians in general. This is not always simple in algebraic geometry since its foundational systems have tended to be more abstract and apparently more idiosyncratic than in other fields such as differential or analytic geometry, and have therefore not become widely known to non-specialists. My idea was to get around this problem by imitating history: i.e., by introducing all the characters simultaneously in their complex analytic and algebraic forms. This did mean that I had to omit discussion of the characteristic p and arithmetic sides. However it also meant that I could immediately compare the strictly analytic constructions (such as Teichmüller Space) with the varieties which we were principally discussing.

When I first started doing research in algebraic geometry, I thought the subject attractive for two reasons: firstly, because it dealt with such down-to-earth and really concrete objects as projective curves and surfaces; secondly, because it was a small, quiet field where a dozen people did not leap on each new idea the minute it became current. As it turned out, the field seems to have acquired the reputation of being esoteric, exclusive and very abstract with adherents who are secretly plotting to take over all the rest of mathematics! In one respect this last point is accurate: algebraic geometry is a subject which relates frequently with a very large number of other fields – analytic and differential geometry, topology, K-theory, commutative algebra, algebraic groups and number theory, for instance – and both gives and receives theorems, techniques and examples with all of them. And certainly Grothendieck's work contributed to the field some very abstract and very powerful ideas which are quite hard to digest. But this subject, like all subjects, has a dual aspect in that all these abstract ideas would collapse of their own weight were it not for the underpinning supplied by concrete classical geometry. For me it has been a real adventure to perceive

[13] Written in 1975.

the interactions of all these aspects, and to learn as much as I could about the theorems both old and new of algebraic geometry.

Dafydd ap Gwilym's *The Lark* seems to me like the muse of mathematics:

> High you soar, Wind's own power,
> And on high you sing each song,
> Bright spell near the wall of stars,
> A far high-turning journey.

If this book entices a few to go on to learn these "spells", I'll be very pleased. I'd like to thank Fred Gehring, Peter Duren and the many people I met at Ann Arbor for their warm hospitality and their willingness to listen. A final point: in order not to interrupt the text, we have omitted almost all references and attributions in the lectures themselves, and instead written a separate section at the end giving references as well as suggestions for good places to learn various topics.

Lecture I: What is a Curve and How Explicitly Can We Describe Them?

In these lectures we shall deal entirely with algebraic geometry over the complex numbers $\mathbb{C}$, leaving aside the fascinating arithmetic and characteristic p side of the subject. In this first lecture, I would like to recapitulate some classical algebraic geometry, giving a leisurely tour of the zoo of curves of low genus, pointing out various features and their generalizations, and leading up to my first main point: the "general" curve of genus g, for g large, is very hard to describe explicitly.

The beginning of the subject is the AMAZING SYNTHESIS, which surely overwhelmed each of us as graduate students and should really not be taken for granted. Starting in 3 distinct fields of mathematics, we can consider 3 types of objects:

a) *Algebra*: consider *field extensions* $K \supset \mathbb{C}$, where K is finitely generated and of transcendence degree 1 over $\mathbb{C}$.

b) *Geometry*: First fix some notations: we denote by $\mathbb{P}^n$ the projective space of complex $(n+1)$-tuples $(X_0, \ldots, X_n)$, not all zero, mod scalars. $X_0, \ldots, X_n$ are called homogeneous coordinates. $\mathbb{P}^n$ is covered by $(n+1)$-affine pieces. $U_i =$ (pts where $X_i \neq 0$) and $x_0 = X_0/X_i, \ldots, x_n = X_n/X_i$ (x_i omitted) are the affine coordinates on U_i. Consider *algebraic curves* $C \subset \mathbb{P}^n$: loci defined by a finite set of homogeneous equations $f_\alpha(X_0, \ldots, X_n) = 0$, and such that for every $x \in X$, C is "locally defined by $n-1$ equations with independent differentials", i.e., $\exists\ f_{\alpha_1}, \ldots f_{\alpha_{n-1}}$ plus g, with $g(x) \neq 0$, such that for all α,

$$g f_\alpha \equiv \sum_{i=1}^{n-1} h_{\alpha,i} f_{\alpha_i}, \quad \text{some polynomials } h_{\alpha,i}$$

and

$$rk\,(\partial f_{\alpha_i} / \partial X_j(x)) = n - 1.$$

c) *Analysis*: consider compact Riemann surfaces[14].

The result is that there are canonical bijections between the set of isomorphism classes of objects of either type. [A word about isomorphism in case (b): the simplest and oldest way to describe isomorphism in the algebraic category is that $C_1 \subset \mathbb{P}^{n_1}$ and $C_2 \subset \mathbb{P}^{n_2}$ are isomorphic if there is a bijective algebraic

[14] Perversely, algebraic geometers persist in talking about curves and analysts about surfaces when they mean essentially the same object!

correspondence between C_1 and C_2, i.e., there is a curve $D \subset C_1 \times C_2$ defined by bihomogeneous equations $g_\alpha(X_0, \ldots X_{n_1}; Y_0, \ldots Y_{n_2}) = 0$ which projects bijectively onto C_1 and onto C_2.] To go back and forth between objects of type a), b), c), for instance, we

1) associate to a curve C the field K of functions $f : C \to \mathbb{C} \cup (\infty)$ given by restricting to C rational functions $p((X_0, \ldots, X_n)/q(X_0, \ldots, X_n)))$, $\deg p = \deg q$; and the Riemann surface just given by C with the induced complex structure from $\mathbb{P}^n$.
2) associate to a Riemann surface X its field of meromorphic functions; and any curve C which is the image of holomorphic embedding of X in $\mathbb{P}^n$.
3) from the field K, we recover C or X as point set just as the set of valuation rings R, $\mathbb{C} \subset R \subset K$.

To X or C or K we can associate a *genus* g as usual:

$$g = \text{ no. of handles of } X$$

or

$$g = \dim \text{ of } \begin{cases} [\text{space of holomorphic differentials } \omega \text{ on } X] \\ \qquad\qquad \| \\ [\text{space of rational differentials } \omega = a\,dx, \\ \quad (a, x \in K,\ x \notin \mathbb{C}) \text{ on } C \text{ with no poles}] \end{cases}$$

or

$$2g - 2 = (\text{no. of zeroes}) - (\text{no. of poles}), \quad \text{of any differential } \omega.$$

For each g, we shall let $\mathfrak{M}_g$ denote the set of isomorphism classes of X or C or K of genus g: we shall discuss the structure of $\mathfrak{M}_g$ in the second lecture.

So much for generalities. Most of what I shall say later is best understood by considering the computable explicit cases of low genus. Let's take these up and see what we have:

$g = 0$: there is only one object here:
X = Riemann sphere $\mathbb{C} \cup (\infty)$
$C = \mathbb{P}^1$ itself
$K = \mathbb{C}(X)$.

$g = 1$: Here we have the famous theory of elliptic curves:
$X = \mathbb{C}/L$, L a lattice which may be taken to be $\mathbb{Z} + \mathbb{Z} \cdot \omega$, $\operatorname{Im} \omega > 0$.
C = any non-singular plane cubic curve, i.e., $C \subset \mathbb{P}^2$ defined by $f(x, y, z) = 0$, f homogeneous of degree 3, with some partial non-zero at each root; in affine coordinates, x, y, C is given as the zeroes of a cubic polynomial $f(x, y) = 0$.
$K = \mathbb{C}\left(X, \sqrt{f(X)}\right)$, where f is a polynomial of degree 3 with distinct roots.

The connections between these are given as follows: given X, form the Weierstrass $\wp$-function:

$$\wp(z) = \frac{1}{z^2} - \sum_{\substack{a \in L \\ a \neq 0}} \left[\frac{1}{(z-a)^2} - \frac{1}{a^2} \right]$$

and map $\mathbb{C}/L$ into $\mathbb{P}^2$ by

$$\begin{array}{lll} z & \longmapsto & (1,\ \wp(z),\ \wp'(z)), \qquad z \notin L \\ z & \longmapsto & (0,\ 0,\ 1), \qquad z \in L, \end{array}$$

(i.e., $(\mathbb{C} - L)/L$ is mapped to the affine piece $X_0 \neq 0$ by $\wp$ and $\wp'$, and the one point L/L is mapped to the "line at infinity" for this affine piece.) Then $\wp$ and $\wp'$ generate the field K of X and since $\wp'^2$ is a cubic polynomial in $\wp$, K is as above. Or starting with any plane cubic C, take affine coordinates x, y so that the line at infinity is a line of inflexion. Then C is readily normalized to the form:

$$y^2 = f(x), \quad \deg f = 3.$$

Therefore the field of rational functions on C is $\mathbb{C}\left(x, \sqrt{f(x)}\right)$. To go back to X, look at the abelian line integral

$$w = \int_{(x_0,y_0)}^{(x,y)} \frac{dx}{y}$$

taken on C; then

$$(x, y) \longmapsto w$$

is well-defined up to a period which lies in a lattice L, hence defines:

$$C \xrightarrow{\approx} \mathbb{C}/L.$$

A few comments on this set-up: $\mathbb{C}/L$ is clearly a group, and hence so is C – here the group law is characterized geometrically by the beautiful:

$$x + y + z = 0 \Longleftrightarrow x, y, z \ \text{ collinear.}$$

For instance, $3x = 0 \Leftrightarrow x$ a point of inflexion. Since $3x = 0 \Leftrightarrow x \in \frac{1}{3}L/L$, there will be 9 of these. Now via $X \approx \mathbb{C}/L$, we get flat metrics on X with curvature $\equiv 0$. But if we instead look at a metric on X induced from the standard metric on $\mathbb{P}^2$ via $X \cong C \subset \mathbb{P}^2$, we get a metric whose curvature at the 9 points of inflexion equals that of $\mathbb{P}^2$, which is positive; and by the Gauss-Bonnet theorem, it must be negative at other points. The wobbly curvature points up the fact that X *does not fit symmetrically in* $\mathbb{P}^2$ – we will discuss this further in Lecture III. Another indication of the antagonism between $\mathbb{C}/\mathbb{Z}+\mathbb{Z}\cdot\omega$ and C is the Gelfand-Schneider result: with a few exceptions for very special ω's, (i.e., $\omega \in \mathbb{Q}(\sqrt{-n})$), ω and the coefficients of any isomorphic cubic C are never simultaneously algebraic.

$\boxed{g=2:}$ Start with the fields K: these are all of the form

$$K = \mathbb{C}\left(X, \sqrt{f(X)}\right), \text{ where degree } f = 5.$$

What this means is that the corresponding curve C admits a $2-1$ mapping onto $\mathbb{P}^1$ ramified at 6 points: the 5 roots of f and the point at infinity. This does not quite give us C *embedded in* $\mathbb{P}^n$ though. We can do 2 things: let

$$\pi : C \longrightarrow \mathbb{P}^1$$

be the above map. Fix x_1, x_2 with $\pi x_1 = \pi x_2$. Then one can prove that C can be mapped to a plane quartic curve C_0 bijectively *except* that x_1 and x_2 are identified to a double point of C_0. This means that at the double point C_0 is given by an equation

$$0 = xy + f_3(x,y) + f_4(x,y)$$

where the double point equals the origin. In this form, $\pi(x,y) = x/y$; or geometrically, $\pi : C \to \mathbb{P}^1$ is defined by "projecting from $(0,0)$." This still doesn't represent C embedded in $\mathbb{P}^n$! In fact, to do this, you need $n = 3$, and at least 3 equations too. You start with a line $\ell \subset \mathbb{P}^3$, then take quadric and cubic surfaces $F, G \subset \mathbb{P}^3$ containing ℓ. Then $F \cap G$ will fall into 2 components – ℓ plus a quintic curve C, and it can be proven that every curve of genus 2 occurs as such a C.

Given such a C, there is only one $2-1$ map $\pi : C \longrightarrow \mathbb{P}^1$ and the most important points on C are the 6 points x_i where it ramifies. They have 2 significances

a) they are the Weierstrass points of C, i.e., the points $x \in C$ such that there is a rational function f on C with a double pole at x and no other poles, (if t is the coordinate on $\mathbb{P}^1$, let $f = (t - t(x))^{-1}$)
b) they represent the "odd theta-characteristics," i.e., look for differentials ω with no poles and zeroes only with even multiplicities: one writes this

$$(\omega) = 2\mathfrak{A}$$

if (ω) = divisor of zeroes and poles of ω. In this case, there are ω_i with one double zero at x_i, i.e.,

$$(\omega_i) = 2x_i$$

and no others (in fact if $a_i = t(x_i)$, $\omega_i = \sqrt{\frac{t-a_i}{\prod_{j\neq i} t-a_j}}\, dt$).

Analytically, C can be represented by a Fuchsian group:

$$C \cong H/\Gamma$$

where:

$$\begin{aligned} H &= \{z \mid \operatorname{Im} z > 0\} \\ \Gamma &= \text{discrete subgroup of } SL(2,\mathbb{R})/(\pm 1) \end{aligned}$$

or by various Kleinian groups:

$$C \cong D/\Gamma$$

where

$$\begin{aligned} D &= \text{open subset of } \mathbb{C} \cup (\infty) \\ \Gamma &= \text{discrete subgroup of } SL(2,\mathbb{C})/(\pm I) \\ & \quad \text{which acts discontinuously on } D. \end{aligned}$$

I want to make only one remark on these representations in connection with my main question of how explicitly one can describe C. Start with a Fuchsian Γ. Choosing a standard basis of $\pi_1(C)$, Γ is generated by hyperbolic transformations A_1, B_1, A_2, B_2 satisfying

$$A_1 B_1 A_1^{-1} B_1^{-1} A_2 B_2 A_2^{-1} B_2^{-1} = e.$$

It is quite clear from the work of Fricke-Klein and of Purzitsky and Keen that there is a small number of inequalities on the traces of small words in $A_1, \ldots, B_2$ which are always satisfied for Fuchsian Γ's, such that *conversely* if $A_1, \ldots, B_2 \in SL(2,\mathbb{R})$ satisfy these inequalities, they generate a Fuchsian Γ. (It would be nice to know these inequalities precisely.) This means that one can actually find all Fuchsian Γ's quite explicitly. For Kleinian Γ's no such simple inequalities are known and presumably do not exist. In the simplest case, the problem arises – describe explicitly the set of pairs $(A, B) \in SL(2,\mathbb{C})^2$ which generate a free group of only hyperbolic elements acting discontinuously at some $z_0 \in \mathbb{C}$, i.e., the Schottky groups. This looks very hard.[15] $\boxed{g = 3:}$ Here we encounter first the phenomenon of not having one easy description of all C's at once: "almost all" C's can be described one way, but some are a special case and must be described a different way. The general type are the C's which are non-singular plane quartic curves. The embedding of C in $\mathbb{P}^2$ is canonical and is given in the following simple way: let $\varphi_1, \varphi_2, \varphi_3$ be a basis of the differentials of first kind (= with no poles) on C. For all $x \in C$, let dt be a differential near x with no zero at x so that $\varphi_i(x) = a_i(x)dt$, a_i a function. Define:

$$C \longrightarrow \mathbb{P}^2$$

by

$$x \longmapsto (a_1(x), a_2(x), a_3(x)) .$$

This is independent of the choice of dt because changing dt multiplies the triple by a scalar. This procedure works in any genus and defines the so-called canonical map

$$\Phi : C \longrightarrow \mathbb{P}^{g-1},$$

given, loosely speaking, by:

[15] Cf. forthcoming book, "Indra's Pearls" by D. Wright, C. Series and myself, Camb. Univ. Press.

$$x \longmapsto (\varphi_1(x), \ldots, \varphi_g(x)),$$

where $\{\varphi_i\}$ is a basis of differentials of 1st kind.

[Note that there is a natural correspondence between linear functions in the homogeneous coordinates on $\mathbb{P}^{g-1}$ and arbitrary differentials $\sum \lambda_i \varphi_i$ of 1st kind on C.] As is well known, there are 2 types of C's: those for which Φ is an embedding (i.e., Φ injective *and* $\Phi(C)$ non-singular), and those for which Φ is $2-1$, and the image is isomorphic to $\mathbb{P}^1$. All C's which admit $2-1$ maps to $\mathbb{P}^1$ fall into the 2nd category and are called *hyperelliptic*. Thus for $g=3$, either $C \cong \Phi(C)$ – then because each φ_i has $2g-2=4$ zeroes, each line in $\mathbb{P}^2$ meets $\Phi(C)$ in 4 points and $\Phi(C)$ is a quartic – or $\Phi(C)$ is a non-singular conic "with multiplicity 2," i.e., Φ is $2-1$. As all non-singular conics are isomorphic to $\mathbb{P}^1$, C is then hyperelliptic. In general, in the non-hyperelliptic case, $\Phi(C)$ will have degree $2g-2$, because the hyperplanes $H \subset \mathbb{P}^{g-1}$ correspond to differentials φ of 1st kind in such a way that:

$$\Phi \;(\text{zeroes of } \varphi) = H \cap \Phi(C).$$

Plane quartic curves C are intricate objects. They have lots of special points on them:

a) their 24 points of inflexion are the Weierstrass points of C: the points x such that there is a function f on C with a triple pole at x and no other[16],
b) their 28 bitangents – lines tangent to C at 2 points – correspond to the odd theta-characteristics. Because if ℓ is tangent to C at x and y, then the differential φ corresponding to ℓ has a double zero at x and y:

$$(\varphi) = 2x + 2y.$$

In fact, projective geometry yields a vast constellation of "*higher Weierstrass points*" too, such as the 108 points x for which there is a conic D touching at x with contact of order 6. More generally, for any degree d, look at the points x for which there is a curve D of degree d touching C at X with contact "one more than is expected," i.e., one more than is possible at most points of D. One can think of this as some kind of analog on C of the finite set of points of order d, $\frac{1}{d}L/L \subset \mathbb{C}/L$ in the genus 1 case. This analogy goes quite far. For

16 You can find the function as follows: let x be a point of inflexion, let ℓ be the tangent line to C at x. Then ℓ meets C at one further point y:

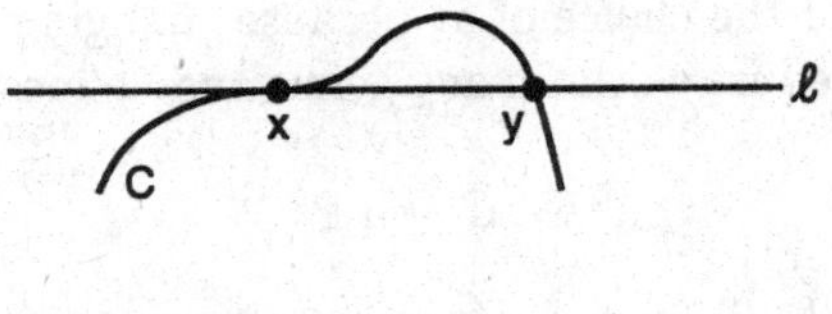

Let u, v be affine coordinates such that y is the origin $u=v=0$, and ℓ is the coordinate axis $u=0$. Consider the function $f=v/u$. Since u and v are zero at y, f is regular at y; but at x, f has a triple pole.

instance, as $d \to \infty$, one can show that these points are dense in C and even fairly evenly distributed in the "Bergman metric," i.e., for any curve C, choose a basis $\varphi_1, \ldots, \varphi_g$ of differentials of 1st kind for which

$$\int_C \varphi_i \wedge \overline{\varphi_j} = \delta_{ij}.$$

Then using such a basis, we can normalize our canonical embedding

$$\Phi : C \longrightarrow \mathbb{P}^{g-1}$$

up to unitary transformations, in which case the standard metric ds^2 on $\mathbb{P}^{g-1}$ has a restriction ds^2_B to C independent of the choice of the φ_i's: this is the Bergman metric.

An interesting question that arises in this connection is the relationship between the Bergman metric ds^2_B and the Poincaré metric ds^2_P of constant negative curvature induced from the standard metric on H:

$$ds^2 = dx^2 + dy^2/y^2, \quad z = x + iy$$

via the Fuchsian uniformization $C = H/\Gamma$. Kazdan suggested that if $\Gamma_n \subset \Gamma$ are subgroups of finite index and cofinal among such subgroups, if $C_n = H/\Gamma_n$, and if $ds^2_{B_n}$ is the Bergman metric on C_n pulled back to H, then with suitable scalars λ_n,

$$\lim_{n\to\infty} \lambda_n \, ds^2_{B_n} = ds^2_p.$$

We won't say much about the hyperelliptic case: in genus g, if $C \to \mathbb{P}^1$ is 2–1, then there are exactly $2g+2$ branch points, and the corresponding fields are just $\mathbb{C}\left(X, \sqrt{f(X)}\right)$, where $\deg f = 2g+1$ or $2g+2$. [If f has degree $2g+2$, by a linear fractional transformation in X, taking some root to ∞, $\mathbb{C}(X, \sqrt{f}) \cong \mathbb{C}(X', \sqrt{f'})$ where $\deg f' = 2g+1$.] These curves are special however in the following precise sense: one can build a big algebraic family of curves of genus g:

$$f : X \to S$$

such that all curves of genus g occur as fibres $X_s = f^{-1}(s)$. Then the set of s such that X_s is not hyperelliptic will form a dense Zariski-open subset of S.

$\boxed{g = 4:}$ Let $\Phi : C \to \mathbb{P}^3$ be the canonical map. If C is not hyperelliptic, we saw that $\Phi(C)$ was a space curve of degree 6. In fact, $\Phi(C)$ is the complete intersection $F \cap G$ of a quadric and cubic surface meeting transversely. One could also ask, however, is C a plane curve or is there a map $\pi : C \to \mathbb{P}^1$ of low degree? The answer to the first question is that C must be given singularities before it can be put in $\mathbb{P}^2$: the simplest way is to identify 2 pairs of points making C into a plane quintic C_0 with 2 double points; as for π, one can always find a π of degree 3.

As the genus g grows, it gets harder and harder to represent the general curve C of genus g either as a plane curve with relatively few singular points, or as a covering of fairly low degree of $\mathbb{P}^1$. For instance, it can be shown that the lowest degree curve representing such a C has degree

$$d = \left[\frac{2g+8}{3}\right].$$

In general, its singularities will only be double points but the number of these will be

$$\delta = \frac{(d-1)(d-2)}{2} - g$$

which is asymptotic to $2/9(g^2)$. If $g \leq 10$, one can work backwards and write down all equations $f(X_0, X_1, X_2)$ defining curves of this degree d and this number δ of double points, hence having genus g. This is because the vector space of such f's has dimension $(d+1)(d+2)/2$ (count the coefficients), and for any point (a_0, a_1, a_2), if we require the coefficients of f to satisfy the 3 linear equations:

$$\frac{\partial f}{\partial X_i}(a_0, a_1, a_2) = 0,$$

then $f = 0$ has a singularity at (a_0, a_1, a_2). Now if $g \leq 10$, then $3\delta < \frac{(d+1)(d+2)}{2}$ (see table below), hence we can pick an *arbitrary* set of δ points $P_i = \left(a_0^{(i)}, a_1^{(i)}, a_2^{(i)}\right)$ in $\mathbb{P}^2$ and always find at least one curve C of degree d with singular points $P_1, \ldots, P_\delta$; in general these will be double points and C will have genus g. However if $g \geq 11$, if we choose the δ singular points generically, there will be no such f, i.e., the coordinates of the δ double points will always satisfy some obscure identities. The upshot is that there is no reasonably explicit way to write down the equations of these plane curves: one is in a realm of unexplicitness almost as bad as with Kleinian groups.

Next, it can be shown that the lowest degree map $\pi : C \to \mathbb{P}^1$ has degree

$$d = \left[\frac{g+3}{2}\right].$$

(This is equivalently the smallest number of poles of any non-constant function on the general curve C.) This also, to my knowledge, does not lead to any explicit polynomial presentation of C, but it does lead to a very explicit topological presentation of C. Namely, assuming the branch points of π are all simple, then one can reconstruct C in 5 steps:

a) Choose the branch points $\{x_i\}$ arbitrarily: there are $2(g+d-1)$ of them.
b) Choose a set of "cuts" joining x_i to a base point z:

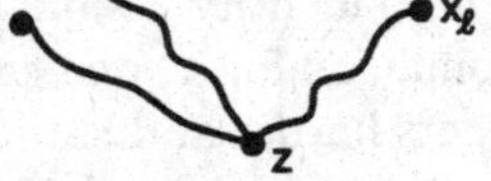

c) Choose $2(g+d-1)$ transpositions σ_i acting on $\{1,\ldots,d\}$ such that $(\sigma_1 \cdot \sigma_2 \cdot \ldots) = e$.
d) Make a topological covering space C_0 of $\mathbb{P}^1$ by glueing together d copies of $\mathbb{P}^1$ via the transposition σ_i on the ith cut.
e) By Riemann's existence theorem, C_0 has a unique algebraic structure, i.e., there is a unique curve C and map $\pi : C \to \mathbb{P}^1$ such that C is homeomorphic to C_0 as covering of $\mathbb{P}^1$.

Unfortunately, step b) is essentially topological and seems very deep from an algebraic point of view. For instance, if you want to algebraize this construction, you are led to ask: given prescribed branch points, cuts and transpositions, find an explicit multi-valued algebraic function with these branch points and transpositions. Thus if $d = 2$, $\sqrt{\prod(x - x_i)}$ is a function; if $d = 3$ or 4, the solvability of S_3, S_4 (the permutation groups) allows one to find such explicit functions too. But I don't know of any general method for larger d. We summarize these discussions in the table on the next page.

For general g, the simplest explicit polynomial presentation of C seems to be one due to K. Petri in a paper that was until recently almost forgotten. He was M. Noether's last student and collaborated with E. Noether and appears to have written only 2 papers. I want to conclude this lecture by describing his results in one of these published in 1922. This is unavoidably a bit messy, but just to be able to brag, I think it is a good idea to be able to say "I have seen *every* curve once."

Let C be a non-hyperelliptic curve of genus g. Petri starts by choosing g points $x_1, \ldots, x_g$ on C in a reasonably general position (we won't worry about this). Let $\varphi_1, \ldots, \varphi_g$ be a dual basis of differential forms, i.e.,

$$\begin{aligned} \varphi_i(x_j) &= 0 \;\text{ if }\; i \neq j \\ &\neq 0 \;\text{ if }\; i = j. \end{aligned}$$

Let $X_1, \ldots, X_g$ be the corresponding homogeneous coordinates in $\mathbb{P}^{g-1}$ for the canonical map $\Phi : C \to \mathbb{P}^{g-1}$. Also, if $3 \le i \le g$, write $\varphi_i = dt_i$, t_i a local coordinate at x_i and then expand

$$\begin{aligned} \varphi_1 &= \lambda_i t_i dt_i + \ldots \\ \varphi_2 &= \mu_i t_i dt_i + \ldots . \end{aligned}$$

(We may assume $\lambda_i \neq 0$ if $3 \le i \le g$.) Then Petri's first step is to write down a *basis* for the vector space of *k-fold* holomorphic differential forms on C for every k: these are differential forms $a(x)(dx)^k$ with no poles. For $k \ge 2$, they form a vector space of dimension $(2k-1)(g-1)$. The table below summarizes his results. Look at it carefully – each column displays a basis for k-fold differentials, $1 \le k \le 5$. Within each column however, we group the differentials in rows according to the multiplicity of their zeroes on $\mathfrak{A} \underset{\text{def}}{=} x_3 + \ldots x_g$. Thus the first row is always $\varphi_3^k, \ldots, \varphi_g^k$ as each of these has no zero at one of the x_i, whereas all other monomials in the φ_i's will be zero at least to 1st order at each point of $\mathfrak{A}$. The second column arises like this:

a) one checks that every quadratic differential which is 0 on $\mathfrak{A}$ is on the form $\varphi_1 \cdot (\,) + \varphi_2 \cdot (\,)$,
b) hence if $3 \leq i < j \leq g$, $\varphi_i \varphi_j$ can be rewritten as $\varphi_1 \cdot (\,) + \varphi_2 \cdot (\,)$.
c) Omitting these $\varphi_i \varphi_j$, the remaining $3g - 3$ monomials form a basis as indicated.

Table of representations of the general curve C of genus g

g	degree of map $\pi : C \to \mathbb{P}^1$	no. of branch points	degree d of plane curve $C_0 \subset \mathbb{P}^2$	no. double points δ of C_0	canonical curve	3δ	vs.	$\frac{(d+1)(d+2)}{2}$	
0	1	0	1	0	–	0	vs.	3	
1	2	4	3	0	–	0	vs.	10	
2	2	6	4	1	–	3	vs.	15	
3	3	10	4	0	$C_4 \subset \mathbb{P}^2$	0	vs.	15	
4	3	12	5	2	$C_6 \subset \mathbb{P}^3$	6	vs.	21	
5	4	16	6	5	$C_8 \subset \mathbb{P}^4$	15	vs.	28	
6	4	18	6	4	$C_{10} \subset \mathbb{P}^5$	12	vs.	28	
⋯	⋯	⋯	⋯						
10	6	30	9	18	$C_{18} \subset \mathbb{P}^9$	54	vs.	55	$\longleftarrow$
11	7	34	10	25	$C_{20} \subset \mathbb{P}^{10}$	75	vs.	66	$\longleftarrow$
⋯	⋯	⋯	⋯						
100	51	300	69	2178	$C_{198} \subset \mathbb{P}^{99}$				

The third column arises like this:

a) one checks the triple differentials $\varphi_1^2(\ )+\varphi_1\varphi_2(\ )+\varphi_2^2(\ )$ are of codimension 1 in the vector space of triple differentials ω, with double zeroes on $\mathfrak{A}$! This is a reflection of the "fundamental class on C": the condition for such an ω to be formed out of $\varphi_1^2, \varphi_1\varphi_2, \varphi_2^2$ alone is that

$$(*) \qquad \sum_{\substack{\text{all zeroes} \\ y \text{ of } \varphi_1 \\ \text{except } x_3,\ldots,x_g}} \operatorname{Res}_y(\omega/\varphi_1\varphi_2) = 0.$$

b) Writing η_i as indicated, this has a double zero on $\mathfrak{A}$ and every difference $\eta_i - \eta_j$ satisfies $(*)$.
c) Hence $\eta_i - \eta_j$ can be rewritten as indicated and this leaves exactly $5g-5$ remaining triple differentials as a basis.

The remaining columns are quite mechanical: the 2 ways of rewriting differentials reduce us to the attached list, and, by counting, leave us with exactly the right number to be a basis!

Let us write out the 2 sets of identities by which these reductions are made. They will be:

$$\begin{aligned} \varphi_i\varphi_j &= \sum_{k=3}^{g} \alpha_{ijk}(\varphi_1,\varphi_2)\,\varphi_k + \nu_{ij}\varphi_1\varphi_2 \\ \eta_i - \eta_j &= \sum_{k=3}^{g} \alpha'_{ijk}(\varphi_1,\varphi_2)\,\varphi_k + \nu'_{ij}\varphi_1^2\varphi_2 + \nu''_{ij}\varphi_1\varphi_2^2 \end{aligned}$$

(here the α is linear, the α' is quadratic, the ν's are scalars, and $3 \le i,j \le g$, $i \ne j$). But what this means in terms of equations in $\mathbb{P}^{g-1}$ is precisely that 2 sets of homogeneous equations:

$$\begin{aligned} f_{ij} &= x_i x_j - \sum_{k=3}^{g} \alpha_{ijk}(x_1,x_2)x_k - \nu_{ij}x_1x_2 \\ g_{ij} &= (\mu_i x_1 - \lambda_i x_2)x_i^2 - (\mu_j x_1 - \lambda_j x_2)x_j^2 \\ &\quad - \sum_{k=3}^{g} \alpha'_{ijk}(x_1,x_2)x_k - \nu'_{ij}x_1^2x_2 - \nu''_{ij}x_1x_2^2 \end{aligned}$$

of degrees 2 and 3 generate the ideal of C!

Petri now goes on to prove 3 beautiful results –

I) These equations are related by syzygies:
   a) $f_{ij} = f_{ji}$, $g_{ij} + g_{jk} = g_{ik}$

b) $$x_k f_{ij} - x_j f_{ik} + \sum_{\substack{\ell=3 \\ \ell \neq k}}^{g} \alpha_{ij\ell} f_{k\ell} - \sum_{\substack{\ell=3 \\ \ell \neq j}}^{g} \alpha_{ik\ell} f_{j\ell} = \rho_{ijk} g_{jk}$$

where $3 \leq i, j, k \leq g$, i, j, k distinct, and the ρ_{ijk}'s are scalars symmetric in i, j, and k.

II) There are 2 possibilities: either $\rho_{ijk} = \alpha_{ijk} = 0$ whenever i, j, k are distinct, and then C is very special – it is a triple covering of $\mathbb{P}^1$ or if $g = 6$ it may also be a non-singular plane quintic; or else most of the ρ's and α's are non-zero (precisely, one can write $\{3, \ldots, g\} = I_1 \cup I_2$ so that for all $j \in I_1$, $k \in I_2$, there exists an i with $\rho_{ijk} \neq 0$, $\alpha_{ijk} \neq 0$), and then the f_{ij}'s alone generate the ideal of C.

III) Given any set of f_{ij}'s, g_{ij}'s as above related by the syzygies in (I), where all $\lambda_i \neq 0$, and at least one $\rho_{ijk} \neq 0$, there exists a curve C of genus g whose canonical image in $\mathbb{P}^{g-1}$ is defined by these equations.

In my mind, (III) is the most remarkable: this means that we have a complete set of identities on the coefficients α, α', ν, ν', ν'', λ, μ, ρ characterizing those that give canonical curves. It would be marvellous to use this formidable and precise machine for applications.

Petri's basis for the canonical ring

	Here, use $\varphi_i\varphi_j = (-)\varphi_1 + (-)\varphi_2$ $3 \le i < j \le g$	Here, use $\eta_i - \eta_j = (-)\varphi_1^2$ $+(-)\varphi_1\varphi_2 + (-)\varphi_2^2$ $3 \le i < j \le g$	$\eta_i = (\mu_i\varphi_1 - \lambda_i\varphi_2)\varphi_i^2$ $3 \le i \le g$		
$\varphi_3, \ldots, \varphi_g$	$\varphi_3^2, \ldots, \varphi_g^2$	$\varphi_3^3, \ldots, \varphi_g^3$	$\varphi_3^4, \ldots, \varphi_g^4$	$\varphi_3^5, \ldots, \varphi_g^5$	*not zero on* $\mathfrak{A}$
φ_1, φ_2	$\varphi_1\varphi_i, \varphi_2\varphi_i$ $3 \le i \le g$	$\varphi_1\varphi_3^2, \ldots, \varphi_1\varphi_g^2$	$\varphi_1\varphi_3^3, \ldots, \varphi_1\varphi_g^3$	$\varphi_1\varphi_3^4, \ldots, \varphi_1\varphi_g^4$	*simple zero on* $\mathfrak{A}$
–	$\varphi_1^2, \varphi_1\varphi_2, \varphi_2^2$	η_3 $\varphi_1^2\varphi_i, \varphi_1\varphi_2\varphi_i, \varphi_2^2\varphi_i$ $3 \le i \le g$	$\varphi_1^2\varphi_3^2, \ldots, \varphi_1^2\varphi_g^2$	$\varphi_1^2\varphi_3^3, \ldots; \varphi_1^2\varphi_g^3$	*double zero on* $\mathfrak{A}$
–	–	$\varphi_1^3, \varphi_1^2\varphi_2, \varphi_1\varphi_2^2, \varphi_2^3$	$\varphi_1\eta_3, \varphi_2\eta_3$ $\varphi_1^3\varphi_i, \varphi_1^2\varphi_2\varphi_i, \varphi_1\varphi_2^2, \varphi_i$ $\varphi_2^3\varphi_i,\ 3 \le i \le g$	$\varphi_1^3\varphi_3^2, \ldots, \varphi_1^3\varphi_g^2$	*triple zero on* $\mathfrak{A}$
–	–	–	$\varphi_1^4, \varphi_1^3\varphi_2, \varphi_1^2\varphi_2^2, \varphi_1\varphi_2^3, \varphi_3^4$	$\varphi_1^2\eta_3, \varphi_1\varphi_2\eta_3, \varphi_2^2\eta_3$ $\varphi_1^4\varphi_i, \varphi_1^3\varphi_2\varphi_i, \varphi_1^2\varphi_2^2\varphi_i$ $\varphi_1\varphi_2^3\varphi_i, \varphi_2^4\varphi_i,\ 3 \le i \le g$	*4-fold zero on* $\mathfrak{A}$
–	–	–	–	$\varphi_1^5, \varphi_1^4\varphi_2, \varphi_1^3\varphi_2^2, \varphi_1^2\varphi_2^3, \varphi_1\varphi_2^4, \varphi_2^5$	*5-fold zero on* $\mathfrak{A}$
g *simple differentials*	$3g - 3$ *quadruple differentials*	$5g - 5$ *triple differentials*	$7g - 7$ *4-tuple differentials*	$9g - 9$ *5-tuple differentials*	

Lecture II: The Moduli Space of Curves: Definition, Coordinatization, and Some Properties

In the previous lecture, we studied each curve separately. We now want to discuss in its own right the space of *all* curves of genus g, which we denote by $\mathfrak{M}_g$. Also very important is the allied space:

$$\mathfrak{M}_{g,n} = \left\{ \begin{array}{l} \text{isomorphism classes of objects } (C, x_1, \ldots, x_n) \\ \text{where } C \text{ is a curve of genus } g \text{, and } x_1, \ldots, x_n \text{ are} \\ \text{distinct ordered points of } C. \end{array} \right\}$$

Let us begin as before by looking first at the simplest cases:

I) $\mathfrak{M}_{0,n} \cong \left[\mathbb{P}^1 - (0, 1, \infty)\right]^{n-3}$ – (all diagonals).

In fact, if we have n distinct points $x_1, \ldots, x_n \in \mathbb{P}^1$, a unique automorphism of $\mathbb{P}^1$ takes x_1 to 0, x_2 to 1, and x_3 to ∞. The remaining $n - 3$ are arbitrary except for being distinct and not equal to 0 or 1 or ∞.

II) $\mathfrak{M}_{1,0} = \mathfrak{M}_{1,1} \cong \mathbb{A}^1_j$ (the affine line[17] with coordinate j).

Because curves of genus 1 are groups, their automorphisms act transitively on them, hence $\mathfrak{M}_{g,0} = \mathfrak{M}_{g,1}$. To determine this space, recall that all such curves are isomorphic to one of the plane cubics C_λ, defined by

$$y^2 = x(x-1)(x-\lambda).$$

Equivalently, C_λ is the double cover of $\mathbb{P}^1$ ramified at $0, 1, \infty, \lambda$. One proves easily that $C_{\lambda_1} \approx C_{\lambda_2}$ if and only if there is an automorphism of $\mathbb{P}^1$ carrying $\{0, 1, \infty, \lambda_1\}$ (unordered sets) to $\{0, 1, \infty, \lambda_2\}$. This happens if and only if

$$\lambda_2 = \lambda_1,\ 1 - \lambda_1,\ 1/\lambda_1,\ (\lambda_1 - 1)/\lambda_1,\ \lambda_1/(\lambda_1 - 1), \text{ or } 1/(1 - \lambda_1)$$

[e.g., note that the map

$$(x, y) \longmapsto (1 - x, y)$$

carries C_λ to $C_{1-\lambda}$; and the map

$$(x, y) \longmapsto (1/x, y/x)$$

[17] Out of habit, I find it more comfortable to call affine n-space $\mathbb{A}^n$ instead of $\mathbb{C}^n$: because $\mathbb{A}^n$ also denotes affine n-space over other ground fields.

carries C_λ to $C_{1/\lambda}$].

One must cook up an expression in λ invariant under these substitutions and no more. It is customary to use:

$$j = 256\frac{(\lambda^2-\lambda+1)^3}{\lambda^2(\lambda-1)^2}.$$

(It is readily checked that this j is invariant under these 6 substitutions and since 6 = max(deg of numerator and denominator), no other λ's give the same j.)

We then get a bijection between the isomorphism classes of genus 1 curves C and the complex numbers $\mathbb{C}$ by taking C to $j(\lambda)$ if $C \approx C_\lambda$.

Analytically, if $C = \mathbb{C}/L$, the j-invariant of C can be calculated from L in the following way: define[18]

$$\begin{aligned} g_2 &= 60\cdot\sum_{\substack{\lambda\in L\\ \lambda\neq 0}} 1/\lambda^4 \\ g_3 &= 140\cdot\sum_{\substack{\lambda\in L\\ \lambda\neq 0}} 1/\lambda^6. \end{aligned}$$

Then it can be shown that:

$$j(C) = 1728\cdot g_2^3/\left(g_2^3-27g_3^2\right).$$

In particular, if $L = \mathbb{Z}+\mathbb{Z}\cdot\omega$, then $j(\omega) = j(\mathbb{C}/\mathbb{Z}+\mathbb{Z}\cdot\omega)$ is the famous elliptic modular function. Its most important property is its invariance under $SL(2,\mathbb{Z})$, which can be explained from a moduli point of view as follows:

$$\left\{\begin{array}{l}\exists\ \alpha\in\mathbb{C}\ \text{such that}\\ \alpha(\mathbb{Z}+\mathbb{Z}\cdot\omega_1) = \mathbb{Z}+\mathbb{Z}\cdot\omega_2\end{array}\right\} \Longleftrightarrow \left\{\begin{array}{l}\exists\ \left(\begin{smallmatrix}a&b\\c&d\end{smallmatrix}\right)\in SL(2,\mathbb{Z})\ \text{such that}\\ \omega_2 = \dfrac{a\omega_1+b}{c\omega_1+d}\end{array}\right\}.$$

(This is trivial to check.) But

$$\begin{array}{c}\left\{\begin{array}{l}\exists\ \text{isomorphism}\\ \mathbb{C}/\mathbb{Z}+\mathbb{Z}\cdot\omega_1\cong\mathbb{C}/\mathbb{Z}+\mathbb{Z}\cdot\omega_2\end{array}\right\} \Longleftrightarrow \left\{\begin{array}{l}\exists\ \alpha\in\mathbb{C}\ \text{such that}\\ \alpha(\mathbb{Z}+\mathbb{Z}\cdot\omega_1)=\mathbb{Z}+\mathbb{Z}\cdot\omega_2\end{array}\right\}\\ \Updownarrow \\ j(\mathbb{C}/\mathbb{Z}+\mathbb{Z}\cdot\omega_1) = j(\mathbb{C}/\mathbb{Z}+\mathbb{Z}\cdot\omega_2).\end{array}$$

Hence:

$$j(\omega_1)=\omega_2 \Longleftrightarrow \left\{\begin{array}{l}\exists\ \left(\begin{smallmatrix}a&b\\c&d\end{smallmatrix}\right)\in SL(2,\mathbb{Z})\ \text{such that}\\ \omega_2 = \dfrac{a\omega_1+b}{c\omega_1+d}\end{array}\right\}.$$

[18] The stream of funny constants can best be explained as making a certain Fourier expansion have *integral*, not just rational, coefficients. This makes the theory work well under "reduction modulo p".

III) $\mathfrak{M}_{2,0}$: this space was studied classically by Bolza among others and in recent years was analyzed completely by Igusa, and was attacked as follows: describe a curve C of genus 2 as a double cover of $\mathbb{P}^1$ ramified in 6 points $\lambda_1, \ldots, \lambda_6$. This sets up a bijection:

$$\left\{ \begin{array}{c} \text{Isom. classes of} \\ C \text{ of genus 2} \end{array} \right\} \cong \left\{ \begin{array}{l} \text{unordered distinct 6-tuples} \\ \lambda_1, \ldots, \lambda_6 \in \mathbb{P}^1 \text{ modulo automorphisms} \\ \text{of } \mathbb{P}^1, \text{ i.e., } PGL(2, \mathbb{C}) \end{array} \right\}.$$

Describe an unordered 6-tuple $\{\lambda_i\}$ by its homogeneous equation $f(X_0, X_1)$ of degree 6, a so-called binary sextic, and we arrive at the problem: find polynomial functions of the coefficients of a binary sextic $f(X_0, X_1)$ invariant under linear substitutions in X_0, X_1 of determinant one. This is a problem worked out by the classical invariant theorists. These invariant functions are then coordinates on $\mathfrak{M}_{2,0}$. Without going into any more detail, suffice it to say that the simplest way to describe the answer you get is:

$$\mathfrak{M}_{2,0} \cong \mathbb{A}^3 \Big/ \underbrace{\begin{array}{l} \text{modulo } \mathbb{Z}/5\mathbb{Z} \text{ acting by} \\ (x, y, z) \mapsto (\zeta^1 x, \zeta^2 y \cdot \zeta^3 z) \\ \text{where } \zeta^5 = 1 \end{array}}_{\begin{array}{l} \text{this, in turn, may be embedded in } \mathbb{A}^8 \\ \text{by the 8 functions} \\ x^5, x^3y, xy^2, y^5, x^2z, xz^3, z^5, yz \end{array}}$$

For all $g \geq 3$, $\mathfrak{M}_{g,0}$ *has never been explicitly described*! This rather discouraging fact does not mean that the other $\mathfrak{M}_{g,n}$'s have not been studied however. The lack of an explicit description is rather a challenge i) to find one and ii) to find the properties of $\mathfrak{M}_{g,n}$ even without such a description!

The first point to be made about $\mathfrak{M}_{g,n}$ in general is why you call it a "space" and expect it to be a variety in the first place. Recall that a *projective variety* $X \subset \mathbb{P}^n$ is defined to be the complete set of zeroes of a set of homogeneous polynomials f_i which generate a prime ideal $\wp \subset \mathbb{C}[X_0, \ldots, X_n]$, and that a *quasi-projective variety* $X \subset \mathbb{P}^n$ is defined to be the difference $\overline{X} - (Y_1 \cup \ldots \cup Y_n)$ where $\overline{X}, Y_i$ are projective varieties. We then say that a normal[19] quasi-projective variety $M_{g,n}$ *is* the moduli space if

i) we are given a bijection between $\mathfrak{M}_{g,n}$ and the set of points of $M_{g,n}$,
ii) for every algebraic family of curves of genus g with n distinct points, i.e., every "proper smooth morphism $\pi : X \to S$ of varieties whose fibres are curves of genus g[20], plus n disjoint sections $\sigma_i : S \to X$," the induced set-theoretic map $\phi : S \to M_{g,n}$ defined by

---

[19] This means that the affine coordinate rings of $M_{g,n}$ are integrally closed in their quotient field. This is a mild condition needed only for technical reasons.

[20] Again it is not essential to know in detail what these terms mean: the idea is to generalize, for instance, the family of curves $y^2 = x(x-1)(x-\lambda_1)(x-\lambda_2)(x-\lambda_3)$, which would represent an algebraic family of curves of genus 2 parametrized by $\mathbb{A}^3$.

$$\phi(s) = \left[\begin{array}{l} \text{pt. of } M_{g,n} \text{ corresponding via (i) to the curve} \\ \pi^{-1}(s), \text{ and points } \sigma_i(s) \end{array} \right]$$

is a morphism of varieties.

It is not hard to show that any 2 such $M_{g,n}$, $M'_{g,n}$ are canonically isomorphic as varieties: hence we may speak of *the* variety $\mathfrak{M}_{g,n}$. It is a non-trivial theorem however that such a variety $M_{g,n}$ exists at all.

The second point is to explain the relationship between $\mathfrak{M}_{g,n}$ and the Teichmüller space $\mathfrak{T}_{g,n}$. Define

$$\Pi = \left\{ \begin{array}{l} \text{free group on } 2g+n \text{ generators } A_1,\ldots,A_g,B_1,\ldots,B_g, \\ C_1,\ldots,C_n \text{ mod one relation} \\ A_1B_1A_1^{-1}B_1^{-1}\ldots A_gB_gA_g^{-1}B_g^{-1}C_1\ldots C_n = e \end{array} \right\}.$$

Define set-theoretically:

$$\mathfrak{T}_{g,n} = \left\{ \begin{array}{l} \text{set of objects } (C,\alpha,x_1,\ldots,x_n), \text{ where } C \text{ is a} \\ \text{curve of genus } g, \ x_1,\ldots,x_n \text{ are distinct points} \\ \text{of } C \text{ and} \\ \qquad \alpha : \Pi \xrightarrow{\approx} \pi_1(C - \{x_1,\ldots,x_n\}) \\ \text{is an isomorphism such that } \alpha(C_i) \text{ is freely} \\ \text{homotopic to a small loop around } x_i \text{ in positive} \\ \text{sense [and if } n=0, \alpha \text{ is ``orientation preserving,''} \\ \text{e.g., via the intersection pairing (.),} \\ (\alpha(A_1)\cdot\alpha(B_1)) = +1], \ \textit{modulo} \ (C,\alpha,x) \sim (C',\alpha',x') \\ \text{if there is an isomorphism } \phi : C \xrightarrow{\approx} C' \text{ such that} \\ \phi(x_i) = x_i' \text{ and such that } (\phi_*)\circ\alpha \text{ differs from } \alpha' \\ \text{by an inner automorphism} \end{array} \right\}.$$

Via the deformation theory of compact complex manifolds, it is easy to put a complex structure on $\mathfrak{T}_{g,n}$: this is the Teichmüller space. It is a deep theorem that $\mathfrak{T}_{g,n}$ is, in fact, a bounded, holomorphically convex domain in $\mathbb{C}^{3g-3+n}$. Let

$$\Gamma_{g,n} = \left\{ \begin{array}{l} \text{automorphisms } \sigma \text{ of } \Pi \text{ such} \\ \text{that } \sigma(C_i) = \text{ conjugate of } C_i \\ \text{(and if } n=0, \sigma \text{ is orientation} \\ \text{preserving in a suitable sense)} \end{array} \right\} \Big/ \begin{array}{l} \text{Inner} \\ \text{automorphisms} \end{array}.$$

Then it follows easily that $\Gamma_{g,n}$ acts discontinuously on $\mathfrak{T}_{g,n}$ via

$$(C,\alpha,x_i) \longmapsto (C,\alpha\circ\sigma,x_i)$$

and that

$$\mathfrak{M}_{g,n} \cong \mathfrak{T}_{g,n}/\Gamma_{g,n}.$$

In the case $g = n = 1$, we just have again the situation mentioned above: viz.

$$\begin{aligned} \mathfrak{J}_{1,1} &\cong \{\omega \in \mathbb{C} \mid \operatorname{Im} \omega > 0\} \\ \Gamma_{1,1} &\cong SL(2, \mathbb{Z})/(\pm I) \\ \mathfrak{M}_{1,1} &\cong \{j \in \mathbb{C}\} = \mathbb{A}^1_j. \end{aligned}$$

In fact, given $\omega \in \mathbb{C}$, define (C, α, x_1) as follows –

$$\begin{aligned} C &= \mathbb{C}/\mathbb{Z} + \mathbb{Z} \cdot \omega \\ x_1 &= \text{image of } 0 \end{aligned}$$

and if we let the image y of $1/2 \in \mathbb{C}$ be the base point C, define $\alpha : \Pi \xrightarrow{\approx} \pi_1(C - x_1, y)$ by

$\alpha(A_1)$ = loop in C obtained by projecting:

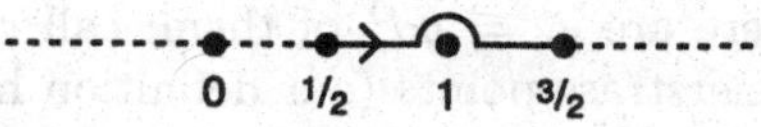

$\alpha(B_1)$ = loop in C obtained by projecting:

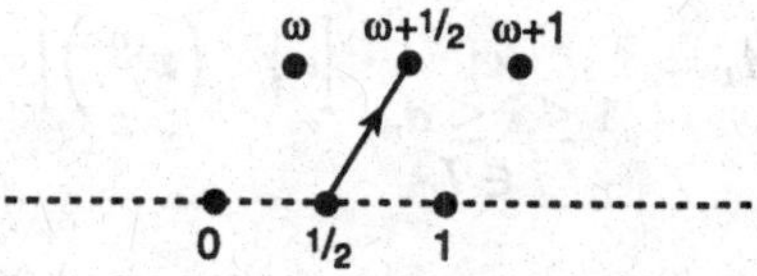

The third point we want to discuss is how one proves that $\mathfrak{M}_{g,n}$ is, in fact, a quasi-projective variety, i.e., how one finds global homogeneous coordinates for $\mathfrak{M}_{g,n}$. To tie this in, for instance, with Petri's approach in Lecture I, one can view his ideas as leading to coordinates on some Zariski-open subset $U \subset \mathfrak{M}_{g,g}$: (i.e., not on all of $\mathfrak{M}_{g,g}$ because the curve C had to be non-hyperelliptic and the g auxiliary points $x_1, \ldots, x_g$ had to be carefully chosen not in too special a position). In general, the hard part of this problem is to make the coordinates work everywhere on $\mathfrak{M}_{g,n}$ and not just on a Zariski-open U however. These coordinates can be viewed as automorphic forms on the Teichmüller space $\mathfrak{J}_{g,n}$ with respect to the Teichmüller modular group $\Gamma_{g,n}$; however this approach to their construction has not been pursued. I know of 3 methods to obtain coordinates:

I. via "theta-null werte,"
II. via the cross-ratios of the higher Weierstrass points,
III. via invariants of the Chow form.

The first method will be discussed in Lecture IV, and we will pass over it for now.

Method II is like this: let C by any curve of genus g. For any $n \geq 3$, let $R_n(C)$ be the vector space of n-fold differential forms with no poles – it has dimension $d_n = (2n-1)(g-1)$ – and let $\omega_i^{(n)}$, $1 \leq i \leq d_n$, be a basis. Define

$$\Phi_n : C \longrightarrow \mathbb{P}^{d_n - 1}$$

by

$$x \longmapsto \left(\omega_1^{(n)}(x), \ldots, \omega_{d_n}^{(n)}(x)\right)$$

just like the usual canonical embedding. Regardless of whether C is hyperelliptic or not, these are all projective embeddings of C. On $\Phi_n(C)$, there is a finite set of points x of hyperosculation, i.e., points where for some hyperplane H, H touches $\Phi_n(C)$ at x with order $\geq d_n$. Allowing these x to be counted with suitable multiplicity, there are $e_n = gd_n^2$ of them: call them $x_i^{(n)}$, $1 \leq i \leq e_n$. These are the n-fold Weierstrass points (our definition here is slightly different from that of Lecture I, but is equivalent). Consider the $e_n \times d_n$-matrix giving the coordinates of the Weierstrass points:

$$\left(\omega_i^{(n)}\left(x_j^{(n)}\right)\right).$$

For every $I \subset \{1, \ldots, e_n\}$, $\#I = d_n$, consider the minor:

$$M_I = \det_{\substack{1 \leq i \leq d_n \\ j \in I}} \left[\omega_i^{(n)}\left(x_j^{(n)}\right)\right].$$

Note that the M_I's are not numbers, but rather products of differential forms at the various points $x_j^{(n)}$, $j \in I$. Now for large N look at monomials in these minors:

$$M_r = \prod_I M_I^{r_I}$$

where $r_I \geq 0$ and $\sum_{i \in I} r_I = N$ for all i. Then these monomials are products over *all* $x_j^{(n)}$ of nN-fold differentials at $x_j^{(n)}$. It follows that although the M_r's are not complex numbers, their *ratios* are! Or if there are μ possible choices of exponents r_I satisfying $r_I \geq 0$ and $\sum_{i \in I} r_I = N$, the set of values M_r, as r varies, is a well-defined point in $\mathbb{P}^{\mu-1}$. Finally we must symmetrize under permutations of the $x_j^{(n)}$ which are not naturally ordered:

$$M'_r = \sum_{\substack{\text{perm. } \sigma \\ \text{of } \{1,\ldots,e_n\}}} \prod_I M_I^{r_{\sigma(I)}}.$$

Then the ratios M'_{r_1}/M'_{r_2} *depend only on* C, and not on the bases $\omega_i^{(n)}$ or on the ordering of the $x_j^{(n)}$'s. Thus we get

$$(\dots, M'_r(C), \dots) \in \mathbb{P}^{\mu-1}$$

depending only on C. One proves a) that not all $M'_r(C)$ are zero, and b) that if $C_1 \not\approx C_2$, $M'_r(C_1)$ is not proportional to $M'_r(C_2)$, all r. Thus we have coordinates on $\mathfrak{M}_{g,0}$. $\mathfrak{M}_{g,n}$ is very similar.

Method III is not so explicit. In general, for any curve $C \subset \mathbb{P}^m$, we can describe C by its "Chow form": let $X_0, \dots, X_m$ be coordinates on $\mathbb{P}^m$ and consider 2 hyperplanes: H_u defined by $\sum u_i X_i = 0$ and H_v defined by $\sum v_i X_i = 0$. Then it turns out that there is *one* equation $F_C(u; v)$ such that

$$F_C(u; v) = 0 \Longleftrightarrow C \cap H_u \cap H_v \neq \emptyset.$$

F_C is called the Chow form of C and it determines C. (For curves in $\mathbb{P}^3$, this idea goes back to Cayley.) Consider the Chow form $F_{\Phi_n(C)}$. This depends on C and on the choice of basis $\omega_i^{(n)}$ of $R_n(C)$. However, changing the basis $\left\{\omega_i^{(n)}\right\}$ changes the Chow form $F_{\Phi_n(C)}(u; v)$ by the contragrediant linear substitution in u and v. Writing out

$$F(u, v) = \sum F_{\alpha\beta} u^\alpha v^\beta,$$

this means that there is a natural representation of $SL(d_n, \mathbb{C})$ in the space of forms F or of the space of coefficients $F_{\alpha\beta}$. One proves that there are "enough" invariant polynomials $c_i(F_{\alpha\beta})$ so that

a) for each curve C, at least one $c_i\left(F_{\Phi_n(C),\alpha\beta}\right)$ is not zero, and
b) if $C_1 \not\approx C_2$, then the set of numbers $c_i\left(F_{\Phi_n(C),\alpha\beta}\right)$ is not proportional to $c_i\left(F_{\Phi_n(C_2),\alpha\beta}\right)$. Thus again the map

$$C \longmapsto \left(\dots, c_i\left(F_{\Phi_n(C),\alpha\beta}\right), \dots\right)$$

embeds $\mathfrak{M}_g$ into projective space.

The fourth point I want to make about $\mathfrak{M}_{g,n}$ is that although it is not compact, because a sequence of curves may "degenerate", $\mathfrak{M}_{g,n}$ has a natural compactification $\overline{\mathfrak{M}}_{g,n}$ obtained by casting out your net further and attempting to make into a moduli space not only the non-singular curves, but also some singular ones too. In fact one looks at curves $C \subset \mathbb{P}^n$ which may have "ordinary double points" and may even have several components. To be precise, we mean either

a) that as an analytic set, C is connected and everywhere is isomorphic locally either to the unit disc Δ, or to 2 copies of the unit disc $\Delta_1 \cup \Delta_2$ crossing transversely

or equivalently

b) that in the Zariski topology, C is connected and everywhere is defined locally
either by $n-1$ equations $f_1, \ldots, f_{n-1}$ with independent differentials df_i or by
$n-1$ equations $g, f_2, \ldots, f_{n-1}$ where g vanishes to second order with leading
term (x, y) and the f_i's vanish only to 1st order, with $dx, dy, df_2, \ldots, df_{n-1}$
all independent.

For instance, we could take 2 non-singular curves and let them cross transversely
at one or more points; or we could take 1 non-singular curve and map it to
$\mathbb{P}^n$ so that it crosses itself transversely at one or more points. Or we combine
both operations! Then $\overline{\mathfrak{M}}_{g,n}$ is to be the space of objects $(C, x_1, \ldots, x_n)$, up to
isomorphism, where C is a projective curve with only ordinary double points as
defined above and the x_i are distinct non-singular points of C and g is the sum of
genuses of the components of C treated as non-singular curves, plus the number
of double points, minus the number of components, plus one: $g = \sum(g_i - 1) + \delta + 1$;
and finally if any component C_0 of C is isomorphic to $\mathbb{P}^1$, then there are at least
3 points of C_0 which are x_i's or where C_0 meets other components of C. It is a
theorem that $\overline{\mathfrak{M}}_{g,n}$ is, in a natural way, a projective variety, esp. it is compact.

The last topic I would like to discuss at some length is the curious ambivalence
in the variety $\mathfrak{M}_{g,n}$ to be in various senses somehow hyperbolic on the one hand,
yet in other senses it wants to be elliptic. To explain this, it's best to go back
first to $\mathfrak{M}_{1,1}$. We can factor the map:

$$
\begin{array}{ccc}
\mathfrak{I}_{1,1} & \longrightarrow & \mathfrak{M}_{1,1} \\
\| & & \| \\
H & & \mathbb{A}_j^1 \cong H/SL(2, \mathbb{Z})
\end{array}
$$

by considering subgroups $\Gamma \subset SL(2, \mathbb{Z})$ of finite index:

$$H \longrightarrow H/\Gamma \longrightarrow H/SL(2, \mathbb{Z}).$$

The curves H/Γ are finite coverings of $\mathfrak{M}_{1,1}$ and are called "higher level" moduli
spaces: I'll denote H/Γ by $\mathfrak{M}_{1,1}^\Gamma$. It too can be naturally compactified by adding
a finite set of points; so we get finally the diagram

$$
\begin{array}{ccc}
& \mathfrak{I}_{1,1} & \\
\alpha & \downarrow & \\
\mathfrak{M}_{1,1}^\Gamma & \subset & \overline{\mathfrak{M}}_{1,1}^\Gamma \\
\downarrow & & \downarrow \beta \\
\mathfrak{M}_{1,1} & \subset & \overline{\mathfrak{M}}_{1,1}
\end{array}
$$

Now of course all curves lie in 3 classes:

Elliptic Class:	$g = 0$;	admits positively curved metric;	no holo. k-forms
Parabolic Class:	$g = 1$;	admits flat metric;	*one* holo. k-form for each k
Hyperbolic Class:	$g \geq 2$;	admits negatively curved metric;	*lots* of holo. k-forms giving proj. embedding

The point is that $\overline{\mathfrak{M}}_{1,1}$ is $\mathbb{P}^1$, hence is elliptic, while if Γ is moderately small, $\overline{\mathfrak{M}}^{\Gamma}_{1,1}$ is hyperbolic. The reason this flip is possible is that β is ramified: in fact there are 2 finite points $j = 0$ and $j = 12^3$ at which $\mathfrak{I}_{1,1} \longrightarrow \mathfrak{M}_{1,1}$ is respectively triply and doubly ramified, and 1 infinite point $j = \infty$ over which the β's are arbitrarily highly ramified. From another point of view, $\mathfrak{I}_{1,1}$ admits a canonical metric with negative curvature, i.e., $ds^2 = dx^2 + dy^2/y^2$, (if $z = x + iy \in H \simeq \mathfrak{I}_{1,1}$ is the coordinate). This induces a negatively curved metric on each $\mathfrak{M}^{\Gamma}_{1,1}$. In this metric, $\mathfrak{M}^{\Gamma}_{1,1}$ has finite volume, but the metric has singularities, a) at points where α is ramified and b) at points of $\overline{\mathfrak{M}}^{\Gamma}_{1,1} - \mathfrak{M}^{\Gamma}_{1,1}$. (If Γ is small enough, α will be unramified and only (b) occurs.)

It is this constellation of facts that to some extent generalizes to $\mathfrak{M}_{g,n}$. In our present state of knowledge, the generalization is very partial. To begin with, we get the same diagram:

$$
\begin{array}{ccc}
\mathfrak{I}_{g,n} & & \\
\alpha \downarrow & & \\
\mathfrak{M}^{\Gamma}_{g,n} & \subset & \overline{\mathfrak{M}}^{\Gamma}_{g,n} \\
\downarrow & & \downarrow \beta \\
\mathfrak{M}_{g,n} & \subset & \overline{\mathfrak{M}}_{g,n}
\end{array}
$$

for each $\Gamma \subset \Gamma_{g,n}$ of finite index. Let me begin with the known elliptic-type properties which are unfortunately weak: we assume $n = 0$ for simplicity.

a) Assume also $g \geq 4$ for simplicity[21]. Then the singular set $S \subset \mathfrak{M}_g$ is the set of points of $\mathfrak{M}_g$ where $\mathfrak{I}_g \to \mathfrak{M}_g$ ramifies and is the set of points corresponding to curves C with automorphisms. Then $B_1(\mathfrak{M}_g - S)$, the first betti number, is zero, hence so is B_1 of $\mathfrak{M}_g$, $\overline{\mathfrak{M}}_g$ and *any non-singular blow-up* $\mathfrak{M}^*_g$ of $\overline{\mathfrak{M}}_g$. This means, e.g., that the so-called Albanese variety of $\mathfrak{M}^*_g$ is trivial.

b) $\mathfrak{M}_g$ has lots of rational curves in it. In fact for any algebraic surface X and rational function f on X, let $C_t \subset X$ be the curve $f(x) = t$, and let $[C_t] \in \mathfrak{M}_g$ denote the corresponding point. Then

$$ t \longmapsto [C_t] $$

---

[21] If $g = 2$ or 3, $\mathrm{Sing}(\mathfrak{M}_g) \subsetneq$ (Ram.Pts. of $\mathfrak{I}_g \to \mathfrak{M}_g) \subseteq \{C$ with automorphisms$\}$. Always $B_1(\mathfrak{M}_g - \mathrm{Sing}\,\mathfrak{M}_g) = 0$, hence $B_1(\mathfrak{M}_g) = B_1(\overline{\mathfrak{M}}_g) = B_1(\mathfrak{M}^*_g) = 0$.

is a morphism

$$\mathbb{P}^1 \longrightarrow \overline{\mathfrak{M}}_g.$$

c) If $g \leq 10$, $\mathfrak{M}_g$ has the much stronger property of being *unirational.* This means equivalently that the field $\mathbb{C}(\mathfrak{M}_g)$ of rational functions is a *subfield* of $\mathbb{C}(t_1, t_2, \ldots, t_n)$ for some n or that there is a Zariski-open set $U \subset \mathbb{A}^n$ and a morphism $f : U \to \mathfrak{M}_g$ with dense image[22]. In terms of moduli, $\mathfrak{M}_g$ being unirational means that one can write down a family of curves of genus g depending on parameters $t_1, \ldots, t_n$ which can be arbitrary complex numbers satisfying some inequalities $f_i(t) \neq 0$, such that "almost all" curves of genus g appear in the family: e.g., if $g = 2$, take the family

$$y^2 = x^5 + t_1 x^4 + t_2 x^3 + t_3 x^2 + t_4 x + t_5$$

and if $g = 3$, take the family

$$y^4 + y^3 (t_1 x + t_2) + y^2 (t_3 x^2 + t_4 x + t_5) + y (t_6 x^3 + t_7 x^2 + t_4 x + t_8) + (t_9 x^4 + t_{10} x^3 + t_{11} x^2 + t_{12} x + t_{13}) = 0.$$

In fact, if $g \leq 10$ we may use the remarks in Lecture I about realizing curves as plane curves with double points to write down a family of plane curves of degree $d = \left[\frac{2g+8}{3}\right]$ with free parameters almost all of which represent curves of genus g and which include almost all curves of genus g.

Whether more $\mathfrak{M}_g$'s, $g \geq 11$, are unirational or not is a very interesting problem, but one which looks very hard too, especially if g is quite large. Now consider the hyperbolic tendencies of $\mathfrak{M}_{g,n}$. First of all, we can put 2 types of metric on $\mathfrak{I}_{g,n}$: one of these is the famous *Teichmüller metric* ρ_T. This is a Finsler metric, so it's a bit messy. However, it equals the Kobayashi metric of $\mathfrak{I}_{g,n}$, so all holomorphic maps $f : \Delta \to \mathfrak{I}_{g,n}$ are distance decreasing for ρ_T and the Poincaré metric on Δ : a hyperbolic property. Its unit balls have been determined and are quite amazingly wrinkled and creased: this led Royden to prove the rigidity theorem that if $\dim \mathfrak{I}_{g,n} > 1$, $\Gamma_{g,n} = \text{Aut}\,(\mathfrak{I}_{g,n})$; esp. $\mathfrak{I}_{g,n}$ is not at all a homogeneous domain in $\mathbb{C}^{3g-3+n}$. On the other hand, in this funny metric, $\mathfrak{I}_{g,n}$ is a straight space in the sense of Busemann, i.e., has unique indefinitely prolongable geodesics, but contrary to a conjecture does not have negative curvature in Busemann's sense (this fly in the ointment shows that my general picture is not entirely accurate!). $\mathfrak{I}_{g,n}$ carries another metric ρ_{P-W}, the *Peterson-Weil* metric[23], which is a Kähler metric, hence locally much nicer. Moreover, it has strictly negative Ricci curvature and holomorphic sectional curvatures. In particular, holomorphic maps $f : \Delta \to \mathfrak{I}_{g,n}$ will also be distance decreasing for ρ_{P-W} (suitably normalized) by the Ahlfors-Pick lemma. All the spaces $\mathfrak{M}_{g,n}$ inherit both metrics (with possible singularities where $\alpha : \mathfrak{I}_{g,n} \to \mathfrak{M}^{\Gamma}_{g,n}$ is ramified), and, esp. with ρ_{P-W}, this makes them rather hyperbolic. A closely related hyperbolic property of $\mathfrak{M}_{g,n}$ is:

[22] If this holds, one can assume $n = 3g - 3$ by restricting f_n to $U \cap L$, L a sufficiently general $3g - 3$-dimensional subspace of $\mathbb{A}^n$; hence $\mathbb{C}(\mathfrak{M}_g) \subset \mathbb{C}(t_1, \ldots, t_{3g-3})$ with finite index too.

[23] If $n > 0$, more precisely, there is a family of $P - W$ metrics depending on assigning branch numbers σ_i, $2 \leq \sigma_i \leq \infty$, to the base points x_i.

The Rigidity Theorem of Arakelov-Paršin-Manin-Grauert. *(also called the "Šafarevitch-Mordell conjecture in the function field case"). Fix $g \geq 2$ and let C be any curve, S a finite set of points of C. Then there are only finitely many families of curves of genus g over $C - S$, i.e.,*

$$\pi : X \longrightarrow C - S$$

which are "non-constant" (i.e., the fibres $\pi^{-1}(s)$ not all isomorphic), and if

$$2(\text{genus } C) - 2 + \#S \leq 0$$

there are none at all; moreover, for each such family there are only finitely many sections, and even for "constant" families, there are only finitely many non-constant sections.

Corollary. *Fix g, n, C, S as above. Then there are only finitely many non-constant morphisms*

$$\phi : C - S \longrightarrow \mathfrak{M}_{g,n}$$

which are locally liftable to $\mathfrak{I}_{g,n}$: i.e., if $x \in C - S$, and $\phi(x)$ is a ramification point for $\mathfrak{I}_{g,n} \to \mathfrak{M}_{g,n}$, one asks that in a small neighborhood of x, ϕ factor through $\mathfrak{I}_{g,n}$.

A sketch of the proof is given in an appendix below. Finally I want to conclude by giving a conjecture which I am hopeful will very soon be a theorem!

Conjecture. For each g, n, there is a $\Gamma_0 \subset \Gamma_{g,n}$ of finite index such that for all $\Gamma \subset \Gamma_0$, $\mathfrak{M}_{g,n}^{\Gamma}$ is a variety of general type in Kodaira's sense.

Here "general type" for a variety X of dimension n means that you compactify X to $\bar{X}$, then blow-up $\bar{X}$ to X^* which is non-singular, and then you look for differential forms of type

$$\omega = a(x) \left(dx_1 \wedge \ldots \wedge dx_n \right)^k$$

on X^*, with no poles. It means that if k is large enough, you can find $n + 2$ such forms whose ratios generate the field of rational functions $\mathbb{C}(X)$ on X. Since on unirational varieties, there are no non-zero differential forms of any type, the conjecture means that for Γ small, $\mathfrak{M}_{g,n}^{\Gamma}$ is more or less the opposite of being unirational.

[Added in 1997 edition] After these notes were written, a remarkable fact was discovered: $\mathfrak{M}_g$ itself, for $g \geq 24$, is of general type. In other words, one finds in the sequence of spaces $\mathfrak{M}_g$ this transition from elliptic to hyperbolic. For $g \leq 13$, $\mathfrak{M}_g$ is unirational, $\mathfrak{M}_{15}$ has Kodaira dimension $-\infty$ (no such ω's, any k), $\mathfrak{M}_{23}$ has positive Kodaira dimension (2 such ω's for some k) and general type thereafter.

Appendix: The idea of the proof of rigidity

The proof has 2 steps. The first consists in showing that the set of all families $\pi : X \to C - S$, and the set of all sections $s : C - S \to X$ of families π, *itself consists in a finite number of families.* The second consists in showing that given one $\pi : X \to C - S$ or one section $s : C - S \to X$ of such a π, then one cannot *deform* π or s, i.e., that the only families π or s lie in are 0-dimensional. Since a finite number of 0-dimensional families is just a finite set, we are done.

To carry out the first step, one can use an explicit projective embedding of $\overline{\mathfrak{M}}_{g,n}$, and for all $\phi : C - S \to \mathfrak{M}_{g,n}$ with $\phi(C - S) \neq$ point, extend ϕ to $\overline{\phi} : C \to \overline{\mathfrak{M}}_{g,n}$ and seek a bound on degree $\overline{\phi}(C)$. Then by general results, the set of morphisms $\phi : C - S \to \mathfrak{M}_{g,n}$ with degree $\overline{\phi}(C)$ bounded can be grouped into a finite number of nice families, the parameter space of each of which is some auxiliary variety. Equivalently, this means take a particular ample line bundle L on $\overline{\mathfrak{M}}_{g,n}$ and seek a bound on $c_1(\overline{\phi}^* L)$. (In fact, the nicest line bundle to pick is not quite ample, but near enough to make the proof go through: we will ignore details like this here.) Choosing a nice L, the next step is to identify $\overline{\phi}^* L$ from the geometry of the family $\pi : X \to C - S$ and the section s. One extends the family π of non-singular curves over $C - S$ to a larger family

$$\overline{\pi} : \overline{X} \longrightarrow C$$

over C of curves, some of which have double points (as in the definition of $\overline{\mathfrak{M}}_{g,n}$). Then it turns out that for the most natural L on $\overline{\mathfrak{M}}_{g,0}$,

$$\overline{\phi}^* L \cong \Lambda^g \overline{\pi}_* \left(\widetilde{\Omega}_{\overline{X}/C} \right)$$

where $\widetilde{\Omega}_{\overline{X}/C}$ denotes the line bundle whose sectinons are differential forms on the curves $\pi^{-1}(s)$, i.e., the cotangent bundle to the fibres of π, except that where $\pi^{-1}(s)$ has a double point, the forms may have simple poles with opposite residues at the 2 branches of $\pi^{-1}(s)$ at this double point. If one is dealing with n sections s_i too, hence a morphism $\phi : C - S \to \mathfrak{M}_{g,n}$ with $n > 0$, then $\overline{\phi}^* L$ is a tensor product of powers of this bundle and the line bundles

$$\overline{s}_i^* \widetilde{\Omega}_{\overline{X}/C}$$

where $\overline{s}_i : C \to \overline{X}$ is the extension of s_i. Now, in fact, by using the theory of algebraic surfaces, one gets a very good bound:

$$c_1 \left(\Lambda^g \overline{\pi}_* \left(\widetilde{\Omega}_{\overline{X}/C} \right) \right) \leq \left(q - 1 + \frac{s}{2} \right) (g - g_0)$$

where

$$\begin{aligned} q &= \text{genus } C \\ s &= \#S \\ g_0 &= \text{dimension of biggest abelian variety} \\ &\quad\ \text{which appears in the Jacobian of every} \\ &\quad\ \text{curve } \pi^{-1}(s) \text{ of the family.} \end{aligned}$$

$c_1\left(\overline{s}^*\widetilde{\Omega}_{\overline{X}/C}\right)$ seems harder to bound: I don't know a nice small explicit bound. However, following Grauert one can show that one exists by showing first that the cotangent bundle $\Omega^1_{\overline{X}}$ (of rank 2) is *ample* on almost all fibres $\pi^{-1}(s)$ of $\overline{X}$ over C and then applying general results on ample vector bundles. A good explicit bound here would be very interesting.

To carry out the second step, one applies Kodaira-Spencer-Grothendieck deformation theory to calculate the vector space of *infinitesimal* deformations of $\pi : X \to C - S$ and of $s : C - S \to X$. More precisely, one looks at deformations of $\overline{X}$ such that the map $\pi : \overline{X} \to C$ extends to this deformation and all singular fibres remain concentrated in $\pi^{-1}(S)$. It turns out that:

$$\left\{\begin{array}{l}\text{Space of}\\ \text{infinitesimal}\\ \text{deformations of}\\ \pi : X \longrightarrow C - S\end{array}\right\} \cong H^1\left(\overline{X}, \widetilde{\Omega}^{-1}_{\overline{X}/S}\right)$$

and

$$\left\{\begin{array}{l}\text{Space of}\\ \text{infinitesimal}\\ \text{deformations of}\\ s : C - S \longrightarrow X\end{array}\right\} \cong H^0\left(C, \overline{s}^*\widetilde{\Omega}^{-1}_{\overline{X}/C}\right).$$

To show these vector spaces are (0), one shows – and this is Arakelov's deepest contribution – that $\widetilde{\Omega}_{\overline{X}/C}$ is an ample line bundle on $\overline{X}$. Then the first space is (0) by Kodaira's Vanishing Theorem, and the second space is (0) because the line bundle has negative degree. Amazingly, Arakelov's proof here involves studying the curve $D \subset \overline{X}$ such that

$$D \cap \pi^{-1}(s) = \text{ the Weierstrass points of } \pi^{-1}(s)$$

and identifying via differentials the line bundle on $\overline{X}$ of which D is a section.

Lecture III: How Jacobians and Theta Functions Arise

I would like to begin by introducing Jacobians in the way that they actually were discovered historically. Unfortunately, my knowledge of 19th-century literature is very scant so this should not be taken too literally. You know the story began with Abel and Jacobi investigating general algebraic integrals

$$I = \int f(x)dx$$

where f was a multi-valued algebraic function of X, i.e., the solution to

$$g(x, f(x)) \equiv 0, \qquad g \text{ polynomial in 2 variables.}$$

So we can write I as

$$I = \int_\gamma y \; dx$$

where γ is a path in plane curve $g(x, y) = 0$; or we may reformulate this as the study of integrals

$$I(a) = \int_{a_0}^{a} \overbrace{\frac{P(x,y)}{Q(x,y)}\, dx}^{\omega}, \qquad \begin{array}{l} P, Q \text{ polynomials} \\ a, a_0 \in \text{ plane curve } C : g(x,y) = 0 \end{array}$$

of *rational* differentials ω on plane curves C. The main result is that such integrals always admit an addition theorem: i.e., there is an integer g such that if a_0 is a base point, and $a_1, \ldots, a_{g+1}$ are any points of C, then one can determine up to permutation $b_1, \ldots, b_g \in C$ rationally in terms of the a's[24] such that

$$\int_{a_0}^{a_1} \omega + \ldots + \int_{a_0}^{a_{g+1}} \omega \equiv \int_{a_0}^{b_1} \omega + \ldots + \int_{a_0}^{b_g} \omega, \text{ mod periods of } \int \omega.$$

For instance, if $C = \mathbb{P}^1$, $\omega = dx/x$, then $g = 1$ and:

$$\int_1^{a_1} \frac{dx}{x} + \int_1^{a_2} \frac{dx}{x} = \int_1^{a_1 a_2} \frac{dx}{x}.$$

Iterating, this implies that for all $a_1, \ldots, a_g, b_1, \ldots, b_g \in C$, there are $c_1, \ldots, c_g \in C$ depending up to permutation rationally on the a's and b's such that

[24] E.g., one can find polynomials $g_i(x, y; a)$ in x, y and the coordinates of the a's such that the b_i's are the set of all $b \in c$ such that $g_i(b; a) = 0$.

$$\sum_{i=1}^{g} \int_{a_0}^{a_i} \omega + \sum_{i=1}^{g} \int_{a_0}^{b_i} \omega \equiv \sum_{i=1}^{g} \int_{a_0}^{c_i} \omega \qquad \text{(mod periods).}$$

Now this looks like a group law! Only a very slight strengthening will lead us to a reformulation in which this most classical of all theorems will suddenly sound very modern. We introduce the concept of an algebraic group G: succinctly, this is a "group object in the category of varieties," i.e., it is simultaneously a variety and a group where the group law $m : G \times G \to G$ and the inverse $i : G \to G$ are morphisms of varieties. Such a G is, of course, automatically a complex analytic Lie group too, hence it has a Lie algebra Lie(G), and an exponential map exp : Lie$(G) \to G$. Now I wish to rephrase Abel's theorem as asserting that if C is a curve, and ω is any rational differential on C, then the multi-valued function

$$a \longmapsto \int_{a_0}^{a} \omega$$

can be factored into a composition of 3 functions:

$$C - (\text{poles of } \omega) \xrightarrow{\phi} J \overset{\exp}{\twoheadleftarrow} \text{Lie } J \xrightarrow{\ell} \mathbb{C}$$

where:

i) J is a commutative algebraic group,
ii) ℓ is a *linear* map from Lie J to $\mathbb{C}$
iii) ϕ is a morphism of varieties; and, in fact, if $g = \dim J$, then if we use addition on J to extend ϕ to

$$\phi^{(g)} : [(C\text{-poles } \omega) \times \ldots \times (C\text{-poles } \omega) / \underset{S_g}{\text{permutations}}] \longrightarrow J$$

then $\phi^{(g)}$ is birational, i.e., is bijective on a Zariski-open set.

In our example

$$C = \mathbb{P}^1, \qquad \omega = dx/x,$$

then $J = \mathbb{P}^1 - (0, \infty)$ which is an algebraic group where the group law is multiplication, and ϕ is the identity. The point is that J is the object that realizes the rule by which 2 g-tuples $(a_1, \ldots, a_g)$, $(b_1, \ldots, b_g)$ are "added" to form a third $(c_1, \ldots, c_g)$, and so that the integral $\sum_{i=1}^{g} I(x_i)$ becomes a homomorphism from J to $\mathbb{C}$. A slightly less fancy way to put it is that there is a $\phi : C$-(poles $\omega) \to J$ and a *translation-invariant* differential η on J such that

$$\phi^* \eta = \omega,$$

hence

$$\int_{\phi(a_0)}^{\phi(a)} \eta \equiv \int_{a_0}^{a_0} \omega \quad \text{(mod periods).}$$

Among the ω's, the most important are those of 1st kind, i.e., without poles, and if we integrate all of them at once, we are led to the most important J of all: the *Jacobian*, which we call Jac. From property (iii), we find that Jac must be a *compact* commutative algebraic group, i.e., a complex torus, and we want that

$$\phi : C \longrightarrow \text{Jac} ,$$

should set up a bijection:

$$\text{iv)} \quad \phi^* : \begin{bmatrix} \text{translation–} \\ \text{invariant 1-forms} \\ \eta \text{ on Jac} \end{bmatrix} \longrightarrow \begin{bmatrix} \text{rational differentials} \\ \omega \text{ on } C \text{ w/o poles} \end{bmatrix} = R_1(C).$$

Thus

$$\begin{aligned} \dim \text{Jac} &= \dim R_1(C) \\ &= \text{genus } g \text{ of } C. \end{aligned}$$

To construct Jac explicitly, there are 2 simple ways:

v) *Analytically:* write Jac $= V/L$, V complex vector space, L a lattice. Define:

$$\begin{aligned} V &= \text{dual of } R_1(C) \\ L &= \left\{ \begin{array}{l} \text{set of } \ell \in V \text{ obtained as periods, i.e.,} \\ \ell(\omega) = \int_\gamma \omega \text{ for some 1-cycle } \gamma \text{ on } C \end{array} \right. . \end{aligned}$$

Fixing a base point $a_0 \in C$, define for all $a \in C$

$$\phi(a) = \left\{ \begin{array}{l} \text{image in } V/L \text{ of any } \ell \in V \text{ defined by} \\ \qquad \ell(\omega) = \int_{a_0}^{a} \omega, \\ \text{where we fix a path from } a_0 \text{ to } a \end{array} \right. .$$

Note that since Jac is a group,

$$V^* \cong \left(\begin{array}{l} \text{translation – invariant} \\ \text{1-forms on Jac} \end{array} \right) \cong \left(\begin{array}{l} \text{cotangent sp. to Jac at } \alpha \\ \text{any } \alpha \in \text{ Jac} \end{array} \right) \cong R_1(C).$$

vi) *Algebraically:* following Weil's original idea, introduce $S^gC = C \times \ldots \times C/S_g$ and construct by the Riemann-Roch theorem, a "group-chunk" structure on S^gC, i.e., a partial group law:

$$\begin{aligned} &m : U_1 \times U_2 \longrightarrow U_3 \\ &U_i \subset S^gC \text{ Zariski-open.} \end{aligned}$$

He then showed that any such algebraic group-chunk prolonged automatically into an algebraic group J with $S^gC \supset U_4 \subset J$ (some Zariski-open U_4).

An important point is that ϕ is an integrated form of the canonical map Φ : $C \to \mathbb{P}^{g-1}$ discussed at length above –

vii) Φ is the Gauss map of ϕ, i.e., for all $x \in C$, $d\phi(T_{x,C})$ is a 1-dimensional subspace of $T_{\phi(x),\text{Jac}}$, and by translation this is isomorphic to Lie(Jac). If $\mathbb{P}^{g-1}$ = [space of 1-dimensional subsp. of Lie(Jac)], then $d\phi : C \to \mathbb{P}^{g-1}$ is just Φ.
(Proof: this is really just a rephrasing of (iv).)

The Jacobian has always been the corner-stone in the analysis of algebraic curves and compact Riemann surfaces. Its power lies in the fact that it *abelianizes* the curve and is a *reification* of H_1, e.g.,

viii) Via $\phi : C \to \text{Jac}$, every abelian covering $\pi : C_1 \to C$ is the "pull-back" of a unique covering $p : G_1 \to \text{Jac}$ (i.e., $C_1 \cong C \underset{\text{Jac}}{\times} G_1$).

Weil's construction in vi) above was the basis of his epoch-making proof of the Riemann Hypothesis for curves over finite fields, which really put characteristic p algebraic geometry on its feet.

There are very close connections between the geometry of the curve C (e.g., whether or not C is hyperelliptic) and Jac. We want to describe these next in order to tie in Jac with the special cases studied in Lecture I, and in order to "see" Jac very concretely in low genus. The main tool we want to use is:

Abel's Theorem. *Given $x_1, \ldots, x_k, y_1, \ldots, y_k \in C$, then*

$$\left\{\begin{array}{l} \exists \text{ rational function } f \\ \text{on } C \text{ with} \\ (f) \underset{\text{def}}{=} (\text{zeroes of } f) - (\text{poles of } f) \\ \quad = \Sigma x_i - \Sigma y_i \end{array}\right\} \iff \sum_{i=1}^{k} \phi(x_i) = \sum_{i=1}^{k} \phi(y_i).$$

When this holds, we say $\Sigma x_i \equiv \Sigma y_i$, or $\Sigma x_i, \Sigma y_i$ are *linearly equivalent.* For instance, when $C = \mathbb{P}^1$, any 2 points a, b are linearly equivalent via the function

$$f(x) = \frac{x-a}{x-b}.$$

For every k, we consider the map:

$$\begin{array}{ccc} \overbrace{C \times \ldots \times C}^{k \text{ times}} & \longrightarrow & \text{Jac} \\ (x_1, \ldots, x_k) & \longmapsto & \displaystyle\sum_{i=1}^{k} \phi(x_i). \end{array}$$

If $S^k C$ denotes C^k divided by permutations, i.e., the kth symmetric power of C, then this map factors via

$$\phi^{(k)} : S^k C \longrightarrow \text{Jac}\ .$$

Define

$$W_k = \text{Im}\ \phi^{(k)}, \qquad 1 \leq k \leq g-1$$

($\phi^{(k)}$ surjective if $k \geq g$). The fibres of this map[25] are called the ***linear systems*** on C of degree k, and by Abel's theorem they are the equivalence classes under linear equivalence and can be constructed as follows:

a) Pick one point $\mathfrak{A} = \sum_{i=1}^{k} x_i \in S^k C$.
b) Let

$$L(\mathfrak{A}) = \left\{ \begin{array}{l} \text{v. sp. of fcns. } f \text{ on } C \text{ with } (f) + \mathfrak{A} \geq 0, \text{ i.e.,} \\ \text{poles only at } x_i, \text{ order bounded by mult. of } x_i \\ \text{in } \mathfrak{A} \end{array} \right\}.$$

c) Let $|\mathfrak{A}| = \{\text{set of divisors } \Sigma y_i = (f) + \Sigma x_i, \ \ f \in L(\mathfrak{A}), \ \ f \neq 0\} \subset S^k C$.
d) Then $|\mathfrak{A}| = \phi^{(k)-1}\left(\phi^{(k)}(\mathfrak{A})\right)$. Note that it follows
$|\mathfrak{A}| \cong$ ***projective space*** of 1-dimensional subspaces of $L(\mathfrak{A})$.
e) We also want to use the Riemann-Roch theorem that tells us that

$$\dim |\mathfrak{A}| = k - g + i$$

where

$$i = \left\{ \begin{array}{l} \dim \ \text{ of v.sp. } R_1(-\mathfrak{A}) \text{ of differentials} \\ \omega \in R_1(C), \ \text{ with } \textit{zeroes} \text{ on } \mathfrak{A} \end{array} \right\}.$$

Now let's look at low genus cases:

$\boxed{g = 0:}$ Jac = (0)

$\boxed{g = 1:}$ (a) $\phi : C \to \text{Jac}$ is an isomorphism, i.e., $C = \text{Jac}$. In fact, for any genus $g \geq 1$,

$$\phi^{(1)} : C \to \text{Jac}$$

is an *embedding*, hence an isomorphism of C with W_1. (*Proof*: the fibres of $\phi^{(k)}$ being $\mathbb{P}^n$'s, $\phi^{(1)}$ would be either an embedding or C itself would be $\mathbb{P}^1$.)

(b) If $k \geq 2$,

$$\phi^{(k)} : S^k C \longrightarrow \text{Jac}$$

makes $S^k C$ into a $\mathbb{P}^{k-1}$-bundle over Jac, whose fibres are the linear systems of degree k. In general, if $k > 2g - 2$,

$$\phi^{(k)} : S^k C \longrightarrow \text{Jac}$$

makes $S^k C$ into a $\mathbb{P}^{k-g}$-bundle over Jac. (*Proof*: This is a consequence of the Riemann-Roch theorem since no differential can have more than $2g - 2$ zeroes.)

---

25. A technical aside: the complete ideal of functions on $S^k C$ vanishing on $\phi^{(k)-1}(\alpha)$ is generated by the functions on Jac vanishing at α – this is needed to make rigorous some of the points made below.

$\boxed{g=2:}$ The interesting case is $1 < k \leq 2g-2$, i.e., $k=2$: the map

$$\phi^{(2)} : S^2C \longrightarrow \text{Jac}.$$

Recall that there is a degree 2 map $\pi : C \to \mathbb{P}^1$. Since the points of $\mathbb{P}^1$ are all linearly equivalent to each other, the degree 2 cycles $\pi^{-1}(x)$ are also linearly equivalent. This gives us a copy E of $\mathbb{P}^1$ inside S^2C. The result is that Jac is isomorphic to the quotient of S^2C after identifying all points of E; i.e., that Jac is obtained by "blowing down" $E \subset S^2C$. Here is a picture:

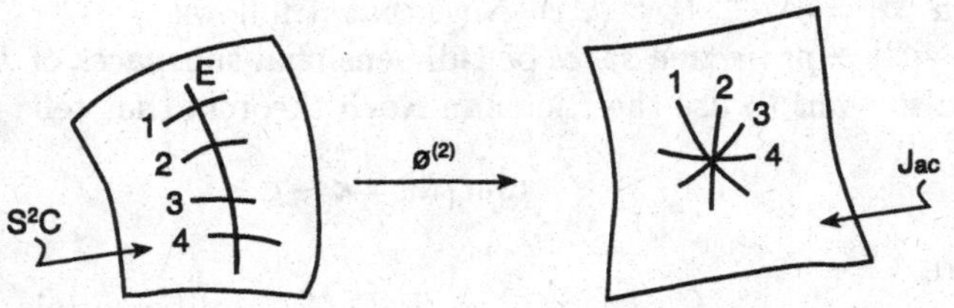

where, as is customary in the theory of algebraic surfaces, we draw real 2-dimensional manifolds in place of manifolds of 2 complex dimensions, which are 4 real-dimensional, hence undrawable! Going backwards, we may say that S^2C is obtained from Jac by "blowing up" $e = \phi^{(2)}(E)$: this is a process applicable to any variety X that replaces one of its points x by the set of tangent lines to X at x, giving you a new variety $B_x(X)$ birational to the first. We see here clearly that if we take the group law $m :$ Jac $\times$ Jac $\to$ Jac and try to transfer it to S^2C, we get merely a group chunk as in Weil's treatment because of E.

$\boxed{g=3:}$ Consider first $k=3$:

$$\phi^{(3)} : S^3C \longrightarrow \text{Jac}.$$

For any $x \in C$, consider the differentials ω on C zero at x: they form a 2-dimensional vector space and have 3 zeroes besides x. These zeroes form a degree 3 cycle, and as ω varies all these are linearly equivalent (use the functions ω_1/ω_2): this gives us a copy E_x of $\mathbb{P}^1$ in S^3C. It turns out:

$$\text{Jac} \cong (S^3C \text{ modulo collapsing each } E_x \text{ to a point }),$$

or putting it backwards if $\gamma =$ locus of points $\phi^{(3)}(E_x)$, then

$$S^3C \cong (\text{ Jac, with a curve } \gamma \subset \text{ Jac, isom. to } C, \text{ blown up}).$$

Most interesting is the case $k=2$:

$$\phi^{(2)} : S^2C \twoheadrightarrow W_2 \subset \text{ Jac}.$$

Then if C is *not hyperelliptic*, there are no non-trivial degree 2 linear systems, so

$$S^2C \xrightarrow{\approx} W_2.$$

But if C is *hyperelliptic*, you get one degree 2 linear system as in the $g = 2$ case, so

$$W_2 \cong (S^2C \text{ with a copy } E \text{ of } \mathbb{P}^1 \text{ blown down}).$$

The image $e \in W_2$ of E is a now double point and it looks like this:

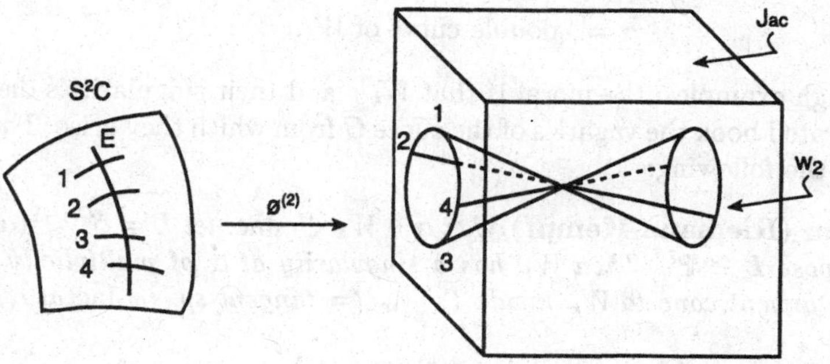

$\boxed{g = 4:}$ In this case, I want to consider because of its importance in Lecture IV only the case $k = 3$:

$$\phi^{(3)} : S^3C \twoheadrightarrow W_3 \subset \text{Jac}.$$

We mentioned briefly in Lecture I that either a) C is hyperelliptic, or b) C was an intersection in $\mathbb{P}^3$ of a quadric F and a cubic G. Now we also distinguish b$_1$) F singular, hence a quadric cone, and b$_2$) F non-singular. b$_2$) is the most common case. Using the 2 rulings on a non-singular quadric, it is a standard fact that such a quadric is isomorphic to $\mathbb{P}^1 \times \mathbb{P}^1$. Thus $C \cong (\mathbb{P}^1 \times \mathbb{P}^1) \cap G$, and since G is a *cubic*, C meets the curves $(\mathbb{P}^1 \times \text{ pt.})$ or $(\text{pt. } \times \mathbb{P}^1)$ in 3 points. Thus the 2 projections of $\mathbb{P}^1 \times \mathbb{P}^1$ to $\mathbb{P}^1$ induce 2 maps π_1, π_2 from C to $\mathbb{P}^1$ of degree 3. The 2 families of degree 3 cycles $\{\pi_1^{-1}(x)\}$ and $\{\pi_2^{-1}(x)\}$ form 2 linear systems $E_1, E_2 \subset S^3C$, with $E_1 \cong E_2 \cong \mathbb{P}^1$. Then:

case b_2: $$W_3 \cong \left(\begin{array}{l} S^3C \text{ with } E_1, E_2 \text{ blown down to} \\ 2 \text{ points } e_1, e_2 \end{array} \right)$$

and

$$e_1, e_2 = \text{ ordinary double points of } W_3.$$

In case b_1, the 2 rulings "come together"; in fact, S^3C contains only one non-trivial linear system E, and *case* b_1: $W_3 \cong \left(S^3C \text{ with } E \text{ blown down to } e\right)$

and

$e =$ higher double point of W_3.

In the hyperelliptic case a, it turns out that there is a whole curve of linear systems $E_x \subset S^3C$ depending on a point $x \in C$: in fact, take the degree 2 linear system, and just add x to each of its members. Thus

$$\text{case } a: \quad W_3 \cong \left(\begin{array}{l} S^3C \text{ with the surface } \cup E_x \text{ blown} \\ \text{down to a curve } \gamma \text{ isomorphic to } C \end{array} \right)$$

and

$\gamma =$ double curve of W_3.

Enough examples: the moral is that W_k's and their singularities display like an illustrated book the vagaries of the curve C from which they arise. The general result is the following:

Theorem (Riemann-Kempf). *Let $\alpha \in W_k \subset$ Jac, let $L = \phi^{(k)-1}(\alpha) \subset S^kC$ and suppose $L \cong \mathbb{P}^\ell$. Then W_k has a singularity at α of multiplicity $\binom{g-k+\ell}{\ell}$, and the tangent cone to W_k inside $T_{\alpha,\,\mathrm{Jac}}$ (= tangent sp. to Jac at α) is equal to:*

$$\bigcup_{\mathfrak{A} \in L} D\phi^{(k)}\left(T_{\mathfrak{A},S^kC}\right).$$

Here $D\phi^{(k)}$ is the differential of $\phi^{(k)}$ and it gives rise to an exact sequence:

$$(*) \qquad 0 \longrightarrow T_{\mathfrak{A},L} \longrightarrow T_{\mathfrak{A},S^kC} \xrightarrow{D\phi^{(k)}} T_{\alpha,\,\mathrm{Jac}}.$$

In fact, this sequence actually "displays" the Riemann-Roch formula in a beautiful way: using the fact that

$$\begin{aligned} R_1(C) &\cong \text{translation-invariant differentials on Jac} \\ &\cong T^*_{\alpha,\,\mathrm{Jac}} \ (= \text{cotangent sp. to Jac at } \alpha), \text{ for all } \alpha, \end{aligned}$$

it is not hard to check that if $\omega \in R_1(C)$ corresponds to $[\omega] \in T^*_{\alpha,\,\mathrm{Jac}}$, then:

$$\left(D\phi^{(k)}\right)^*[\omega] = 0 \text{ in } T_{\mathfrak{A},S^kC} \Longleftrightarrow \omega \text{ is zero on } \mathfrak{A}.$$

Therefore

$$\operatorname{Coker} D\phi^{(k)} \cong \text{dual of } R_1(-\mathfrak{A}), \text{ the space of differentials zero on } \mathfrak{A}.$$

Therefore counting the dimensions of the vector spaces in $(*)$:

$$\begin{aligned} \dim L &= \dim S^kC - \dim \mathrm{Jac} + \dim \operatorname{Coker} D\phi^{(k)} \\ &= k - g + \dim R_1(-\mathfrak{A}), \end{aligned}$$

which is the Riemann-Roch theorem! What comes next is going to be harder to follow, but we can go much further: let $\{\omega_i\}$ be a basis of $R_1(-\mathfrak{A})$ and let $\{f_j\}$ be a basis of $L(\mathfrak{A})$. Then a general member of L is given by

$$\mathfrak{A}_t = \mathfrak{A} + \left(\sum_{j=0}^{\ell} t_j f_j\right)$$

and a basis of $R_1(-\mathfrak{A}_t)$ is given by $\left(\sum_{j=0}^{\ell} t_j f_j\right) \cdot \omega_i$. Therefore $\sum_{j=0}^{\ell} t_j[f_j\omega_i]$ span the dual of the cokernel of

$$T_{\mathfrak{A}_t, S^k C} \xrightarrow{D\phi^{(k)}} T_{\alpha,\mathrm{Jac}},$$

or $\sum_{j=0}^{\ell} t_j[f_j\omega_i] = 0$ are linear equations on $T_{\alpha,\mathrm{Jac}}$ which define the subspace $D\phi^{(k)}\left(T_{\mathfrak{A}_t,S^kC}\right)$. It follows that if we put together a big $(\ell+1)\times(g-k+\ell)$-matrix of linear functions on $T_{\alpha,\mathrm{Jac}}$ out of $[f_j\omega_i]$, then all its $(\ell+1)\times(\ell+1)$ minors vanish on each $D\phi^{(k)}\left(T_{\mathfrak{A}_t,S^kC}\right)$, hence vanish on the whole tangent cone to W_k. Kempf proved that these equations suffice, and that W_k itself has equations of this type:

Theorem (Kempf). *There is a $(\ell+1)\times(g-k+\ell)$-matrix of holomorphic functions (f_{ij}) on* Jac *near α such that W_k is the set of zeroes of all its $(\ell+1)\times(\ell+1)$ minors: i.e., W_k is a* determinantal variety. *Moreover, $[f_j\omega_i]$ = linear term of f_{ij} and the tangent cone to W_k is the set of zeroes of the $(\ell+1)\times(\ell+1)$-minors of the matrix $[f_j\omega_i]$ of linear functions.*

The feature of the Jacobian, however, which really gives it its punch is the theta function. There are 3 very good reasons to look next at the function theory of Jac –

a) to define projective embeddings of Jac, hence understand better the algebraic structure, moduli, etc.
b) because Jac is a group, one hopes that its function theory will reflect this in interesting ways,
c) by pull-back, functions on Jac will define functions on S^gC, hence on C, and may give a good way to expand functions on C, prove the Riemann-Roch theorem, etc.

So write

$$\mathrm{Jac} = \mathbb{C}^g/L.$$

Instead of constructing L-periodic meromorphic functions f on $\mathbb{C}^g$, one seeks L-automorphic entire functions f, i.e.,

$$f(x+\alpha) = e_\alpha(x) \cdot f(x), \qquad \alpha \in L,\ x \in \mathbb{C}^g$$
$$\{e_\alpha\} = \text{"automorphy factor"}^{26}$$

Equivalently, such f are holomorphic sections of a line bundle $L_{\{e_\alpha\}}$ on Jac and clearly the quotient of 2 such f is always L-periodic. The simplest choice of $\{e_\alpha\}$ is something in the general form:

$$e_\alpha(x) = e^{\beta B(x,\alpha)+c(\alpha)}, \qquad \beta \text{ bilinear.}$$

($e_\alpha =$ (constant) is too simple, because no f's will exist.)

Now if $g \geq 2$, most complex tori $\mathbb{C}^g/L$ have no non-constant meromorphic functions on them at all, and are not algebraic varieties, and do not carry any but "trivial" $\{e_\alpha\}$'s[27]. In the case of a curve C, however, special things happen; let's look for bilinear forms as candidates for B. We saw above that on $R_1(C)$ one has a positive definite Hermitian form:

$$(\omega_1, \omega_2) = \int_C \omega_1 \wedge \overline{\omega}_2$$

hence its dual, which is the universal covering space $\mathbb{C}^g$ of Jac, gets a Hermitian form that we will write H. But also $H_1(C,\mathbb{Z})$ carries an integral skew-symmetric form

$$E : H_1(C,\mathbb{Z}) \times H_1(C,\mathbb{Z}) \longrightarrow \mathbb{Z}$$

given by intersection pairing. As we saw in (v) above, there is an isomorphism $H_1(C,\mathbb{Z}) \cong L$, hence L carries such an E. It is not hard to show that H and E are connected by:

$$(*) \qquad E(x_1,x_2) = \operatorname{Im} H(x_1,x_2), \qquad \text{all } x_1, x_2 \in L,$$

and that when $(*)$ holds there is a (nearly canonical[28]) choice of $\{e_\alpha\}$, viz.

$$e_\alpha(x) = \pm e^{\pi[H(x,\alpha)+\frac{1}{2}H(\alpha,\alpha)]}.$$

Moreover, one has the beautiful theorem:

---

[26] i.e., entire functions on $\mathbb{C}^g$, nowhere zero, such that

$$e_{\alpha+\beta}(x) \equiv e_\alpha(x+\beta) \cdot e_\beta(x).$$

[27] $\{e_\alpha\}$ is trivial if $e_\alpha(x) \equiv e(x+\alpha)/e(x)$ for some e.

[28] The sign $\pm$ is not canonical; it satisfies some funny identities that I don't want to discuss; any 2 choices, however, are related by a transformation

$$e'_\alpha(x) = \ell(\alpha) e_\alpha(x), \qquad \ell \in \operatorname{Hom}\,(L/2L,\ \pm 1).$$

Theorem. *The existence of a positive definite Hermitian H on $\mathbb{C}^g$ and an integral skew-symmetric E on L satisfying $E = \operatorname{Im} H$ is necessary and sufficient for a complex torus $\mathbb{C}^g/L$ to carry g algebraically independent meromorphic functions and if it has such functions, it admits an embedding into $\mathbb{P}^n$, some n, hence is a projective variety*[29].

Here we see the principle emerging that a complex torus does not fit easily in $\mathbb{P}^n$: non-trivial identities $(*)$ are required before it will fit at all. Now define a *theta-function*[30] *of order n* to be an entire function f on $\mathbb{C}^g$ such that

$$f(x+\alpha) = \left(\pm e^{\pi[H(x,\alpha)+\frac{1}{2}H(\alpha,\alpha)]}\right)^n \cdot f(x),$$

and let S_n be the space of such f. Then $S = \sum S_n$ is a graded ring. Elementary Fourier analysis combined with the fact that E is a unimodular pairing leads to

$$(**) \qquad \dim S_n = n^g, \qquad (n \geq 1).$$

In particular, there is exactly one first order theta-function, up to scalars. This important function, written $\vartheta(x)$, is called *Riemann's theta function*[31]. If, instead, we take any $n \geq 3$, and let $\psi_1, \ldots, \psi_{n^g}$ be a basis of S_n, we get:

Lefschetz's embedding theorem. *$\mathbb{C}^g/L$ is embedded in $\mathbb{P}^{n^g-1}$ by*

$$x \longmapsto (\psi_1(x), \ldots, \psi_{n^g}(x)) = \Psi_n(x).$$

This makes sense because ψ_i/ψ_j are single-valued functions on $\mathbb{C}^g/L$. This solves problem (a) raised above. (b) however is even more remarkable. In fact, to introduce the group structure into the picture, for all $\beta \in \mathbb{C}^g$, define

$$(T_\beta f)(x) = f(x+\beta).$$

For all nowhere zero holomorphic functions e on $\mathbb{C}^g$, define

---

29 By Chow's theorem, if you embed it in projective space at all, the image is projective a variety; and if you embed it in 2 ways, the 2 projective varieties are isomorphic algebraically as well as analytically.

30 Since this is not *exactly* the classical definition, let me indicate the connection. Classically, one splits $L = L_1 + L_2$, when $L_i \cong \mathbb{Z}^g$ and $E(x_1, x_2) = 0$ if x_1, x_2 are *both* in L_1 or *both* in L_2. For all $\alpha \in L_2$, define a *complex* linear $\ell_\alpha : \mathbb{C}^g \to \mathbb{C}$ by $\ell_\alpha(x) = E(x,\alpha)$ if $x \in L_1$. Require instead

$$\begin{aligned} f(x+\alpha) &= f(x), \qquad \alpha \in L_1 \\ f(x+\alpha) &= e^{2\pi i n(\ell_\alpha(x)+\frac{1}{2}\ell_\alpha(\alpha))} f(x), \qquad \forall \alpha \in L_2. \end{aligned}$$

Then these f's differ from the other f's by an elementary factor.

31 In the classical normalization, it is

$$\vartheta(x) = \sum_{\alpha \in L_2} e^{-2\pi i(\ell_\alpha(x)+\frac{1}{2}\ell_\alpha(\alpha))}.$$

$$(U_e f)(x) = e(x) f(x).$$

Then refining the analysis leading to (**), one finds

Lemma. *i)* $\forall \beta \in \mathbb{C}^g$, *there exists* e *such that* $U_e T_\beta S_n = S_n$ *if and only if* $\beta \in \frac{1}{n} L$.

ii) Choosing such an $e(\beta)$ *for each* $\beta \in \frac{1}{n}L$, $\beta \mapsto U_{e(\beta)} \cdot T_\beta$ *defines a projective representation*[32] *of* $\frac{1}{n}L/L$ *on* S_n*: this representation is* irreducible.

It seems to me remarkable that although $\mathbb{C}^g/L$ is an abelian group, its function-theory is full of irreducible representations of dimensions bigger than one: in fact, these are ordinary representations of a finite 2-step nilpotent group $\mathfrak{G}_n$:

$$1 \longrightarrow \mathbb{Z}/n\mathbb{Z} \longrightarrow \mathfrak{G}_n \longrightarrow \frac{1}{n}L/L \longrightarrow 1$$

analogous to the nilpotent Lie group:

$$1 \longrightarrow \mathbb{R} \longrightarrow \mathfrak{D} \longrightarrow V \oplus \hat{V} \longrightarrow 1 \qquad (V = \text{real vector space})$$

whose Lie algebra is the Heisenberg commutation relations[33]. This rather easy lemma has lots of consequences.

Corollary. *i) In the embedding* Ψ_n*, translation by* β *on* $\mathbb{C}^g/L$ *extends to a linear transformation* $\mathbb{P}^{n^g-1} \to \mathbb{P}^{n^g-1}$ *if and only if* $\beta \in \frac{1}{n}L/L$.

ii) Modulo the choice of distinguished generators for the associated finite group $\mathfrak{G}_n$*, we get, up to scalars, a distinguished basis of* S_n*, hence a normalization of*

$$\Psi_n : \mathbb{C}^g/L \longrightarrow \mathbb{P}^{n^g-1}$$

under projective transformations. In this normalization, translations by $\beta \in \frac{1}{n}L/L$ *acts on* $\Psi_n(\mathbb{C}^g/L)$ *by a simple set of explicit* $n^g \times n^g$*-matrices.*

To be more explicit, start with $\vartheta \in S_1$. Choose $\phi : \mathbb{Z}^g \times \mathbb{Z}^g \xrightarrow{\approx} L$ such that

$$E(\phi(n,m), \phi(n',m')) = \langle n, m' \rangle - \langle m, n' \rangle,$$

and extend ϕ to $\mathbb{Q}^g \times \mathbb{Q}^g \xrightarrow{\approx} L \otimes \mathbb{Q}$. Then if $n = m^2$, a typical distinguished basis of S_{m^2} is of the form

[32] I.e., $g \mapsto U_g$ is a projective representation of G if

$$U_{g_1 g_2} = \text{const.}\ U_{g_1} \cdot U_{g_2},$$

all $g_1, g_2 \in G$.

[33] $\mathfrak{D}$ = set of triples (α, x, ξ), $\alpha \in \mathbb{R}$, $x \in V$, $\xi \in \hat{V}$, with group law:

$$(\alpha, x, \xi) \cdot (\alpha', x', \xi') = (\alpha + \alpha' + \langle x, \xi' \rangle, x + x', \xi + \xi').$$

$$\vartheta\begin{bmatrix}\alpha\\ \beta\end{bmatrix}(x) = \begin{bmatrix}\text{exponential}\\ \text{of suitable}\\ \text{linear fcn.}\end{bmatrix}\cdot\vartheta(mx+\phi(\alpha,\beta))$$

where α, β range over coset representatives of $\frac{1}{m}\mathbb{Z}^g$ modulo $\mathbb{Z}^g$. Thus

$$x \longmapsto \left(\ldots\ldots, \vartheta\begin{bmatrix}\alpha\\ \beta\end{bmatrix}(x), \ldots\ldots\right)$$

is the normalized projective embedding of $\mathbb{C}^g/L$. The most important point here is that while translations by $\beta \in \frac{1}{n}L/L$ are normalized, $\Psi_n(0)$ is *not normalized.* Hence $\Psi_n(0) = \left(\ldots, \vartheta\left[\begin{smallmatrix}\alpha\\ \beta\end{smallmatrix}\right](0), \ldots\right)$ is an *invariant* of the torus $\mathbb{C}^g/L$ and the distinguished generators of $\mathfrak{G}_n$. These $\vartheta\left[\begin{smallmatrix}\alpha\\ \beta\end{smallmatrix}\right](0)$ are classically called the *theta-null werte.* We will discuss their role as moduli at more length in Lecture IV.

Summarizing this discussion, you can say that you can take a) $\mathbb{C}^g/L$, and b) $\mathbb{P}^n$: both innocent *homogeneous* complex manifolds. You marry them via Ψ_n and the children they produce are these highly unsymmetric and intricate functions $\vartheta\left[\begin{smallmatrix}\alpha\\ \beta\end{smallmatrix}\right]$:

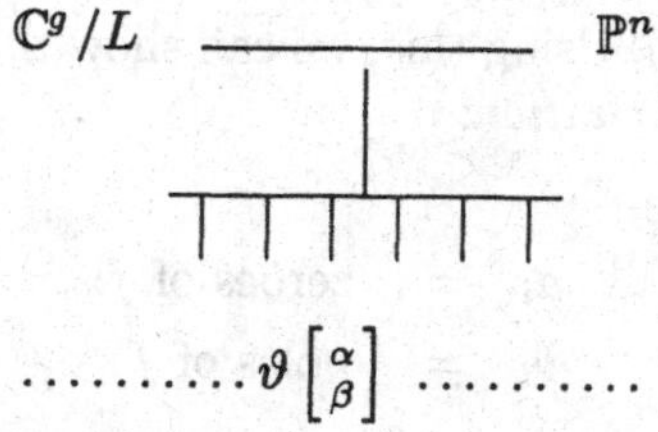

We pass on now to problem (c): when $\mathbb{C}^g/L$ is the Jacobian of C, pull back functions on $\mathbb{C}^g/L$ to C and see what you get. We follow Riemann and consider the basic functions:

$$E_e(x,y) = \vartheta\left(\int_y^x \omega - \mathbf{e}\right)$$

where $\mathbf{e} \in \mathbb{C}^g$, and $\boldsymbol{\omega} = (\omega_1, \ldots, \omega_g)$, ω_i the basis of $R_1(C)$ (recall that ϑ is naturally a function on the dual space to $R_1(C)$, and we have identified this space with $\mathbb{C}^g$; so $\{\omega_i\}$ is the dual basis of $R_1(C)$). For fixed y and $\mathbf{e}$, this is a multi-valued function on C that changes by a multiplicative factor

$$\mathrm{e}^{\left(\int_y^x \omega + \text{const.}\right)}$$

when analytically continued around a cycle. Riemann showed that when not identically zero it had exactly g zeroes $z_1, \ldots, z_g$ and that there was a point $\Delta \in$ Jac (depending on the choice of sign $\pm$ in the definition of the automorphy factor $\{e_\alpha\}$ for ϑ) such that in Jac:

$$\sum_{i=1}^{g} \phi(z_i) = \phi(y) + \mathbf{e} + \Delta.$$

In fact, we saw that

$$\phi^{(g)} : S^g C \longrightarrow \text{Jac}$$

was birational, and this shows that if $\mathbf{e}_0 = \phi(y) + \Delta$ we get an *inverse* to $\phi^{(g)}$ almost everywhere by

$$\mathbf{e} \longmapsto g\text{-tuple of zeroes of } \vartheta\left(\int_y^x \omega - \mathbf{e} + \mathbf{e}_0\right).$$

Moreover, we find:

$$\vartheta(\mathbf{e}) = 0 \Longleftrightarrow E_e(y) = 0 \Longleftrightarrow \begin{array}{c}\text{some } z_i \\ \text{equals } y\end{array} \Longleftrightarrow \begin{array}{c}\exists z_1, \ldots, z_{g-1} \in C \\ \text{such that } \sum_{i=1}^{g-1} \phi(z_i) = \Delta + \mathbf{e}\end{array}.$$

This means that if we define the codimension 1 subset $\Theta \subset \text{Jac}$ by

$$\Theta = \{x \in \text{ Jac} \mid \vartheta(x) = 0\},$$

then, up to a translation, Θ is just W_{g-1}! This is the basic link between theta-functions and the geometry discussed earlier. Moreover, if we fix $\mathbf{e}$ such that $\vartheta(\mathbf{e}) = 0$, and fix $z_1, \ldots, z_{g-1} \in C$ such that $\sum \phi(z_i) = \Delta + \mathbf{e}$, then consider the function $E_e(x, y)$ for variables x and y. It follows that so long as $x \mapsto E_e(x, y)$ is not identically 0, its zeroes are $x = y$ and $x = z_1, \ldots, z_{g-1}$, i.e., ignoring certain bad points z_i independent of y, $E_e(x, y)$ is a *Prime Form* as a function of x: has a unique variable zero at y. Using this, we can show that every rational function f on C has a *unique factorization*:

If

$$\begin{aligned} a_i &= \text{zeroes of } f \\ b_i &= \text{poles of } f \end{aligned}$$

then

$$f(x) = e^{\int^x \omega} \cdot \frac{\prod_i E_e(x, a_i)}{\prod_i E_e(x, b_i)}$$

(for some $\omega \in R_1(C)$).

This beautiful decomposition is the higher genus analog of the factorization:

$$f(x) = C \cdot \frac{\prod (x - a_i)}{\prod (x - b_i)}$$

of rational functions on $\mathbb{P}^1$. Nor do these factorizations depend much on $\mathbf{e}$, because if $\vartheta(\mathbf{e}_1) = \vartheta(\mathbf{e}_2) = 0$, then

$$E_{e_1}(x, y) = (\text{fcn of } x \text{ alone})(\text{fcn of } y \text{ alone}) E_{e_2}(x, y).$$

Using the E_e's, we also get beautiful expressions for differentials on C with various poles too, e.g.,

$$\left(\frac{\partial}{\partial x} \log \frac{E_e(x, a)}{E_e(x, b)}\right) dx$$

is a rational 1-form on C, with simple poles at a, b only, residues ± 1 respectively; and

$$\left(\frac{\partial^2}{\partial x \partial y} \log E_e(x, y)\right)\bigg|_{y=a} dx$$

is a rational 1-form on C, with a double pole at $x = a$ and no others.

Lecture IV: The Torelli Theorem and the Schottky Problem

The purpose of this lecture is to consider the map carrying C to its Jacobian Jac from a moduli point of view. Jac is a particular kind of complex torus and the Schottky problem is simply the problem of characterizing the complex tori that arise as Jacobians. The Torelli theorem says that Jac, plus the form H on its universal covering space, determine the curve C up to isomorphism.

First of all, we saw that if $g \geq 2$, not all complex tori $X = \mathbb{C}^g/L$ are even projective varieties: in fact, necessary and sufficient for X to be a projective variety is that there exists a positive definite Hermitian form H on $\mathbb{C}^g$, such that $E \underset{\text{def}}{=} \operatorname{Im} H$ is integral on $L \times L$. The varieties that arise this way are called *abelian varieties*. The forms H are called *polarizations* of X. Since $rk\ L = 2g$, and E is skew-symmetric and integral on L, $\det E = (-1)^g d^2$, for some $d \in \mathbb{Z}, d \geq 1$: d is called the *degree* of the polarization. A polarization of degree 1 is called a *principal polarization*. Jacobians come with a natural polarization in which E is just the intersection form on $L \cong H_1(C, \mathbb{Z})$: this form is unimodular, so this is in fact a principal polarization. In general, if $(\mathbb{C}^g/L, H)$ is any polarized abelian variety, one can find $L_1 \subset L$ of finite index and $n \geq 1$ such that $(\mathbb{C}^g/L_1, \frac{1}{n}H)$ is a principally polarized abelian variety – so in studying all abelian varieties, the principally polarized ones play a central role.

Secondly, we saw in Lecture III that starting with any principal polarized abelian variety $(\mathbb{C}^g/L, H)$, we get Riemann's theta function $\vartheta : \mathbb{C}^g \to \mathbb{C}$, hence $\Theta = (\text{zeroes of } \vartheta) \subset \mathbb{C}^g/L$. A more succinct way to describe how $\mathbb{C}^g/L$ and H canonically determine the codimension 1 subvariety Θ up to translation[34] is the following:

Θ = any codim. 1 subvariety D of $\mathbb{C}^g/L$ whose fundamental class $[D] \in H^2(\mathbb{C}^g/L, \mathbb{Z})$ is represented by E, under the canonical identification:

$$H^2(\mathbb{C}^g/L, \mathbb{Z}) \cong (\text{skew-symmetric, integral forms on } L).$$

Up to a translation, the only such Θ is the set of zeroes of ϑ. This shows that H, or Θ (up to translation) are equivalent data. Moreover it is also possible to describe which codimension 1 subvarieties $D \subset \mathbb{C}^g/L$ arise from H and a ϑ: for

[34] From Lecture III, Θ looks like it is unique even without a possible translation: however, remember the annoying ambiguity of sign in $\{e_\alpha\}$ – this means we actually only found Θ up to translation by a point $x \in \frac{1}{2}L/L$.

any $a \in \mathbb{C}^g/L$, let D_a = translate of D by a. For any D, choose $a_1, \ldots, a_g$ so that $D_{a_1}, \ldots, D_{a_g}$ meet transversely and consider the number of intersections:

$$D_{a_1} \bigcap \cdots \bigcap D_{a_g}.$$

This is denoted (D^g) and is always divisible by $g!$ Then

$$D = \text{some } \Theta \qquad \text{iff} \qquad (D^g)/g! = 1.$$

We say such D's are of degree one. Therefore, instead of describing principal polarizations on $\mathbb{C}^g/L$ as forms H with Im E integral and unimodular on L, we can describe them as codimension 1 subvarieties $\Theta \subset \mathbb{C}^g/L$ with $(\Theta^g) = g!$ given up to translation.

This gives a completely algebraic way to describe such polarizations. There are also quite simple ways to describe algebraically polarizations of higher degree, but we do not need to know these[35]. We can now introduce the moduli space of principally polarized abelian varieties:

$$\mathcal{A}_g = \left\{ \begin{array}{l} \text{set of pairs } (X, \Theta), \text{ where } X \\ \text{is an abelian variety and} \\ \Theta \subset X \text{ is a codimension 1} \\ \text{subvariety such that} \\ (\Theta^g) = g! \end{array} \right\} \Bigg/ \left\{ \begin{array}{l} \text{isomorphisms} \\ f: X_1 \to X_2 \\ \text{such that} \\ f\Theta_1 = \Theta_2 \text{ -} \\ \text{but } f \text{ need not} \\ \text{take the identity} \\ 0 \in X_1 \text{ to } 0 \in X_2 \end{array} \right\}.$$

As in Lecture II, it turns out that $\mathcal{A}_g$ has a natural structure of normal quasi-projective variety. Moreover, we obtain a morphism:

$$t: \mathfrak{M}_g \longrightarrow \mathcal{A}_g$$

by defining $t(C) = (\text{Jac}, W_{g-1})$. The Torelli theorem simply says that t is injective and the Schottky problem can be rephrased as asking for a characterization of the image $t(\mathfrak{M}_g)$. Before studying these in more detail, I would like, in parallel with the treatment in Lecture II, to i) indicate the analytic description of $\mathcal{A}_g$ via an infinite covering, ii) indicate how to explicitly coordinatize $\mathcal{A}_g$, and iii) describe the closure of $t(\mathfrak{M}_g)$ in $\mathcal{A}_g$.

In regards to (i), we consider set-theoretically:

$$\mathfrak{H}_g = \left\{ \begin{array}{l} \text{set of 4-tuples } (V, L, H, \alpha), \text{ where} \\ V = \text{ a complex vector space} \\ L = \text{ a lattice in } V \\ H = \text{ a positive definite Hermitian form on } V \\ \alpha = \text{ an isomorphism } \mathbb{Z}^g \times \mathbb{Z}^g \to L \\ \text{with} \\ \text{Im } H\big(\alpha(n,m), \alpha(n',m')\big) = \langle n, m' \rangle - \langle n', m \rangle \end{array} \right\} \Big/ \text{isomorphism}$$

[35] The 2 methods are i) by a suitable line bundle $\mathfrak{L}$ on $\mathbb{C}^g/L$ given up to translation, or ii) by a suitable homomorphism $\phi: \mathbb{C}^g/L \to (\mathbb{C}^g/L)^\wedge$ where $(\mathbb{C}^g/L)^\wedge$ is the "dual" abelian variety.

$$\cong \left\{ \begin{array}{l} \text{set of 3-tuples } (X, \Theta, \alpha), \text{ where} \\ X = \text{ an abelian variety} \\ \Theta = \text{ codim 1 subvariety with } (\Theta^g) = g! \\ \alpha = \text{ an isomorphism } \mathbb{Z}^g \times \mathbb{Z}^g \to H_1(X, \mathbb{Z}) \\ \text{where if } [\Theta] = \text{ fundamental class of } \Theta, \\ \text{then} \\ [\Theta]\big(\alpha(u,m), \alpha(u,m')\big) = \langle n, m' \rangle - \langle n', m \rangle \end{array} \right\} \Big/ \text{isomorphism}$$

(The connection being given by $X = V/L$, $\Theta \rightleftarrows H$ as above). Clearly, forgetting α defines a map

$$\mathfrak{H}_g \longrightarrow \mathcal{A}_g$$

and for all $\sigma \in Sp(2g, \mathbb{Z}) =$ [group of $2g \times 2g$ integral symplectic matrices],

$$(X, \Theta, \alpha) \longmapsto (X, \Theta, \alpha \cdot \sigma)$$

defines an action of $Sp(2g, \mathbb{Z})$ on $\mathfrak{H}_g$ such that

$$\mathcal{A}_g \cong \mathfrak{H}_g / Sp(2g, \mathbb{Z}).$$

On the other hand, given (V, L, H, α), there is a unique isomorphism $\Psi : V \cong \mathbb{C}^g$ such that $\Psi\big(\alpha(n, 0)\big) = n$, i.e., such that the first g generators of L are just the unit vectors in $\mathbb{C}^g$. Define the $g \times g$ complex matrix Ω by $\Psi\big(\alpha, (0, m)\big) = \Omega \cdot m$, i.e., the second g generators of L are just the g columns of Ω. Write H via a $g \times g$ Hermitian matrix h via

$$H(x, y) = {}^t\Psi(x) \cdot h \cdot \overline{\Psi(y)}.$$

Then the condition on Im H written out is:

$$\left. \begin{array}{l} \text{Im } {}^t n \cdot h \cdot m = 0 \\ \text{Im } {}^t n {}^t \Omega \cdot h \cdot \overline{\Omega} \cdot m = 0 \\ \text{Im } {}^t n {}^t \Omega \cdot h \cdot m = {}^t n \cdot m \end{array} \right\} \forall n, m \in \mathbb{Z}^g$$

which works out as simply:

$$\Omega = {}^t\Omega, \qquad h = (\text{Im } \Omega)^{-1}.$$

On the other hand, if Ω is any symmetric $g \times g$ complex matrix with Im Ω positive definite, then

$$\begin{aligned} V &= \mathbb{C}^g \\ L &= \mathbb{Z}^g + \Omega \cdot \mathbb{Z}^g \\ H(x, y) &= {}^t x \cdot (\text{Im } \Omega)^{-1} \cdot y \\ \alpha(n, m) &= n + \Omega \cdot m \end{aligned}$$

is an element of $\mathfrak{H}_g$. This proves that

$$\mathfrak{H}_g \cong \left\{ \begin{array}{l} \text{open subset of } \mathbb{C}^{g(g+1)/2} \text{ of } g \times g \text{ complex symmetric} \\ \text{matrices } \Omega, \text{ with Im } \Omega \text{ positive definite} \end{array} \right\}$$

which is called the "Siegel upper half-space". Bringing in Teichmüller space again, it is not hard to see that we get the big diagram:

$$\begin{array}{ccc} \mathbb{C}^{3g-3} & & \mathbb{C}^{g(g+1)/2} \\ \cup & & \cup \\ \mathfrak{T}_g & \xrightarrow{\tilde{t}} & \mathfrak{H}_g \\ \downarrow & & \downarrow \\ \mathfrak{M}_g & \xrightarrow{t} & \mathcal{A}_g \end{array}$$

Bounded Domains on this line ($\mathfrak{T}_g \to \mathfrak{H}_g$)

Quasi-Projective Varieties on this line ($\mathfrak{M}_g \to \mathcal{A}_g$)

where $\tilde{t}$ is even an equivariant holomorphic map for a homomorphism

$$\tau : \Gamma_g \longrightarrow Sp(2g, \mathbb{Z})$$
$$\Gamma_g = \text{ Teichmüller modular group.}$$

In regards to (ii), we want to mention how to use theta-null werte to explicitly embed $\mathcal{A}_g$ in a big projective space $\mathbb{P}^N$. To be precise, the ideas of Lecture III lead to the following: there is a subgroup $\Gamma_m \subset Sp(2g, \mathbb{Z})$ of finite index such that for all $m \geq 2$, the functions $\vartheta \left[\begin{smallmatrix} \alpha \\ \beta \end{smallmatrix}\right](0)$, $\alpha, \beta \in \frac{1}{m}\mathbb{Z}^g$ running over cosets mod $\mathbb{Z}^g$, which are called the *theta-nulls* of the abelian variety X, are global homogeneous coordinates not on $\mathcal{A}_g$ but on the covering space

$$\mathcal{A}_g^{\Gamma_m} \underset{\text{def}}{=} \mathfrak{H}_g / \Gamma_m,$$

i.e., define an embedding

$$\mathcal{A}_g^{\Gamma_m} \hookrightarrow \mathbb{P}^{(m^{2g}-1)}.$$

Suitable polynomials in the theta-nulls $\vartheta \left[\begin{smallmatrix} \alpha \\ \beta \end{smallmatrix}\right](0)$, invariant under the finite group $Sp(2g, \mathbb{Z})/\Gamma_m$, will then be coordinates on $\mathcal{A}_g$ itself. Assuming the injectivity of t, this gives by composition coordinates once more on $\mathfrak{M}_g$: this is method I alluded to in Lecture II. Other ways to get coordinates on $\mathcal{A}_g$ are to use other modular functions on $\mathcal{A}_g$, i.e., holomorphic functions automorphic with respect to $Sp(2g, \mathbb{Z})$, such as Poincaré series or Eisenstein series. The coordinates gotten in this way seem harder to interpret algebraically: in particular, finding an algebraic interpretation of the Eisenstein series in terms of the definition of $\mathcal{A}_g$ via moduli seems to be a very interesting problem.

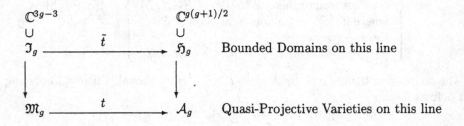

In regards to (iii), although unfortunately $t(\mathfrak{M}_g)$ is not closed in $\mathcal{A}_g$, it is very nearly so. We can look at the compactification $\overline{\mathfrak{M}}_g$ of $\mathfrak{M}_g$ mentioned in Lecture II and study the "limit" of the Jacobian of a non-singular curve C as C approaches a singular curve C_0 representing a point of $\overline{\mathfrak{M}}_g - \mathfrak{M}_g$. It turns out these Jacobians have limits which are still abelian varieties if and only if C_0 is made up of a set of non-singular components $\{D_i\}$ connected together like a tree, and that in this case the limit of the Jacobian of C is the product of the Jacobians of the D_i. From this one proves:

$$\overline{t(\mathfrak{M}_g)} = \left\{ \begin{array}{l} \text{set of pairs } (X, \Theta) \text{ of the following type:} \\ X = \mathrm{Jac}(D_1) \times \ldots \times \mathrm{Jac}(D_k), \\ \quad D_i \ \text{ non-singular curves of genus } g_i \\ \quad \sum g_i = g \\ \Theta = \bigcup_{i=1}^{k} \mathrm{Jac}\,(D_1) \times \ldots \times \Theta_i \times \ldots \times \mathrm{Jac}(D_k) \end{array} \right\}.$$

Note that inside $\overline{t(\mathfrak{M}_g)}$, $t(\mathfrak{M}_g)$ is readily characterized as the set of (X, Θ) with *irreducible* Θ.

We now come to the final and most fascinating point (for me): exploration of the special properties that Jacobians have and that general principally polarized abelian varieties do not have. I would like to thank Harry Rauch, as well as Alan Mayer and John Fay, who introduced me to these questions and helped me see what a subtle thing was going on. The first step is to reconstruct C from Jac, i.e., prove t is injective (Torelli's theorem). Once this is proven, it follows that

$$\begin{aligned} \dim \overline{t(\mathfrak{M}_g)} &= 3g - 3 \\ \dim \mathcal{A}_g &= g(g+1)/2, \end{aligned}$$

hence:

$$g \geq 4 \Longrightarrow \overline{t(\mathfrak{M}_g)} \overset{\subset}{\neq} \mathcal{A}_g.$$

The second point is to try to characterize $\overline{t(\mathfrak{M}_g)}$ by some special properties (Schottky's problem). I know of $\underline{\underline{4}}$ essentially different approaches to these closely related questions. At the outset, however, let me say that none of them seems to me to be a definitive solution to the second question and that I strongly suspect that although many special things about Jac are known, there is much more to be discovered in this direction.

Approach I: Reducibility of $\Theta \cap \Theta_a$

Recall that Θ_a denotes the translate of Θ by a. Almost all classical work on Torelli's theorem is closely related in some way to the lemma:

Lemma. *Let* Jac *be a Jacobian,* Θ *its theta-divisor. Then given* $a \in$ Jac, $a \neq 0$

$$\left\{ \begin{array}{l} \Theta \cap \Theta_a \subset \Theta_b \cup \Theta_c \\ \text{for some } b, c \in \ \mathrm{Jac} \\ \text{distinct from } 0, a \end{array} \right\} \Longleftrightarrow \left\{ \begin{array}{l} \text{for some } x, y \in C, \\ a = \phi(x) - \phi(y) \end{array} \right\}.$$

In fact, what this means is that if $a = \phi(x) - \phi(y)$, then $\Theta \cap \Theta_a$ breaks up into 2 components W_1, W_2 of dimension $g-2$, and $W_1 \subset \Theta_b$, $W_2 \subset \Theta_c$. This lemma is fairly elementary: let's check the easiest implication "$\Longleftarrow$". Using lecture 3, we recall that $\Theta = W_{g-1}$, esp. Θ is the image of:

$$\phi^{(g-1)} : S^{g-1}C \longrightarrow \text{Jac}\,.$$

Then $\Theta \cap \Theta_a$ is the image under $\phi^{(g-1)}$ of:

$$W = \left\{ \mathfrak{A} \in S^{g-1}C \,\middle|\, \begin{array}{l} \exists \mathfrak{A}' \in S^{g-1}C \text{ with} \\ \phi^{(g-1)}(\mathfrak{A}') = \phi^{(g-1)}(\mathfrak{A}) - a \end{array} \right\}$$

hence if $a = \phi(x) - \phi(y)$, by Abel's theorem:

$$W = \left\{ \mathfrak{A} \in S^{g-1}C \,\middle|\, \begin{array}{l} \exists \mathfrak{A}' \in S^{g-1}C \text{ with} \\ \mathfrak{A}' \equiv \mathfrak{A} - x + y \end{array} \right\}$$

Clearly one way for $\mathfrak{A}'$ to exist is if x is one of the points in the divisor $\mathfrak{A}$: i.e., $\mathfrak{A} = \mathfrak{A}_0 + x$, $\mathfrak{A}' = \mathfrak{A}_0 + y$: thus

$$W \supset W_y = \{ \text{ set of divisors } \mathfrak{A}_0 + x,\ \mathfrak{A}_0 \in S^{g-2}C \}\,.$$

The only other way is if $\mathfrak{A} + y$, $\mathfrak{A}' + x$ are distinct linearly equivalent divisors of degree g; but by Riemann-Roch, $\dim |\mathfrak{A} + y| \geq 1$ if and only if there is a 1-form ω, zero on $\mathfrak{A} + y$. Such an ω must have $g - 2$ more zeroes: call these $\mathfrak{A}_0$. Then

$$W \supset W'_y = \left\{ \begin{array}{l} \text{set of divisors } \mathfrak{A}, \text{ where } \mathfrak{A} + \mathfrak{A}_0 + y = \text{ zeroes} \\ \text{of some } \omega \in R_1(C),\ \ \mathfrak{A}_0 \in S^{g-2}C \end{array} \right\}$$

and

$$W = W_x \cup W'_y.$$

Therefore

$$\Theta \cap \Theta_a = \left(\phi^{(g-1)} W_x\right) \cup \left(\phi^{(g-1)} W'_y\right)$$

and it is not hard to see that:

$$\begin{aligned} \phi^{(g-1)}(W_x) &= (W_{g-2})_{\phi(x)} \\ \phi^{(g-1)}(W'_y) &= (-W_{g-2})_{k-\phi(y)} \end{aligned}$$

(where $-W_{g-2}$ is the set of points $-x$, $x \in W_{g-2}$; and $k = \sum_{i=1}^{2g-2} \phi(x_i)$, $\{x_i\}$ the zeroes of some $\omega \in R_1(C)$). Finally, if $b = \phi(x) - \phi(z)$, $c = \phi(w) - \phi(y)$, then the same argument shows:

$$\begin{aligned} \Theta \cap \Theta_b &= (W_{g-2})_{\phi(x)} \cup (-W_{g-2})_{k-\phi(z)} \\ \Theta \cap \Theta_c &= (W_{g-2})_{\phi(w)} \cup (-W_{g-2})_{k-\phi(y)} \end{aligned}$$

hence

$$\Theta \cap \Theta_a \subset \Theta_b \cup \Theta_c.$$

Weil investigated the deeper problem of classifying *all* cases where $\Theta \cap \Theta_a$ was reducible: it appears that for most curves, this only happens if $a = \phi(x) - \phi(y)$ again. But for some curves of genus 3 or 4 or for curves C which are double coverings of elliptic curves, there are other a's for which $\Theta \cap \Theta_a$ is reducible.

However, for *general principally polarized abelian varieties X, it seems very likely that $\Theta \cap \Theta_a$ is irreducible for all $a \in X$.*

There are various ways to use variants of the lemma to prove Torelli's theorem: one can stick to the implication "$\Longleftarrow$", and generalize it substantially, playing an elaborate Boolean algebra game with all the translates of all the $W_r \subset$ Jac. This leads eventually to the conclusion that there are only two possible ways to set up this whole Boolean configuration inside Jac, given the divisor Θ: one being obtained from the other by reflection in the origin. Or using the full strength of the lemma, one sees that Θ determines the surface

$$V = \{\phi(x) - \phi(y) \big| x, y \in C\} \subset \text{ Jac} .$$

If C is not hyperelliptic, it turns out that the tangent cone $T_{V,0}$ to V at $0 \in$ Jac is just a cone over the curve C itself: more precisely, if P is the projective space of 1-dimensional subspaces of $T_{\mathbf{Jac},0}$, then the canonical curve $\Phi(X)$ sits in P, and

$$T_{V,0} = \left(\begin{array}{l} \text{union of lines } \ell, \quad \ell \text{ corresponding to} \\ \text{points of } \Phi(C) \end{array} \right).$$

If C is hyperelliptic, other arguments are needed. Or one may use variants of the lemma where a is infinitesimal. The geometric meaning of $\Theta \cap \Theta_a$, a infinitesimal, is the following: let $\check{P}$ be the projective space of $(g-1)$-dimensional subspaces of $T_{\mathbf{Jac},0}$. Then we get the so-called Gauss map:

$$\pi : \Theta\text{-(singular pts. of } \Theta) \longrightarrow \check{P}$$

defined by

$$\pi(x) = \left(\begin{array}{l} \text{tangent space } T_{\Theta,x}, \text{ translated} \\ \text{to a subspace of } T_{\mathbf{Jac},0} \end{array} \right).$$

The divisors $\pi^{-1}(H)$, $H \subset \check{P}$ a hyperplane, are the limits of the intersections $\Theta \cap \Theta_a$ as $a \to 0$. Thus the lemma says that at least a 1-dimensional family of divisors $\pi^{-1}(H)$ is reducible: in fact, note that hyperplanes in $\check{P}$ are points in P, and if we let $H_x \subset \check{P}$ denote the hyperplane corresponding to $x \in P$, the lemma says that $\pi^{-1}(H_x)$ is reducible for $x \in \Phi(C)$. Andreotti showed that one could say more: let $B \subset \check{P}$ be the branch locus of π, then

$$B \quad = \quad \text{"envelope" of the family of hyperplanes } \{H_x\},$$

i.e.,

$$\begin{aligned} B &= \bigcup_{x \in \Phi(C)} (H_x \cap H_{x+\delta x}) \quad (\delta x = \text{ infinitesimal change of } x) \\ &= \bigcup_{x \in C} \left(\begin{array}{l} \text{linear } (g-2)\text{-space of } \check{P} \text{ dual to the line} \\ \text{in } P \text{ tangent to } \Phi(C) \text{ at } \Phi(x) \end{array} \right). \end{aligned}$$

Again for non-hyperelliptic C's, this enables one to reconstruct $\Phi(C)$ immediately from (Jac, Θ).

Fay has given an analytic form of the lemma: if Jac $= \mathbb{C}^g/L$ and $\vartheta : \mathbb{C}^g \longrightarrow \mathbb{C}$ is the theta function whose zeroes are Θ, then he shows:

$$\begin{aligned} & E(x,v)\cdot E(u,y)\cdot\vartheta\left(\mathbf{z}+\int_u^x\omega\right)\cdot\vartheta\left(\mathbf{z}+\int_v^y\omega\right)\\ +\ & E(x,u)\cdot E(v,y)\cdot\vartheta\left(\mathbf{z}+\int_v^x\omega\right)\cdot\vartheta\left(\mathbf{z}+\int_u^y\omega\right)\\ =\ & E(x,y)\cdot E(u,v)\cdot\vartheta\left(\mathbf{z}+\int_{u+v}^{x+y}\omega\right)\cdot\vartheta(\mathbf{z}) \end{aligned}$$

where $E(x,y)$ is a certain "Prime form" on $C\times C$. In particular, it follows that if $x\neq u$:

$$\vartheta(\mathbf{z})=\vartheta\left(\mathbf{z}+\int_u^x\omega\right)=0\Longrightarrow\vartheta\left(\mathbf{z}+\int_v^x\omega\right)=0\ \text{ or }\ \vartheta\left(\mathbf{z}+\int_u^y\omega\right)=0$$

which is the "$\Longleftarrow$" of the lemma. Another pretty way to interpret this half of the lemma is via the Kummer variety: one uses the set of theta-functions of order 2 to map Jac to a projective space. All these functions are *even*, so the map factors through $K=\text{Jac}/(\pm 1)$ (here $-1=$ inverse of group law on Jac), and defines:

$$\Psi: K\hookrightarrow\mathbb{P}^k,\qquad k=2^g-1.$$

Then we find that $\Psi(K)$ has *trisecants*; more precisely, for any $x,y,u,v\in C$, fix a point $a\in$ Jac such that:

$$2a=\phi(x)+\phi(y)-\phi(u)-\phi(v).$$

Write $a=\frac{1}{2}(x+y-u-v)$ for clarity. Define $\frac{1}{2}(x-y+u-v)$ and $\frac{1}{2}(x-y-u+v)$ as $a-\phi(y)+\phi(u)$ and as $a-\phi(y)+\phi(v)$. Then:

$$\Psi\left(\frac{1}{2}(x+y-u-v)\right),\qquad\Psi\left(\frac{1}{2}(x-y+u-v)\right),\qquad\Psi\left(\frac{1}{2}(x-y-u+v)\right)$$

are collinear. Contrast this with the situation for generic principally polarized abelian varieties: because $\dim K\ll\dim\mathbb{P}^k$ for g large, it seems very likely that $\Psi(K)$ has no trisecants.

In connection with the Schottky problem, I would like to raise the following questions: given a principally polarized abelian variety (X,Θ), suppose there is a 2-dimensional set of points $a\in X$ such that $\Theta\cap\Theta_a$ is contained in $\Theta_b\cup\Theta_c(\{b,c\}\cap\{0,a\}=\phi)$. Then is X a Jacobian? Or if not, are there some small extra conditions that suffice to characterize Jacobians?

Approach II: Θ of translation type

Since $\Theta=W_{g-1}$, Θ is just the sum, using the group law of Jac, of the curve $\phi(C)=W_1$ with itself $(g-1)$-times. One can localize this property and come up with the following:

– Let H be a germ of a hypersurface at $0\in\mathbb{C}^n$.

– Then H is of *translation type* if there are $(n-1)$ germs of analytic curves γ_i at $0 \in \mathbb{C}^n$ such that $H = \gamma_1 + \ldots + \gamma_{n-1}$. (+ represents vector addition pointwise of these subsets of $\mathbb{C}^n$.)

In fact, since up to translation Θ is symmetric, i.e., $-\Theta + k = \Theta$ for some $k \in$ Jac, whereas $\phi(C)$ is symmetric only when C is hyperelliptic, we find that for non-hyperelliptic C, Θ is *doubly of translation type*: i.e., for all $x \in \Theta$, represent Θ near x as a sum:

$$\begin{aligned}(\text{germ of } \Theta \text{ at } x) &= \gamma_1^{(x)} + \ldots + \gamma_{g-1}^{(x)} + x\\ \gamma_i^{(x)} &= \text{a 1-dimensional germ at } 0.\end{aligned}$$

Then if $-\Theta + k = \Theta$, so that $k - x \in \Theta$

$$(\text{germ of } \Theta \text{ at } x) = -\gamma_1^{(k-x)} - \ldots - \gamma_{g-1}^{(k-x)} + x$$

gives a 2nd representation of Θ as a hypersurface of translation type. The beautiful fact, which was conjectured by Sophus Lie and proven by W. Wirtinger, is that the only hypersurfaces doubly of translation type are the theta divisors in Jacobians and certain degenerate limits. Moreover, the theta divisor is never of translation type in a *third* way, which then proves Torelli's theorem! The following sort of answer to the Schottky problem is presumably a consequence – although the details have not been written down: given a principally polarized abelian variety (X, Θ),

$$\begin{pmatrix} \Theta \text{ is doubly of translation} \\ \text{type at some point } x \in \Theta \end{pmatrix} \Longleftrightarrow \begin{pmatrix} (X,\Theta) \text{ is the Jacobian of a} \\ \text{non-hyperelliptic curve} \end{pmatrix}.$$

The only thing lacking here is a nice differential-geometric criterion for a hypersurface to be singly or doubly of translation type. Also, since Θ is symmetric, one would hope that "usually" being *simply* of translation type by analytic prolongation would force it to be *doubly* so: but this is not clear.

Recently, Saint-Donat discovered a very beautiful proof of the Lie-Wirtinger results that I want to sketch. It is based on the following beautiful criterion:

Theorem. *Let $H \subset \mathbb{P}^n$ be a hyperplane, let $x_1, \ldots, x_d \in H$ and let $\gamma_1, \ldots, \gamma_d$ be germs of analytic curves at $x_1, \ldots, x_d$ which cross H transversely. Suppose $t_1, \ldots, t_d$ are coordinate functions on $\gamma_1, \ldots, \gamma_d$ such that for all hyperplanes H' near H:*

$$\sum_{i=1}^{d} t_i\,(H' \cap \gamma_i) = 0.$$

Then by analytic continuation, the $\gamma_1, \ldots, \gamma_d$ are part of an algebraic curve $\Gamma \subset \mathbb{P}^n$ of degree d (possibly reducible), so that in some neighbourhood U of H, $\Gamma \cap U = \bigcup_{i=1}^{d} \gamma_i \cap U$.

This can be proven, e.g., by reducing to $n = 2$ and in this case showing that in some neighbourhood U of H, there is a meromorphic 2-form ω on U, with simple poles on $\cup\gamma_i \cap U$, such that

$$t_i(z) = \int_{x_i}^{z} \mathrm{Res}_{\gamma_i}\omega,$$

and finally using the pseudoconcavity of U to extend ω to a rational 2-form on $\mathbb{P}^2$, whose poles will be Γ. To apply the theorem, let

$$H \subset \mathbb{C}^n$$

be a germ of a hypersurface such that

$$H = \sum_{i=1}^{n-1} \gamma_i = \sum_{i=1}^{n-1} \delta_i, \qquad \gamma_i, \delta_i \text{ germs of analytic curves.}$$

Let P denote the projective space of lines in $\mathbb{C}^n$ through 0. Associating to each point x of γ_i or δ_i the tangent line to γ_i or δ_i at x and translating to the origin, we get germs of analytic curves $\dot{\gamma}_i, \dot{\delta}_i \subset P$. For each $z \in H$, write

$$(*) \qquad z = \sum_{i=1}^{n-1} \gamma_i(x_i) = \sum_{i=1}^{n-1} \delta_i(y_i).$$

Then $T_{H,z}$ defines a hyperplane $H(z) \subset P$, and since $T_{H,z} \supset T_{\gamma_i,x_i} \cup T_{\delta_i,y_i}$, it follows that $\dot{\gamma}_i(x_i), \dot{\delta}_i(x_i) \in H(z)$. Now parametrize the branches γ_i and δ_i by any linear function L on $\mathbb{C}^n$, i.e.,

$$L(\gamma_i(x)) = x, \qquad L(\delta_i(y)) = y.$$

Then it follows from $(*)$ that:

$$\sum_{i=1}^{n-1} x_i = L\left(\sum \gamma_i(x_i)\right) = L(z) = L\left(\sum \delta_i(y_i)\right) = \sum_{i=1}^{n-1} y_i.$$

Assuming that all hyperplanes H' near $H(0)$ are of the form $H(z)$, this proves that the $2n-2$ branches $\dot{\gamma}_i, \dot{\delta}_i$ with coordinate functions $x_i, -y_i$ satisfy the condition of the theorem! Analytically prolonging, the $\dot{\gamma}_i, \dot{\delta}_i$ therefore are part of a curve $C \subset P$ and one goes on to prove that C is a canonically embedded curve of genus n (or a singular limit of such) and H is its theta divisor.

Approach III: Singularities of Θ

We use the fact that $\Theta = W_{g-1}$ and apply the results of Lecture III: it follows that every $\alpha \in \Theta$ is equal to $\phi^{(g-1)}(\mathfrak{A})$ for some divisor $\mathfrak{A} = \sum_{i=1}^{g-1} x_i$ in C. If $\ell = \dim |\mathfrak{A}|$, i.e.,

$$\left(\phi^{(g-1)}\right)^{-1}(\alpha) \cong \mathbb{P}^\ell,$$

then Kempf's results show in this case:

1) Θ is defined near α by an equation $\det(f_{ij}) = 0$, where f_{ij} is an $(\ell+1)\times(\ell+1)$ matrix of holomorphic functions at α,
2) the multiplicity of Θ at α is $\ell+1$ and in fact the tangent cone to Θ is defined by the polynomial equation $\det(df_{ij}) = 0$ on $T_{\alpha,\mathrm{Jac}}$.

Let Sing Θ denote the set of singular points on Θ. Then it is not hard to see that $\mathrm{Sing}_2\Theta$, the set of double points, is dense in Sing Θ; also, by the results quoted in Lecture I,

$$\begin{aligned} g \geq 4 \quad &\Longrightarrow \quad \exists \text{ at least one map } \pi : C \longrightarrow \mathbb{P}^1 \text{ of degree } d \leq g-1 \\ &\Longrightarrow \quad \exists \text{ at least one } \mathfrak{A} \in S^{g-1}C \text{ with } \dim|\mathfrak{A}| \geq 1 \\ &\Longrightarrow \quad \mathrm{Sing}\ \Theta \neq \emptyset. \end{aligned}$$

But if α is a double point of Θ, Θ is defined near α by an equation

$$f_{11}f_{22} - f_{12}f_{21} = 0.$$

Therefore Θ is also singular at any point x where $f_{11}(x) = f_{12}(x) = f_{21}(x) = f_{22}(x) = 0$. But 4 equations define a set of points of codimension at most 4, hence

3) Sing $\Theta \neq \emptyset$ and all components have $\dim \geq g-4$[36].

In fact, a closer analysis shows that:

3′) (C hyperelliptic) $\Longrightarrow$ (Sing Θ irreducible of dim. exactly $g-3$)
(C not hyperelliptic) $\Longrightarrow$ $\left(\begin{array}{l}\text{all comp. of Sing } \Theta \text{ have dim.} \\ \text{exactly } g-4\end{array}\right)$.

Notice, for instance, that this was exactly what we found in Lecture III if $g = 3$ or $g = 4$. This immediately distinguishes Jacobians from generic abelian varieties when $g \geq 4$, because for almost all $(X, \Theta) \in a_g$, Θ will be non-singular! In fact, Andreotti and Mayer prove:

---

[36] A heuristic argument for "proving" this is to count the dimension of the space S of coverings C of $\mathbb{P}^1$, of degree $g-1$, simple branching, and genus g. Simple topology shows that there must be $4g-4$ branch points, so we get $\dim S = 4g-4$. Looking at the curve C, we get a morphism $p : S \to \mathfrak{M}_g$, hence almost all fibres of p have $\dim g-1$, i.e., almost all curves C admit of $g-1$-dimensional family of maps $\pi : C \to \mathbb{P}^1$. Allowing for the 3-dimensional automorphism group of $\mathbb{P}^1$, this gives a $g-4$-dimensional family of 1-dimensional linear systems $L \subset S^{g-1}C$, hence $\dim$ Sing $\Theta \geq g-4$.

Theorem. *Let*

$$\mathcal{A}_g^{(n)} = \{(X,\Theta) \in \mathcal{A}_g \mid \dim\ \mathrm{Sing}\ \Theta \geq n\}.$$

Then $\overline{t(\mathfrak{M}_g)}$ is a component of $\mathcal{A}_g^{(g-4)}$.

The proof of the former fact and one ingredient of the proof of the theorem is the *heat equation* that ϑ satisfies: if we describe a principally polarized abelian variety X as above

$$X = \mathbb{C}^g / \mathbb{Z}^g + \Omega \cdot \mathbb{Z}^g$$

then in the classical normalization:

$$\vartheta(z) = \sum_{n \in \mathbb{Z}^g} e^{2\pi i({}^t n z + \frac{1}{2}{}^t n \Omega n)}$$

and it is immediate that considering ϑ as a function of z *and* Ω:

$$\frac{1}{2\pi i} \cdot \frac{\partial^2 \vartheta}{\partial z_i \partial z_j} = (1 + \delta_{ij}) \cdot \frac{\partial \vartheta}{\partial \Omega_{ij}}.$$

If $\alpha \in \Theta$ is a double point, and H_α is the hyperplane $\sum_{i,j=1}^g \frac{\partial^2 \vartheta}{\partial z_i \partial z_j}(\alpha) \cdot d\Omega_{ij} = 0$ in the tangent space to $\mathcal{A}_g$ at (X,Θ), then this shows that the singularity α "disappears" if you move (X,Θ) in a direction transversal to H_α. The idea of Andreotti and Mayer's proof is to show that for almost all curves C, corresponding to points $\gamma \in \mathfrak{I}_g$, $\tilde{t}(\mathfrak{I}_g)$ is non-singular at $\tilde{t}(\gamma)$ and its tangent space is the intersection of these H_α's, hence upstairs in $\mathfrak{H}_g$, $t(\mathfrak{I}_g)$ and the inverse image of $\mathcal{A}_g^{(g-4)}$ are *both* non-singular with the *same* tangent space at $\tilde{t}(\gamma)$. This will prove their theorem.

Now if

$$X = \mathbb{C}^g / \mathbb{Z}^g + \Omega \cdot \mathbb{Z}^g \cong \mathrm{Jac}\ (C),$$

let ω_i be the differential on C gotten by restricting dz_i to C; then one shows that if C is not hyperelliptic, $\tilde{t}(\mathfrak{I}_g)$ is non-singular at $\tilde{t}(\gamma)$ and its tangent space is defined by equations $\sum \lambda_{ij} d\Omega_{ij} = 0$ for all $\{\lambda_{ij}\}$ such that the quadratic differential $\sum \lambda_{ij}\omega_i\omega_j$ vanishes identically on C. More canonically, the point here is that the cotangent spaces to $\mathfrak{I}_g$ and $\mathfrak{H}_g$ can be identified as follows:

$$T^*_{\mathfrak{I}_g,\gamma} \cong R_2(C), \quad \text{quadratic differentials on } C$$

$$T^*_{\mathfrak{H}_g,\Omega} \cong \mathrm{Symm}^2 \begin{pmatrix} \text{space of transl.-inv.} \\ \text{1-forms on } X \end{pmatrix}$$

and when $X = \mathrm{Jac}$, $\tilde{t}^*$ is multiplication taking quadratic expressions in the $\omega \in R_1(C)$ to the corresponding quadratic differentials; the kernel is thus the quadratic forms in $R_1(C)$ which vanish identically on C: call this $\mathrm{Ker}(\mathrm{Symm}^2 R_1 C \to R_2 C)$ or I_2. So what Adreotti and Mayer need is that for almost all curves C, I_2 is spanned by the forms:

$$q_\alpha = \sum \frac{\partial^2 \vartheta}{\partial z_i \partial z_j}(\alpha) \cdot [\omega_i] \cdot [\omega_j], \qquad \alpha = \text{ double pt. of } \Theta.$$

Looking back at Lecture III, we can see what these special quadratic forms are: we take $\mathfrak{A}$ of degree $g-1$ such that $\dim|\mathfrak{A}| = 1$, i.e., $L(\mathfrak{A})$ has a basis $\{1, f\}$. By Riemann-Roch, $R_1(-\mathfrak{A})$ is 2-dimensional: let ω_1, ω_2 be a basis. It follows that $\eta_i = f\omega_i$, $i = 1, 2$, have no poles, hence are in $R_1(C)$. Then the 2 quadratic differentials $\eta_1\omega_2$ and $\eta_2\omega_1$ are equal, i.e.,

$$q_{\mathfrak{A}} = [\eta_1] \cdot [\omega_2] - [\eta_2] \cdot [\omega_1]$$

is a quadratic form in $R_1(C)$ which vanishes on C (equivalently represents a quadric in $\mathbb{P}^{g-1}$ vanishing on $\Phi(C)$). According to the results of Lecture III, if $\alpha = \phi^{(g-1)}(\mathfrak{A})$, then $q_\alpha = \text{constant} \cdot q_{\mathfrak{A}}$. It appears to be an open question whether or not for *every* non-hyperelliptic C, these $q_{\mathfrak{A}}$'s span $I_2 = \operatorname{Ker}\left(\operatorname{Symm}^2 R_1 C \to R_2 C\right)$. However, Andreotti and Mayer were able to check this for C which were triple covers of $\mathbb{P}^1$, hence it does hold for *almost all* C and their theorem is proven.

This approach does not establish Torelli's theorem for all curves C, but it does show that for *almost all* C's, $t^{-1}(t(C)) = \{C\}$. In fact, for almost all C, 2 good things happen – i) in the canonical embedding $\Phi : C \to \mathbb{P}^{g-1}$, $\phi(C)$ is the intersection of the quadrics containing it, and ii) the space I_2 of quadrics through $\Phi(C)$ is generated by the tangent cones q_α to Θ at its double points. Thus we have a simple prescription for recovering C from (Jac, Θ) when C is "good": i) take the tangent cone to Θ at all its double points, ii) translate to the origin of Jac and projectivize to get a quadric in $\mathbb{P}^{g-1}$, iii) intersect all these quadrics: this is generally the C you started with!

These ideas, though very elegant, do not seem to work without exception – e.g., $\mathcal{A}_g^{(g-4)}$ has other components besides $\overline{t(\mathfrak{M}_g)}$.

Approach IV: Prym varieties

The final approach to the Schottky problem is due to Schottky himself, in collaboration with Jung. One may start like this: since the curve C has a non-abelian π_1, can one use the *non-abelian coverings* of C to derive additional invariants of C which will be related by certain identities to the natural invariants of the "abelian part of C", i.e., to the theta-nulls of the Jacobian? And then, perhaps, use this whole set of identities to show that the theta-nulls of the Jacobian alone satisfy non-trivial identities? Now the simplest non-abelian groups are the dihedral groups, and this leads us to consider unramified covering spaces:

$$\begin{array}{c} C_2 \\ \big\downarrow \text{ degree } n, \text{ abelian covering, group } A \\ C_1 \\ \big\downarrow \text{ degree 2} \\ C \end{array}$$

where the involution $\iota : C_1 \to C_1$ of C_1 over C lifts to an involution on C_2 and $\iota\alpha\iota^{-1} = -\alpha$, all $\alpha \in A$. These, in turn, may be constructed by starting with the degree 2 covering C_1, taking its Jacobian Jac_1, and taking the "odd" part of Jac, when it is decomposed into a product of even and odd pieces under ι. More precisely, we define:

$$\begin{aligned} \mathrm{Prym}\,(C_1/C) &= \left(\begin{array}{l}\text{subabelian variety of } \mathrm{Jac}_1 \text{ of all} \\ \text{points } x - \iota(X), \quad x \in \mathrm{Jac}_1\end{array}\right) \\ &= \left(\begin{array}{l}\text{connected component of the set of} \\ x \in \mathrm{Jac}_1 \text{ such that } \iota(x) = -x.\end{array}\right) \end{aligned}$$

There is a natural map

$$\phi_- : C_1 \longrightarrow \mathrm{Prym}$$

given by

$$\begin{aligned} &\phi_-(x) = \phi_1(x) - \iota\left(\phi_1(x)\right), \quad \text{with} \\ &\phi_1 : C_1 \longrightarrow \mathrm{Jac}_1 \quad \text{the canonical map.} \end{aligned}$$

Then the coverings C_2 in question are pull-backs of abelian coverings of Prym via ϕ_-.

Now Jac_1 is very nearly the product of Jac and Prym: in fact there are homomorphisms

$$\mathrm{Jac} \times \mathrm{Prym} \underset{\beta}{\overset{\alpha}{\rightleftarrows}} \mathrm{Jac}_1$$

such that $\alpha \circ \beta$, $\beta \circ \alpha$ are multiplication by 2, and $\ker\alpha$, $\ker\beta$ are finite abelian groups isomorphic to $(\mathbb{Z}/2\mathbb{Z})^{2g-1}$ (g = genus of C). The genus of C_1 will be $2g-1$, hence

$$\begin{aligned} \dim \mathrm{Jac}_1 &= 2g-1 \\ \dim \mathrm{Jac} &= g \\ \dim \mathrm{Prym} &= g-1. \end{aligned}$$

The beautiful and surprising fact is that the new abelian variety Prym carries a canonical principal polarization too. In fact, if $\Theta \subset \mathrm{Jac}$, $\Theta_1 \subset \mathrm{Jac}_1$ and $\Xi \subset \mathrm{Prym}$ are the 3 theta-divisors, Ξ is characterized by either of the properties:

$$\alpha^{-1}(\Theta_1) \sim 2\Theta + 2\Xi$$
$$\beta^{-1}(\Theta + \Xi) \sim 2\Theta_1$$

(where $\sim$ means the fundamental classes of the divisors are cohomologous; or equivalently that suitable translates of the divisors are linearly equivalent). So far, these facts tie Jac, Jac_1 and Prym into a tight but quite elementary configuration of abelian varieties, but one that does not impose any restriction on Jac itself. Thus if ϑ, ϑ_1 and ξ are the theta functions of these three abelian varieties, one can calculate ϑ_1 from ϑ and ξ and vice-versa, but ϑ and ξ can be arbitrary theta-functions of g and $g-1$ variables respectively. But now the underlying configuration of curves comes in and tells us:

$$(*) \qquad (\mathrm{Jac} \times (0)) \cap \alpha^{-1}\Theta_1 = \Theta + \Theta_\eta$$

where $\eta \in \mathrm{Jac}$ is the one non-trivial point of order 2 such that $\alpha(\eta) = 0$ (i.e., the original double cover C_1/C "corresponds" to η). This follows in fact directly from the interpretations $\Theta = W_{g-1} \subset \mathrm{Jac}$ and $\Theta_1 = W_{2g} \subset \mathrm{Jac}_1$. Now ϑ and ξ cannot be arbitrary any longer: $(*)$ turns out to be equivalent to asserting that the squares of the theta-nulls of Prym:

$$\xi^2 \begin{bmatrix} \alpha \\ \beta \end{bmatrix} (0), \qquad \alpha, \beta \in \frac{1}{2}\mathbb{Z}^{g-1}$$

are *proportional* to certain monomials in the theta-nulls of Jac:

$$\vartheta \begin{bmatrix} \alpha\, 0 \\ \beta\, 0 \end{bmatrix} (0) \cdot \vartheta \begin{bmatrix} \alpha\, 0 \\ \beta\, 1 \end{bmatrix} (0), \qquad \alpha, \beta \in \frac{1}{2}\mathbb{Z}^{g-1}$$

(after one makes the correct simultaneous coordinatization of Prym and Jac). A third way to interpret $(*)$ is via the Kummer variety: embed $\mathrm{Jac}/\pm 1$ and $\mathrm{Prym}/\pm 1$ in projective spaces.

$$\begin{aligned} \Phi: &\quad \mathrm{Jac}/\pm 1 \hookrightarrow \mathbb{P}^{2^g-1} = \mathbb{P}_j \\ \Psi: &\quad \mathrm{Prym}/\pm 1 \hookrightarrow \mathbb{P}^{2^{g-1}-1} = \mathbb{P}_p. \end{aligned}$$

Via suitable normalized coordinate systems as in Lecture III, there is a canonical way to identify $\mathbb{P}_p$ with a linear subspace of $\mathbb{P}_j$. Then $(*)$ says:

$$(**) \qquad \Psi(0) = \Phi\left(\frac{\eta}{2}\right).$$

Since Im Φ, Im Ψ have such large codimension, one certainly expects that for most g and $(g-1)$-dimensional principally polarized abelian varieties X, Y, $\Psi(Y)$ and $\Phi(X)$ would be disjoint.

The case when $g = 4$ is the first one where $\overline{t(\mathfrak{M}_g)} \subsetneq \mathcal{A}_g$ and in this case:

$$\dim \mathcal{A}_4 = 10$$
$$\dim \overline{t(\mathfrak{M}_4)} = 9.$$

Schottky was able to show that the above identities on ϑ and ξ implied one identity on ϑ alone, of degree 8, and Igusa has asserted that this identity holds only on $\overline{t(\mathfrak{M}_4)}$! Moreover, when $g > 4$, no efficient method of eliminating ξ from the above identities is known and the ultimate problem of characterizing $\overline{t(\mathfrak{M}_g)}$ by simple identities in the theta-nulls remains open. I am confident that Schottky's approach has not been exhausted, however, and a full theta-function theoretic analysis of the dihedral (or even higher non-abelian) coverings of C remains to be carried out.

I hope these lectures have perhaps convinced the patient reader that nature's secrets in this corner of existence are fascinating and subtle and worthy of his time!

Survey of Work on the Schottky Problem up to 1996

Enrico Arbarello

Approach I.

The reducibility condition for the theta divisor on a p.p.a.v. X can be written as

$$\Theta_a \cap \Theta_b \subset \Theta_c \cup \Theta_{-c}$$

(here $\pm c \neq a, b$). If $a \neq -b$, this condition expresses the fact that $\psi(a), \psi(b), \psi(c)$ belong to a trisecant of the Kummer variety $K = K(X)$. Gunning [Gu] defines the locus

$$V_{a,b,c} = \{2x \mid \Theta_{x+a} \cap \Theta_{x+b} \subset \Theta_{x+c} \cup \Theta_{-x-c}\} \subset X$$

($V_{a,b,c}$ has a natural scheme structure) and shows that if $V_{a,b,c}$ is positive dimensional, and if some mild condition is satisfied, then $V_{a,b,c}$ is a smooth curve and X is its Jacobian. So, in this sense,

"infinitely many trisecants" $\Rightarrow$ "X is a Jacobian".

In [W3], Welters defines a subscheme $V_Y \subset X$, for any length 3 artinian subscheme $Y \subset X$ (except for the case $Y = \operatorname{Spec} \mathbb{C}[\epsilon_1, \epsilon_2] / (\epsilon_1, \epsilon_2)^2$) by letting

$$V_Y = \{2x \mid \exists \text{ a line } \mathrm{l} \subset \mathbb{P}^{2^g-1}, \text{ such that } \psi^{-1}(\mathrm{l}) \supset \mathrm{Y} + 2\mathrm{x}\},$$

and proves, without extra conditions, that

"V_Y is positive dimensional" $\Rightarrow$ "V_Y is a smooth curve and X is its Jacobian".

It is easy to see that V_Y always contains $Y - s$, where $\mathrm{Supp Y} = \{\mathrm{a, b, c}\}$, and $s = a + b + c$ (with a, b, c possibly coincident). Welters asks the question: is the condition $V_Y \not\supseteq Y - s$ sufficient to insure that V_Y is a smooth curve and X is its Jacobian? This is the so-called trisecant conjecture.

Particular cases of this question are: is the existence of a trisecant or of an inflectionary tangent to the Kummer variety sufficient to conclude that X is a Jacobian? These questions are analyzed and partially answered in [D2] and [M2], but they remain open.

Another particular case of this question is the one in which $Y = \{a + \epsilon D_1 + \epsilon^2 D_2\}$, $2x \in Y - s$, $2x + \epsilon D_1 + \epsilon^2 D_2 + \epsilon^3 D_3 \in V_Y$. This translates into the K.P. equation

$$D_1^4\theta\cdot\theta - 4D_1^3\theta\cdot D_1\theta + 3(D_1^2\theta)^2$$
$$-3(D_2\theta)^2 + 3D_2^2\theta\cdot\theta + 3D_1\theta\cdot D_3\theta - 3D_1D_3\theta\cdot\theta + d\theta\cdot\theta = 0,$$

where d is a suitable constant. This case is Novikov's conjecture solved by T. Shiota [S].

The relationship between the K.P. hierarchy of differential equations and the scheme V_Y is established in [A-D2]. Following the point of view of [A], [A-D2], a geometrical proof of Shiota's Theorem is given in [M1]. There, to overcome the difficulties coming from the singularities of the theta divisor, an important role is played by the following fundamental result of Ein and Lazarsfeld [E-L]:

If X is a g-dimensional, indecomposable p.p.a.v. and Θ its theta divisor, then

$$\dim(\Theta_{\text{sing}}) < g - 2.$$

Approach II.

One can read a complete survey of the classical point of view in the papers by J. Little [Li1], [Li2]. There, one can find the developments of the ideas in Saint-Donat's paper [S-D]. A discussion of Darboux's criterion [D] for the algebraicity of local analytic arcs can be found in [Gr].

In [Li3] it is proved that the existence of a curve of flexes of the Kummer variety is equivalent to the double translation property for the Θ-divisor, thus relating Approaches I and II.

Approach III.

The loci $\mathcal{A}_g^{(k)}$ (often denoted by $\mathcal{N}_k$) have been studied especially in the case $k \le g-4$.

In [B 1], Beauville proves that $\mathcal{A}_4^{(0)}$ is the union of $\mathrm{t}(\mathcal{M}_4)$ and the locus θ_{null} of 4-dimensional p.p.a.v. with a vanishing theta-null.

$\mathcal{A}_5^{(1)}$ has five components that are described in [De1] and [Do4]. In [De1] a list of components for $\mathcal{A}_g^{(g-4)}$ is described.

In [B-D1], a bridge between Approaches I and III is established by showing that the locus of g-dimensional p.p.a.v. whose associated Kummer has a trisecant line is contained in $\mathcal{A}_g^{(g-4)}$.

As it was mentioned above, the case $k = g-2$ has been settled by Ein and Lazarsfeld:

$$\mathcal{A}_g^{(g-2)} = \text{decomposable p.p.a.v.}$$

The study of tangent cones at singular points of the theta-divisor has been carried out by M. Green in [G] where he proves the following theorem: *For* *any non-hyperelliptic curve C of genusg ≥ 4, the space of quadrics through the canonical curve is spanned by tangent cones at double points of its theta-divisor.*

When the curve in question is non-trigonal, and not a smooth plane quintic, this gives a new proof of Torelli's theorem. Green's proof relies on a subtle theorem by Kempf [K] which asserts that, when C is not hyperrelliptic, the Kodaira-Spencer map:

$$H^1(C, \mathcal{T}_C) \to H^1(C^{(g-1)}, \mathcal{T}_{C^{(g-1)}})$$

is an isomorphism.

A detailed analysis of the local Torelli theorem including the case of hyperelliptic curves is contained in [O-S].

The heat equation satisfied by the theta function of an abelian variety, which plays a central role in the paper of Andreotti and Mayer, has been studied by Welters in [W1]. This paper was instrumental in Hitchin's laying the foundation of a parallel theory in the case of higher rank vector bundles on curves [H].

Approach IV.

Using Kummer varieties and suitable normalized coordinate systems one can define the Schottky loci $\mathcal{S}_g$ and $\mathcal{S}_g^{(\text{big})}$ to be

$$\mathcal{S}_g^{(\text{big})} = \{X \in \mathcal{A}_g \mid \exists\, \eta \in X\backslash\{0\} \text{ of order } 2\text{, and } P \in \mathcal{A}_{g-1},\ \text{s.t. } \Psi(0) = \Phi(\tfrac{\eta}{2})\}$$

$$\mathcal{S} = \{X \in \mathcal{A}_g \mid \forall\, \eta \in X\backslash\{0\} \text{ of order } 2,\ \exists\, P \in \mathcal{A}_{g-1},\ \text{s.t. } \Psi(0) = \Phi(\tfrac{\eta}{2})$$

By virtue of Prym varieties one has

$$t(\mathcal{M}_g) \subseteq \mathcal{S}_g \subseteq \mathcal{S}_g^{(\text{big})}\,.$$

In [vG1], van Geemen proves that $t(\mathcal{M}_g)$ is an irreducible component of $\mathcal{S}_g$ and Donagi [Do1] proves that $t(\mathcal{M}_g)$ is an irreducible component of $\mathcal{S}_g^{(\text{big})}$.

An important tool in the proofs is Welter's characterization (for $g \geq 5$) of the difference $C - C \subset J(C)$ as the support of the base locus of the linear system $\Gamma_{00} = |\{s \in \mathcal{O}_X(2\Theta) \mid \text{mult}_0(s) \geq 4\}|$. This result was conjectured in [vG-vG]. There it is also conjectured that this equality is yet another way to characterize Jacobians, to the point that what should happen is: *if X is not a Jacobian, the support of Γ_{00} is* $\{0\} \in X$. In the same paper an infinitesimal version of this conjecture is also discussed.

In [B-D3] it is proved that for a generic p.p.a.v. of genus $g \geq 4$ the base locus of Γ_{00} is finite, and that the infinitesimal version of the conjecture holds. Moreover, the conjecture itself is related with the trisecant conjecture. In [Iz] the conjecture is proved for $g = 4$.

Analogous to the Torelli map t is the Prym map

$$p : \mathcal{R}_{g+1} \to \mathcal{A}_g\ ,$$

where $\mathcal{R}_{g+1}$ is the moduli space of unramified 2-sheeted coverings of genus $g+1$ curves, or equivalently, of pairs (C, η) where C is a smooth curve and $\eta \in H^1(C, \mathbb{Z}_2)$ a non-zero half period.

The Prym map is generically injective for $g \geq 6$ [F-S] but, unlike the Torelli map, it is never injective because of Donagi's tetragonal construction [Do 4].

Turning the attention to low genera, Igusa [I] and Freitag [Fr] prove that $t(\mathcal{M}_4) = \mathcal{S}_4 = \mathcal{S}_4^{(\mathrm{big})}$.

Also, the Prym map is generically surjective for $g \leq 5$ and its structure has been extensively analyzed in [D-S] and [Do5].

Donagi, in an unpublished work, establishes the equality $t(\mathcal{M}_5) = \mathcal{S}_5$. On the other hand, $\mathcal{S}_5 \neq \mathcal{S}_5^{(\mathrm{big})}$, as the latter contains intermediate Jacobians of cubic threefolds [Do2].

The conjecture $t(\mathcal{M}_g) = \mathcal{S}_g$ is still open.

The Schottky problem for Prym varieties is studied in [B-D2] where it is proved that the moduli space of Prym varieties of dimension ≥ 7 is an irreducible components of the locus of p.p.a.v. whose associated Kummer varieties admit a *quadrisecant* plane.

Finally, two words about new ideas. Let $(X = \mathbb{C}^g/\Lambda, \Theta)$ be p.p.a.v.; Buser and Sarnak defined a metric invariant $m(X) = m(X, \Theta)$ by setting it equal to the minimum square length, with respect to the Hermitian form, of a non-zero lattice vector. Then, on the one hand, they show that for a general p.p.a.v. X,

$$m(X) \geq \frac{1}{\pi}(2g!)^{1/g} \approx \frac{g}{\pi e},$$

while on the other hand, they show that for a Jacobian $J(C)$,

$$m(J(C)) \leq \frac{3}{\pi} \log(4g+3) .$$

As a consequence, for $g \gg 0$, the Jacobian locus $t(\mathcal{M}_g)$ lies in a small neighbourhood of the boundary of $\mathcal{A}_g$.

Lazarsfeld [L] takes a different point of view. He considers the Seshadri constant

$$\epsilon(X) = \epsilon(X, \Theta) = \sup\{\epsilon \geq 0 \mid f^* c_1(\Theta) - \epsilon[E] \text{ is nef}\},$$

where f is the blow-up of X at one of its points x, and E is the exceptional divisor of f (by homogeneity the definition does not depend on x). He proves that

$$\epsilon(X) \geq \frac{\pi}{4} m(X).$$

In particular, for a general p.p.a.v. X,

$$\epsilon(X) \geq \frac{1}{4}(2g!)^{1/g} \approx \frac{g}{4e}.$$

He then gets upper bounds for Jacobians of curves

$$\epsilon(J(C)) \leq g^{1/g},$$

and, even more precisely, for Jacobians of d-gonal curves

$$\epsilon(J(C)) \leq \frac{4d}{\pi}.$$

In conclusion, the minimal length of lattice vectors and the Seshadri constant behave in a similar way and both tend to distinguish Jacobians of curves among all principally polarized abelian varieties.

References

The Red Book of Varieties and Schemes

1. Auslander, L., Mackenzie, R.E.: Introduction to Differentiable Manifolds. McGraw-Hill: New York 1963
2. Atiyah, M.: K-Theory. W.A. Benjamin: New York 1967
3. Bourbaki, N.: Commutative Algebra 1–7. Springer: Heidelberg 1988. Orig. publ. by Addison-Wesley, Reading, MA, 1972
4. Bourbaki, N.: Algèbre Ch. 8: Modules et Anneaux Semi-Simples. Hermann: Paris 1958
5. Bourbaki, N.: Algèbre Commutative 1–4. Reprint Masson: Paris 1985
6. Grothendieck, A., Dieudonné, J.A.: Eléments de Géométrie Algébrique I. Springer: Berlin-Heidelberg 1971
7. Grothendieck, A., Dieudonné, J.A.: Eléments de Géométrie Algébrique III/1. Publ. Math. IHES **11**: Paris 1961
8. Gunning, R.C., Rossi, H.: Analytic Functions of Several Complex Variables. Prentice Hall 1965
9. Klein, F.: Vorlesungen über die Entwicklung der Mathematik im 19. Jahrhundert. Springer: Berlin 1926
10. Lang, S.: Introduction to Algebraic Geometry. Wiley-Interscience: New York 1958
11. Semple, J.G., Roth, L.: Introduction to Algebraic Geometry. Clarendon Press: Oxford 1985
12. Serre, J.-P.: Algèbre Locale. Multiplicités. Lect. Notes in Math. vol. 11. Springer: Berlin-Heidelberg 1975
13. Zariski, O., Samuel, P.: Commutative Algebra 1. Springer New York 1975
14. Zariski, O., Samuel, P.: Commutative Algebra 2. Springer New York 1976

Guide to the Literature and References

Curves and Their Jacobians

Lecture I:

The best beginning book on the theory of curves is:

R. Walker, *Algebraic Curves*, (Dover reprint 1962).

With a little more background, the best modern book is:

J.-P. Serre, *Groupes algébriques et corps de classes*, (Hermann, 1959).

A more analytical treatment can be found in:

R.C. Gunning, *Lectures on Riemann Surfaces*, (Math. Notes, Numbers 2, 6, 12, Princeton Univ. Press).

C.L. Siegel, *Topics in Complex Function Theory* (3 volumes, Wiley-Interscience, 1969–1973).

There is a huge classical literature, of which the following are best known to me:

K. Hensel and G. Landsberg, *Theorie der Algebraischen Funktionen einer Variablen*, (Chelsea reprint, 1965).

F. Severi, *Vorlesungen über Algebraische Geometrie* (Johnson reprint, 1968).

J.L. Coolidge, *A treatise on algebraic plane curves* (Oxford, 1931).

The most famous book which drew together clearly the topological, analytical and algebraic strands is:

H. Weyl, *Die Idee der Riemannschen Fläche* (Teubner, 3rd ed., 1955).

For elliptic curves specifically, one can look for instance in:

Ch. 7 of L. Ahlfors, *Complex Analysis* (2nd ed., McGraw-Hill, 1966): a good quick introduction to the analysis.

A. Hurwitz, R. Courant, *Allgemeine Funktionentheorie und Elliptische Funktionen* (Springer, 1964).

S. Lang, *Elliptic Functions*, (Addison-Wesley 1973).

J. Tate, Notes from Phillips Lectures at Haverford, to be published

J.W.S. Cassels, *Diophantine equations with special reference to elliptic curves* (Survey article: J. London, Math. Soc., **41** (1966), pp. 193–291).

The Gelfand-Schneider theorem quoted in Lecture I can be found in:

S. Lang, *Introduction to transcendental numbers* (Addison-Wesley, 1966), p. 22.

For higher Weierstrass points, see:

B. Olsen, *On higher Weierstrass points* (Annals of Math., **95** (1972), pp. 357–364).

J. Hubbard, *Sur les sections analytique de la courbe universelle de Teichmüller*, (to appear in Memoirs of AMS).

For the theory of theta-characteristics, some references are:

D. Mumford, *Theta characteristics of an algebraic curve* (Annales Ecole Norm. Sup., **4** (1971), p. 181).

Ch. 13, Vol. 2 of H. Weber, *Lehrbuch der Algebra*, (Chelsea reprint) – for the case of $g = 3$.

The details of the hyperelliptic vs. non-hyperelliptic story can be found in most of the books cited above, e.g., Walker, Hensel-Landsberg or Severi. Concerning the explicit description of the inequalities on generation of Fuchsian groups, see:

R. Fricke, F. Klein, *Vorlesungen über die Theorie der Automorphen Funktionen* (Johnson reprint 1965), esp. Vol. 1, Part 2, Ch. 2.

N. Purzitsky, *2 generator discrete free products*, (Math. Zeits., **126**, p. 209 and other papers in Math. Zeit., Ill. J.)

L. Keen, *On Fricke Moduli*, in Annals of Math. Studies 66, 1971.

Kazhdan's ideas on the metrics of varieties D/Γ are contained in:

D. Kazhdan, *Arithmetic varieties and their fields of quasi-definition*, (Actes du Cong. Int., Nice, Vol. 2).

Proofs of the facts concerning representation of a general curve C of genus g as a plane curve of lowest possible degree or as a covering of $\mathbb{P}^1$ of lowest possible degree can be found in:

S. Kleiman, D. Laksov, *Another proof of the existence of special divisors*, (Acta Math., **132** (1974)).

Petri's work can be found in:

K. Petri, *Über die invariante Darstellung algebraischer Funktionen einer Veränderlichen*, (Math. Ann. **88** (1922)).

B. Saint-Donat, *On Petri's Analysis of the Linear System of Quadrics through a Canonical Curve*, (Math. Ann., **206** (1973)).

K. Petri, *Über Specialkurven* I, (Math. Ann., **93** (1924)).

Lecture II:

I gave some introductory talks on moduli problems in general and on $\mathfrak{M}_{1,0}$, i.e., the elliptic curve case, in particular, in:

D. Mumford, K. Suominen, *Introduction to the Theory of Moduli*, in "Algebraic Geometry, Oslo, 1970", (Wolters-Noordhoff, 1971).

For $g = 1$, cf. also references given above for elliptic curves. The genus 2 case is in:

J.-I. Igusa, *Arithmetic variety of moduli for genus two*, (Annals of Math. **72** (1960)).

The precise definition of $\mathfrak{M}_g$ and proof that it is a quasi-projective variety are in:

W. Baily, *On the theory of Θ-functions, the moduli of abelian varieties and the moduli of curves*, (Annals of Math., **75** (1962)).

D. Mumford, *Geometric Invariant Theory*, (Springer 1965).

In particular, the first reference coordinatizes $\mathfrak{M}_g$ essentially by theta-nulls; the second coordinatizes $\mathfrak{M}_g$ both by a variant of theta-nulls, using cross-ratios of points of finite order on the Jacobian, and also by invariants of the Chow form as in the text of the Lecture. The final method of using cross-ratios of higher Weierstrass points has been suggested by Lipman Bers. The method as described in the text has never been published, but details can be filled in as follows: (1) one must first assign multiplicities to the higher Weierstrass points so that the divisor $\mathfrak{A}_k$ of all of them varies algebraically with C – cf. Hubbard (op. cit.), (2) one proves that no $x \in C$ occurs with multiplicity $> g^2$ in $\mathfrak{A}_k$; (3) using this one deduces that $\Phi_k(\mathfrak{A}_k)$ is stable in this sense of "Geometric Invariant Theory", Ch. 3; (4) check that if a quadric contains $\Phi_k(\mathfrak{A}_k)$, then it contains $\Phi_k(C)$, and since $\Phi_k(C)$ is an intersection of quadrics, $\Phi_k(\mathfrak{A}_k)$ determines C up to isomorphism; (5) apply the results of "Geometric Invariant Theory", Ch. 3.

Concerning Teichmüller space, the best reference is the survey article of Bers:

L. Bers, *Uniformization, Moduli and Kleinian groups*, (Bull. Lond. Math. Soc. **4** (1972)).

For the compactification $\overline{\mathfrak{M}}_g$ of $\mathfrak{M}_g$, see:

P. Deligne, D. Mumford, *The irreducibility of the space of curves of given genus*, (Publ. IHES, **36** (1969)).

L. Bers, *Spaces of degenerating Riemann Surfaces* (in Annals of Math. Studies **79** (1974)),

as well as forthcoming articles by F. Knudsen and myself. The references for the "positive curvature" assertions on $\mathfrak{M}_g$ are as follows:

E. Arbarello, *Weierstrass points and moduli of curves*, (thesis; to appear in Comp. Math.).

D. Mumford, *Abelian quotients of the Teichmüller modular group*, (J. d'Anal. Math., **18** (1967)).

H. Rauch, *The singularities of the modulus space*, (Bull. AMS, **68** (1962)).

B. Segre, *Sui moduli delle curve algebriche*, (Annali di Mat. **7** (1930)).

Concerning the Teichmüller metric and Petersson-Weil metrics on $\mathfrak{I}_{g,n}$, see:

L. Ahlfors, *Curvature properties of Teichmüller's space*, (J. d'Analyze, **9** (1961)).

L. Ahlfors, *Some remarks on Teichmüller's space of Riemann surfaces* (Annals of Math., **74** (1961)).

H. Masur, *On a class of geodesics in Teichmüller space* (to appear)

H. Royden, *Automorphisms and isometries in Teichmüller space* (in Annals of Math. Studies 66 (1971)).

H. Royden, *Metrics on Teichmüller space* (in Springer Lecture Notes, 400).

For the Ahlfors-Pick lemma, cf.

M. Cornalba, P. Griffiths, *Some transcendental aspects of algebraic geometry*, §6 (in Proc. Summer Institute on Alg. Geom. 1974, AMS).

The rigidity theorem of $A-P-M-G$ has usually been considered separately in 2 parts: one on the finiteness of the set of sections of a fixed family $\pi : X \to C - S$; the other on the finiteness of the set of families π. Both have much deeper, still unsolved number-theoretic analogs – viz. given a number field K, then one conjectures that 1) given a curve D defined over K of genus $g \geq 2$, D has only a finite set of K-rational points and 2) given a finite set S of primes of K and $g \geq 2$, there are only finitely many curves D defined over K with "good reduction outside S" and of genus g. The first is called Mordell's conjecture and the second is called Šafarevitch's conjecture (cf. his talk at the Stockholm International Congress of 1962). If one replaces K by the field of rational functions on a curve C over $\mathbb{C}$, these conjectures are equivalent to the Rigidity Theorem of the lecture. The Mordell part was proven first, independently by Manin and Grauert:

H. Grauert, *Mordells Vermutung über Punkte auf algebraischen Kurven und Funktionenkörper* (Publ. IHES **25** (1965)).

Y. Manin, *Rational points on algebraic curves over function fields*, (Izvestija Akad. Nauk, **27** (1963)).

P. Samuel, *La conjecture de Mordell pour les corps de fonctions*, (Sem. Bourbaki, exp. 287, 1964–65).

The Šafarevitch part was proven by Arakelov using earlier partial results of Paršin:

S. Ju. Arakelov, *Families of algebraic curves with fixed degeneracies*, (Izvest, Akad. Nauk **35** (1971)).

A.N. Paršin, *Algebraic curves over function fields*, (Izvest. Akad. Nauk, **32** (1968)).

References for recent results on $\mathfrak{M}_g$ added in proof are:

E. Arbarello, E. Sernesi, *The equation of a plane curve*, Duke Math. J. **46** (1979).

E. Sernesi, *Unirationality of moduli of curves of genus* 12, Ann. Sc. Norm. Pisa **8** (1981).

M. Chang, Z. Ran, *Unirationality of moduli of curves of genus* 11, 13, Inv. Math. **76** (1984).

M. Chang, Z. Ran, *The Kodaira dimension of moduli of curves of genus* 15, J. Diff. Geom. **24** (1986).

J. Harris, D. Mumford, *On the Kodaira dimension of moduli of curves*, (1982).

D. Eisenbud, J. Harris, *The Kodaira dimension of moduli of curves*, $g \geq 23$, Inv. Math. **90** (1987)

Lecture III:

The Jacobian is introduced in the standard books referred to in our notes to Lecture I: esp. Serre (op. cit.) shows that all rational differentials ω on curves C are pull-backs of translation-invariant differentials η on algebraic groups J via some rational map $\phi : C \to J$. Moreover Gunning (op. cit., esp. Math. Notes 12, subtitled "Jacobi Varieties") treats in detail many of the topics of this and the next Lecture. For Weil's original algebraic construction of the Jacobian and its application to the Riemann hypothesis, see:

A. Weil, *Variétés abéliennes et courbes algébriques*, (Hermann, 1948).

The result that I called the theorem of Riemann and Kempf on the singularities of W_k was proved for $k = g - 1$ by Riemann. Kempf's results appear in:

G. Kempf, *On the geometry of a theorem of Riemann* (Annals of Math., **98** (1973)).

G. Kempf, *Schubert methods with an application to algebraic curves* (Stichting Math. Centrom, Amsterdam, 1971).

The elegant proof of Riemann-Roch using the differential of $\phi^{(k)}$ is worked out in detail in:

A. Mattuck, A. Mayer, *The Riemann-Roch theorem of algebraic curves*, (Annali Sc. Norm. Pisa, **17** (1963)).

Here are references for the theory of theta functions; and more generally, function theory on abelian varieties:

W. Baily, *Classical theory of theta functions* (In AMS Proc. of Symp. in pure Math., vol. 9).

F. Conforto, *Abelsche Funktionen und algebraische Geometrie* (Springer-Verlag, 1956).

J. Fay, *Theta functions on Riemann Surfaces* (Springer Lecture Notes 352).

J.-I. Igusa, *Theta functions* (Springer-Verlag, 1972).

A. Krazer, *Lehrbuch der Thetafunktionen* (Teubner, 1903).

D. Mumford, *On the equations defining abelian varieties* (Inv. Math., **1** and **3** (1966–67)).

D. Mumford, *Abelian varieties* (Tata studies in Math. 5, Oxford, 2nd ed. 1974).

H. Rauch, H. Farkas, *Theta functions with applications to Riemann Surfaces*, (Wiliams and Wilkins, 1974).

My treatment in the lecture follows more or less my book *Abelian Varieties*; e.g., the 2 embedding theorems above are proven in Ch. 1 of this book. The group-theoretic aspects discussed in the lecture are in my book, §23, as well as in my paper and Igusa's book. The prime form E_e and its applications are due to Riemann and are discussed at length with many more applications in Fay.

Lecture IV:

Besides the references given above on abelian varieties and moduli problems, one can find further material in:

J.-I. Igusa, *On the graded ring of theta-constants* (Am. J. Math., **86** (1964) and **88** (1966)).

D. Mumford, *The structure of the moduli spaces of curves and abelian varieties* (in Actes du Cong. Int., Nice, 1971).

G. Shimura, *Moduli of abelian varieties and number theory*, (in AMS Proc. of Symp. in Pure Math., vol. 9).

Because of their arithmetic and representation-theoretic importance, there is a huge literature on various types of modular forms on $\mathfrak{H}_g$ and on other bounded symmetric domains. I am not competent even to begin to list references on these topics, but I would like to emphasize here as in the Lecture one big gap: the lack of a moduli-theoretic interpretation of the Eisenstein series.

Concerning Torelli and Schottky, good general references are the books of Gunning (op. cit., notes to Lecture I, part III), Fay (op. cit.) and Rauch-Farkas (op. cit.). In more detail, for Approach I, see:

A. Andreotti, *On a theorem of Torelli* (Am. J. Math., **80** (1958)).

H. Martens, *A new proof of Torelli's theorem* (Annals of Math., **78** (1963)).

T. Matsusaka, *On a theorem of Torelli*, (Am. J. Math., **80** (1958)).

A. Weil, *Zum Beweis des Torellischen Satzes*, (Nachr. Akad. Wiss. Göttingen (1957)).

Fay's formula is in his Springer Lecture Notes (op. cit.), p. 34 (formula (45); the interpretation via trisecants follows from the Proposition on p. 335 of my paper:

D. Mumford, *Prym Varieties I* (in: Contributions to Analysis, Academic Press, 1974).

Approach II can be found in:

W. Wirtinger, *Lie's Translationsmannigfaltigkeiten und Abelsche Integrale*, (Monatshefte für Math. und Physik, **46** (1938)).

S. Lie, *Werke*, Band 2, Abt. 2, Teil 2, p. 481.

Aproach III is in:

A. Andreotti, A. Mayer, *On period relations for Abelian integrals on algebraic curves* (Ann. Scu. Norm. Sup. Pisa (1967)).

My discussion of Approach IV follows my paper "Prym Varieties I" mentioned above. A more analytic approach is in Fay (op. cit.) or Rauch-Farkas (op. cit.) as well as in the many papers of Rauch and Farkas cited in their book. The original works of Schottky and Jung are:

F. Schottky, *Über die Moduln der Thetafunktionen* (Acta Math. **27** (1903)).

F. Schottky, H. Jung, *Neue Sätze über Symmetralfunktionen und die Abel'schen Funktionen* (2 parts, Sitzungsber. Berlin Akad. Wiss., **1** (1909)).

Supplementary Bibliography on the Schottky Problem

Enrico Arbarello

Note:

– The papers [A-D 2], [B3] and [Do3] are survey papers and contain more bibliographical references.

– The paper [B3] is in the Proceedings of the Summer Research Institute on Theta Functions held at Bowdoin College in 1987 which is altogether another source of references.

– Also the Proceedings of an AMS-IMS-SIAM Joint Summer Research Conference on the Schottky Problem 1992, is a good source of references. This is where [Do5] is.

– [B2] is a Séminaire Bourbaki and [De3] provides a continuation of it.

[A] E. Arbarello, *Fay's trisecant Formula and a Characterisation of Jacobian Varieties*, Proceedings of Symposia in Pure Mathematics Vol. **46**, (1987) 49–61.

[A-D 1] E. Arbarello and C. De Concini, *On a set of equations characterizing Riemann matrices*, Annals of Math. **120**, (1984), 119–140.

[A-D 2] E. Arbarello and C. De Concini, *Geometrical aspects of the Kadomtsev-Petviashvili equation*, L.N.M. **1451**, 95–137, Springer 1990.

[B1] A. Beauville, *Prym varieties and the Schottky problem*, Inv. Math. **41**, (1977) 149–196.

[B2] A. Beauville, *Le problème de Schottky et la conjecture de Novikov*, Séminaire Bourbaki no. 675, Astérisque 1988.

[B3] A. Beauville, *Prym varieties: a survey*, Proceedings of Symposia in Pure Mathematics Vol. **49** (1989), 607–620.

[B-D 1] A. Beauville and O. Debarre, *Une relation entre deux approches du problème de Schottky*, Inv. Math. **86**, 195–207, (1986).

[B-D 2] A. Beauville and O. Debarre, *Sur le problème de Schottky pour les variétés de Prym*, Annali Scuola Norm. Sup. Pisa, Cl. Sci. (4) **14** , (1987), 613–623.

[B-D 3] A. Beauville and O. Debarre, *Sur les function thêta du second ordre*, Arithmetic of Complex manifolds, Proceedings, Erlangen 1988, Lecture Notes in Mathematics **1399** Springer 1989.

[B-S] P. Buser and P. Sarnak, *On the period matrix of a Riemann surface of large genus*, Invent. Math. **117**, (1994), 27–56.

[C] A. Collino, *A new proof of the Ran-Matsusaka criterion for Jacobians*, Proc. Amer. Math. Soc. **92**, (1984), 329–331.

[D] G. Darboux, *Leçons sur la Théorie Générale des Surfaces*, 2nd Ed., Paris Gauthier-Villars, 1914.

[De1] O. Debarre, *Sur les variétés abéliennes dont le diviseur thêta est singulier en codimension 3*, Duke Math. Journ. **56** (1988), 221–273.

[De2] O. Debarre, *Trisecant lines and Jacobians*, J. Alg. Geom. **1**, (1992), 5–14.

[De3] O. Debarre, *The Schottky Problem: an Update*, Complex Algebraic Geometry, MSRI Publications, Volume *28*, 1995.

[De4] O. Debarre, *The trisecant conjecture for Pryms*, Proceedings of Symposia in Pure Mathematics Vol. **49**, (1989).

[Do1] R. Donagi, *Big Schottky*, Inv. Math. **89** (1987), 569–599.

[Do2] R. Donagi, *Non-Jacobians in the Schottky locus*, Annals of Math. **126** (1987), 193–217.

[Do3] R. Donagi, *The Schottky Problem*, in "Theory of Moduli" (E. Sernesi ed.), L.N.M. **1337**, 82–137, Springer 1988.

[Do4] R. Donagi, *The tetragonal connstruction*, Bull. Amer. Math. Soc. **4** (1981), 181–185.

[Do5] R. Donagi, *The fibers of the Prym map*, Proceedings of an AMS-IMS-SIAM Joint Summer Research Conference on the Schottky Problem 1992.

[D-S] R. Donagi and R. Smith, *The structure of the Prym map*, Acta Math **146**, (1981), 25–102.

[Du] B.A. Dubrovin, *Theta functions and non-linear equations*, Russian Math. Surveys 36, **2**, (1981), 11–92.

[E-L] L. Ein and R. Lazarsfeld, *Singularities of Theta Divisors and the Birational Geometry of Irregular Varieties*, Journal of the AMS 10, (1997) pp. 243–258.

[F] H.M. Farkas, *On Fay's trisecant formula*, Journal d'Analyse Mathematique (**44**), (1984-85), 205–217.

[Fr] E. Freitag, *Die Irreduzibilität der Schottkyrelation*, Arch. Math **40**, (1983), 255–259.

[F-S] R. Friedman and R. Smith, *The generic Torelli Theorem for the Prym map*, Inv. Math. (1982), 473–490.

[vG1] B. van Geemen, *Siegel modular forms vanishing on the moduli space of curves*, Inv. Math. **78** 1984, 329–349.

[vG2] B. van Geemen, *Schotty-Jung relations and vector bundles on hyperelliptic curves*, Math Ann. **281**, (1988), 431–449.

[vG-vG] B. van Geemen and G. van der Geer, *Kummer Varieties and the Moduli Spaces of Abelian Varieties*, American Journal of Mathematics **108** (1986), 615–642.

[vG-P] B. van Geemen and E. Previato, *Prym varieties and the Verlinde formula*, Math. Annalen **294**, 1992, 741–754.

[G] M. Green, *Quadrics of rank four in the ideal of the canonical curve*, Invent. Math. **75**, (1984), 85–104.

[Gr] P.A. Griffiths, *Variations on a Theorem of Abel*, Inventiones Math. **35**, (1976), 321–390.

[Gu] R.C. Gunning, *Some Curves in Abelian Varieties*, Invent. Math. **66** (1982) 285–296.

[H] N. Hitchin, *Flat connections and geometric quantization*, Comm. Math. Physics **131**, (1990), 347–380.

[I] J. Igusa, *On the irreducibility of Schottky's divisor*, Journal of the Faculty of Science Tokyo **28**, (1981), 531–545.

[Iz] E. Izadi, *The geometric structure of $\mathcal{A}_4$, the structure of the Prym map, double solids and Γ_{00}-divisors*, Journal für die Reine und Angewandte Mathematik, (1995).

[K] V. Kanev, *The global Torelli theorem for Prym varieties at a generic point*, Math. U.S.S.R. Izvestija **20** (1983), 235–258.

[Ke] G. Kempf, *Deformations of symmetric products*, Riemann Surfaces and related topics: Proceedings of the 1978 Stony Brook Conference. Princeton (1981), 319–341.

[L] R. Lazarsfeld, *Length of periods and Seshadri constants of abelian varieties*, Math Research Letters **3**, (1996), 439–447.

[Li1] J. Little, *Translation manifolds and the converse of Abel's theorem*, Compositio Math. **49** (1983), 147–171.

[Li2] J. Little, *Translation manifolds and the Schottky problem*, Proceedings of Symposia in Pure Mathematics Vol. **49** (1989), 517–529.

[Li3] J. Little, *Another Relation between approaches to the Schottky Problem.*

[M1] G. Marini, *A geometrical proof of Shiota's theorem on a conjecture of S.P. Novikov*, Compositio Mathematica 111: (1998), 305–322.

[M2] G. Marini, *An inflectionary tangent to the Kummer variety and the Jacobian condition*, Math. Ann. 309, (1997), 483–490.

[Mu] M. Mulase, *Cohomological structure in soliton equations and Jacobian varieties*, J. Diff. Geo. **19**, (1984), 403–430.

[O-S] F. Oort and J. Steenbrink, *The local Torelli problem for algebraic curves*, Journées de Géométrie Algébrique d'Angers (A. Beauville ed.) Süthoff & Noordhoff, 1980, 157–204.

[R] Z. Ran, *On subvarieties of abelian varieties*, Invent. Math. **62**, (1981), 459–479.

[Re] S. Recillas, *Jacobians of curves with g_4^1's are Pryms of trigonal curves*, Bol. Soc. Mat. Mexicana **19** (1974), 9–13.

[S] T. Shiota, *Characterization of Jacobian varieties in terms of soliton equations*, Inv. Math. **83**, (1986), 333–382.

[S-V] R. Smith and R. Varley, *Components of the locus of singular theta divisors in genus 5*, LNM 1124, Springer-Verlag, (1983) 338–416.

[W1] G.E. Welters, *Polarized Abelian Varieties and the Heat Equations*, Compositio Mathematica **49** (1983) 173–194.

[W2] G.E. Welters, *On flexes of the Kummer variety*, Indagationes Math. **45**, (1983), 501–520.

[W3] G.E. Welters, *A Criterion for Jacobian varieties*, Annals of Math. **120**, (1984), 497–504.

[W4] G.E. Welters, *Recovering the curve data from a general Prym variety*, Amer. J. of Math. **109** (1987), 165–182.

[W5] G.E. Welters, *The surface $C - C$ on Jacobi varieties and 2nd order theta functions*, Acta math. **109** (1987), 165–182.

Druck: Strauss Offsetdruck, Mörlenbach
Verarbeitung: Schäffer, Grünstadt

Lecture Notes in Mathematics

For information about Vols. 1–1520 please contact your bookseller or Springer-Verlag

Vol. 1521: S. Busenberg, B. Forte, H. K. Kuiken, Mathematical Modelling of Industrial Process. Bari, 1990. Editors: V. Capasso, A. Fasano. VII, 162 pages. 1992.

Vol. 1522: J.-M. Delort, F. B. I. Transformation. VII, 101 pages. 1992.

Vol. 1523: W. Xue, Rings with Morita Duality. X, 168 pages. 1992.

Vol. 1524: M. Coste, L. Mahé, M.-F. Roy (Eds.), Real Algebraic Geometry. Proceedings, 1991. VIII, 418 pages. 1992.

Vol. 1525: C. Casacuberta, M. Castellet (Eds.), Mathematical Research Today and Tomorrow. VII, 112 pages. 1992.

Vol. 1526: J. Azéma, P. A. Meyer, M. Yor (Eds.), Séminaire de Probabilités XXVI. X, 633 pages. 1992.

Vol. 1527: M. I. Freidlin, J.-F. Le Gall, Ecole d'Eté de Probabilités de Saint-Flour XX – 1990. Editor: P. L. Hennequin. VIII, 244 pages. 1992.

Vol. 1528: G. Isac, Complementarity Problems. VI, 297 pages. 1992.

Vol. 1529: J. van Neerven, The Adjoint of a Semigroup of Linear Operators. X, 195 pages. 1992.

Vol. 1530: J. G. Heywood, K. Masuda, R. Rautmann, S. A. Solonnikov (Eds.), The Navier-Stokes Equations II – Theory and Numerical Methods. IX, 322 pages. 1992.

Vol. 1531: M. Stoer, Design of Survivable Networks. IV, 206 pages. 1992.

Vol. 1532: J. F. Colombeau, Multiplication of Distributions. X, 184 pages. 1992.

Vol. 1533: P. Jipsen, H. Rose, Varieties of Lattices. X, 162 pages. 1992.

Vol. 1534: C. Greither, Cyclic Galois Extensions of Commutative Rings. X, 145 pages. 1992.

Vol. 1535: A. B. Evans, Orthomorphism Graphs of Groups. VIII, 114 pages. 1992.

Vol. 1536: M. K. Kwong, A. Zettl, Norm Inequalities for Derivatives and Differences. VII, 150 pages. 1992.

Vol. 1537: P. Fitzpatrick, M. Martelli, J. Mawhin, R. Nussbaum, Topological Methods for Ordinary Differential Equations. Montecatini Terme, 1991. Editors: M. Furi, P. Zecca. VII, 218 pages. 1993.

Vol. 1538: P.-A. Meyer, Quantum Probability for Probabilists. X, 287 pages. 1993.

Vol. 1539: M. Coornaert, A. Papadopoulos, Symbolic Dynamics and Hyperbolic Groups. VIII, 138 pages. 1993.

Vol. 1540: H. Komatsu (Ed.), Functional Analysis and Related Topics, 1991. Proceedings. XXI, 413 pages. 1993.

Vol. 1541: D. A. Dawson, B. Maisonneuve, J. Spencer, Ecole d' Eté de Probabilités de Saint-Flour XXI - 1991. Editor: P. L. Hennequin. VIII, 356 pages. 1993.

Vol. 1542: J.Fröhlich, Th.Kerler, Quantum Groups, Quantum Categories and Quantum Field Theory. VII, 431 pages. 1993.

Vol. 1543: A. L. Dontchev, T. Zolezzi, Well-Posed Optimization Problems. XII, 421 pages. 1993.

Vol. 1544: M.Schürmann, White Noise on Bialgebras. VII, 146 pages. 1993.

Vol. 1545: J. Morgan, K. O'Grady, Differential Topology of Complex Surfaces. VIII, 224 pages. 1993.

Vol. 1546: V. V. Kalashnikov, V. M. Zolotarev (Eds.), Stability Problems for Stochastic Models. Proceedings, 1991. VIII, 229 pages. 1993.

Vol. 1547: P. Harmand, D. Werner, W. Werner, M-ideals in Banach Spaces and Banach Algebras. VIII, 387 pages. 1993.

Vol. 1548: T. Urabe, Dynkin Graphs and Quadrilateral Singularities. VI, 233 pages. 1993.

Vol. 1549: G. Vainikko, Multidimensional Weakly Singular Integral Equations. XI, 159 pages. 1993.

Vol. 1550: A. A. Gonchar, E. B. Saff (Eds.), Methods of Approximation Theory in Complex Analysis and Mathematical Physics IV, 222 pages, 1993.

Vol. 1551: L. Arkeryd, P. L. Lions, P.A. Markowich, S.R. S. Varadhan. Nonequilibrium Problems in Many-Particle Systems. Montecatini, 1992. Editors: C. Cercignani, M. Pulvirenti. VII, 158 pages 1993.

Vol. 1552: J. Hilgert, K.-H. Neeb, Lie Semigroups and their Applications. XII, 315 pages. 1993.

Vol. 1553: J.-L- Colliot-Thélène, J. Kato, P. Vojta. Arithmetic Algebraic Geometry. Trento, 1991. Editor: E. Ballico. VII, 223 pages. 1993.

Vol. 1554: A. K. Lenstra, H. W. Lenstra, Jr. (Eds.), The Development of the Number Field Sieve. VIII, 131 pages. 1993.

Vol. 1555: O. Liess, Conical Refraction and Higher Microlocalization. X, 389 pages. 1993.

Vol. 1556: S. B. Kuksin, Nearly Integrable Infinite-Dimensional Hamiltonian Systems. XXVII, 101 pages. 1993.

Vol. 1557: J. Azéma, P. A. Meyer, M. Yor (Eds.), Séminaire de Probabilités XXVII. VI, 327 pages. 1993.

Vol. 1558: T. J. Bridges, J. E. Furter, Singularity Theory and Equivariant Symplectic Maps. VI, 226 pages. 1993.

Vol. 1559: V. G. Sprindžuk, Classical Diophantine Equations. XII, 228 pages. 1993.

Vol. 1560: T. Bartsch, Topological Methods for Variational Problems with Symmetries. X, 152 pages. 1993.

Vol. 1561: I. S. Molchanov, Limit Theorems for Unions of Random Closed Sets. X, 157 pages. 1993.

Vol. 1562: G. Harder, Eisensteinkohomologie und die Konstruktion gemischter Motive. XX, 184 pages. 1993.

Vol. 1563: E. Fabes, M. Fukushima, L. Gross, C. Kenig, M. Röckner, D. W. Stroock, Dirichlet Forms. Varenna, 1992. Editors: G. Dell'Antonio, U. Mosco. VII, 245 pages. 1993.

Vol. 1564: J. Jorgenson, S. Lang, Basic Analysis of Regularized Series and Products. IX, 122 pages. 1993.

Vol. 1565: L. Boutet de Monvel, C. De Concini, C. Procesi, P. Schapira, M. Vergne. D-modules, Representation Theory, and Quantum Groups. Venezia, 1992. Editors: G. Zampieri, A. D'Agnolo. VII, 217 pages. 1993.

Vol. 1566: B. Edixhoven, J.-H. Evertse (Eds.), Diophantine Approximation and Abelian Varieties. XIII, 127 pages. 1993.

Vol. 1567: R. L. Dobrushin, S. Kusuoka, Statistical Mechanics and Fractals. VII, 98 pages. 1993.

Vol. 1568: F. Weisz, Martingale Hardy Spaces and their Application in Fourier Analysis. VIII, 217 pages. 1994.

Vol. 1569: V. Totik, Weighted Approximation with Varying Weight. VI, 117 pages. 1994.

Vol. 1570: R. deLaubenfels, Existence Families, Functional Calculi and Evolution Equations. XV, 234 pages. 1994.

Vol. 1571: S. Yu. Pilyugin, The Space of Dynamical Systems with the C^0-Topology. X, 188 pages. 1994.

Vol. 1572: L. Göttsche, Hilbert Schemes of Zero-Dimensional Subschemes of Smooth Varieties. IX, 196 pages. 1994.

Vol. 1573: V. P. Havin, N. K. Nikolski (Eds.), Linear and Complex Analysis – Problem Book 3 – Part I. XXII, 489 pages. 1994.

Vol. 1574: V. P. Havin, N. K. Nikolski (Eds.), Linear and Complex Analysis – Problem Book 3 – Part II. XXII, 507 pages. 1994.

Vol. 1575: M. Mitrea, Clifford Wavelets, Singular Integrals, and Hardy Spaces. XI, 116 pages. 1994.

Vol. 1576: K. Kitahara, Spaces of Approximating Functions with Haar-Like Conditions. X, 110 pages. 1994.

Vol. 1577: N. Obata, White Noise Calculus and Fock Space. X, 183 pages. 1994.

Vol. 1578: J. Bernstein, V. Lunts, Equivariant Sheaves and Functors. V, 139 pages. 1994.

Vol. 1579: N. Kazamaki, Continuous Exponential Martingales and *BMO*. VII, 91 pages. 1994.

Vol. 1580: M. Milman, Extrapolation and Optimal Decompositions with Applications to Analysis. XI, 161 pages. 1994.

Vol. 1581: D. Bakry, R. D. Gill, S. A. Molchanov, Lectures on Probability Theory. Editor: P. Bernard. VIII, 420 pages. 1994.

Vol. 1582: W. Balser, From Divergent Power Series to Analytic Functions. X, 108 pages. 1994.

Vol. 1583: J. Azéma, P. A. Meyer, M. Yor (Eds.), Séminaire de Probabilités XXVIII. VI, 334 pages. 1994.

Vol. 1584: M. Brokate, N. Kenmochi, I. Müller, J. F. Rodriguez, C. Verdi, Phase Transitions and Hysteresis. Montecatini Terme, 1993. Editor: A. Visintin. VII. 291 pages. 1994.

Vol. 1585: G. Frey (Ed.), On Artin's Conjecture for Odd 2-dimensional Representations. VIII, 148 pages. 1994.

Vol. 1586: R. Nillsen, Difference Spaces and Invariant Linear Forms. XII, 186 pages. 1994.

Vol. 1587: N. Xi, Representations of Affine Hecke Algebras. VIII, 137 pages. 1994.

Vol. 1588: C. Scheiderer, Real and Étale Cohomology. XXIV, 273 pages. 1994.

Vol. 1589: J. Bellissard, M. Degli Esposti, G. Forni, S. Graffi, S. Isola, J. N. Mather, Transition to Chaos in Classical and Quantum Mechanics. Montecatini Terme, 1991. Editor: 2S. Graffi. VII, 192 pages. 1994.

Vol. 1590: P. M. Soardi, Potential Theory on Infinite Networks. VIII, 187 pages. 1994.

Vol. 1591: M. Abate, G. Patrizio, Finsler Metrics – A Global Approach. IX, 180 pages. 1994.

Vol. 1592: K. W. Breitung, Asymptotic Approximations for Probability Integrals. IX, 146 pages. 1994.

Vol. 1593: J. Jorgenson & S. Lang, D. Goldfeld, Explicit Formulas for Regularized Products and Series. VIII, 154 pages. 1994.

Vol. 1594: M. Green, J. Murre, C. Voisin, Algebraic Cycles and Hodge Theory. Torino, 1993. Editors: A. Albano, F. Bardelli. VII, 275 pages. 1994.

Vol. 1595: R.D.M. Accola, Topics in the Theory of Riemann Surfaces. IX, 105 pages. 1994.

Vol. 1596: L. Heindorf, L. B. Shapiro, Nearly Projective Boolean Algebras. X, 202 pages. 1994.

Vol. 1597: B. Herzog, Kodaira-Spencer Maps in Local Algebra. XVII, 176 pages. 1994.

Vol. 1598: J. Berndt, F. Tricerri, L. Vanhecke, Generalized Heisenberg Groups and Damek-Ricci Harmonic Spaces. VIII, 125 pages. 1995.

Vol. 1599: K. Johannson, Topology and Combinatorics of 3-Manifolds. XVIII, 446 pages. 1995.

Vol. 1600: W. Narkiewicz, Polynomial Mappings. VII, 130 pages. 1995.

Vol. 1601: A. Pott, Finite Geometry and Character Theory. VII, 181 pages. 1995.

Vol. 1602: J. Winkelmann, The Classification of Three-dimensional Homogeneous Complex Manifolds. XI, 230 pages. 1995.

Vol. 1603: V. Ene, Real Functions – Current Topics. XIII, 310 pages. 1995.

Vol. 1604: A. Huber, Mixed Motives and their Realization in Derived Categories. XV, 207 pages. 1995.

Vol. 1605: L. B. Wahlbin, Superconvergence in Galerkin Finite Element Methods. XI, 166 pages. 1995.

Vol. 1606: P.-D. Liu, M. Qian, Smooth Ergodic Theory of Random Dynamical Systems. XI, 221 pages. 1995.

Vol. 1607: G. Schwarz, Hodge Decomposition – A Method for Solving Boundary Value Problems. VII, 155 pages. 1995.

Vol. 1608: P. Biane, R. Durrett, Lectures on Probability Theory. Editor: P. Bernard. VII, 210 pages. 1995.

Vol. 1609: L. Arnold, C. Jones, K. Mischaikow, G. Raugel, Dynamical Systems. Montecatini Terme, 1994. Editor: R. Johnson. VIII, 329 pages. 1995.

Vol. 1610: A. S. Üstünel, An Introduction to Analysis on Wiener Space. X, 95 pages. 1995.

Vol. 1611: N. Knarr, Translation Planes. VI, 112 pages. 1995.

Vol. 1612: W. Kühnel, Tight Polyhedral Submanifolds and Tight Triangulations. VII, 122 pages. 1995.

Vol. 1613: J. Azéma, M. Emery, P. A. Meyer, M. Yor (Eds.), Séminaire de Probabilités XXIX. VI, 326 pages. 1995.

Vol. 1614: A. Koshelev, Regularity Problem for Quasilinear Elliptic and Parabolic Systems. XXI, 255 pages. 1995.

Vol. 1615: D. B. Massey, Le Cycles and Hypersurface Singularities. XI, 131 pages. 1995.

Vol. 1616: I. Moerdijk, Classifying Spaces and Classifying Topoi. VII, 94 pages. 1995.

Vol. 1617: V. Yurinsky, Sums and Gaussian Vectors. XI, 305 pages. 1995.

Vol. 1618: G. Pisier, Similarity Problems and Completely Bounded Maps. VII, 156 pages. 1996.

Vol. 1619: E. Landvogt, A Compactification of the Bruhat-Tits Building. VII, 152 pages. 1996.

Vol. 1620: R. Donagi, B. Dubrovin, E. Frenkel, E. Previato, Integrable Systems and Quantum Groups. Montecatini Terme, 1993. Editors:M. Francaviglia, S. Greco. VIII, 488 pages. 1996.

Vol. 1621: H. Bass, M. V. Otero-Espinar, D. N. Rockmore, C. P. L. Tresser, Cyclic Renormalization and Auto-morphism Groups of Rooted Trees. XXI, 136 pages. 1996.

Vol. 1622: E. D. Farjoun, Cellular Spaces, Null Spaces and Homotopy Localization. XIV, 199 pages. 1996.

Vol. 1623: H.P. Yap, Total Colourings of Graphs. VIII, 131 pages. 1996.

Vol. 1624: V. Brınzanescu, Holomorphic Vector Bundles over Compact Complex Surfaces. X, 170 pages. 1996.

Vol.1625: S. Lang, Topics in Cohomology of Groups. VII, 226 pages. 1996.

Vol. 1626: J. Azéma, M. Emery, M. Yor (Eds.), Séminaire de Probabilités XXX. VIII, 382 pages. 1996.

Vol. 1627: C. Graham, Th. G. Kurtz, S. Méléard, Ph. E. Protter, M. Pulvirenti, D. Talay, Probabilistic Models for Nonlinear Partial Differential Equations. Montecatini Terme, 1995. Editors: D. Talay, L. Tubaro. X, 301 pages. 1996.

Vol. 1628: P.-H. Zieschang, An Algebraic Approach to Association Schemes. XII, 189 pages. 1996.

Vol. 1629: J. D. Moore, Lectures on Seiberg-Witten Invariants. VII, 105 pages. 1996.

Vol. 1630: D. Neuenschwander, Probabilities on the Heisenberg Group: Limit Theorems and Brownian Motion. VIII, 139 pages. 1996.

Vol. 1631: K. Nishioka, Mahler Functions and Transcendence.VIII, 185 pages.1996.

Vol. 1632: A. Kushkuley, Z. Balanov, Geometric Methods in Degree Theory for Equivariant Maps. VII, 136 pages. 1996.

Vol.1633: H. Aikawa, M. Essén, Potential Theory – Selected Topics. IX, 200 pages.1996.

Vol. 1634: J. Xu, Flat Covers of Modules. IX, 161 pages. 1996.

Vol. 1635: E. Hebey, Sobolev Spaces on Riemannian Manifolds. X, 116 pages. 1996.

Vol. 1636: M. A. Marshall, Spaces of Orderings and Abstract Real Spectra. VI, 190 pages. 1996.

Vol. 1637: B. Hunt, The Geometry of some special Arithmetic Quotients. XIII, 332 pages. 1996.

Vol. 1638: P. Vanhaecke, Integrable Systems in the realm of Algebraic Geometry. VIII, 218 pages. 1996.

Vol. 1639: K. Dekimpe, Almost-Bieberbach Groups: Affine and Polynomial Structures. X, 259 pages. 1996.

Vol. 1640: G. Boillat, C. M. Dafermos, P. D. Lax, T. P. Liu, Recent Mathematical Methods in Nonlinear Wave Propagation. Montecatini Terme, 1994. Editor: T. Ruggeri. VII, 142 pages. 1996.

Vol. 1641: P. Abramenko, Twin Buildings and Applications to S-Arithmetic Groups. IX, 123 pages. 1996.

Vol. 1642: M. Puschnigg, Asymptotic Cyclic Cohomology. XXII, 138 pages. 1996.

Vol. 1643: J. Richter-Gebert, Realization Spaces of Polytopes. XI, 187 pages. 1996.

Vol. 1644: A. Adler, S. Ramanan, Moduli of Abelian Varieties. VI, 196 pages. 1996.

Vol. 1645: H. W. Broer, G. B. Huitema, M. B. Sevryuk, Quasi-Periodic Motions in Families of Dynamical Systems. XI, 195 pages. 1996.

Vol. 1646: J.-P. Demailly, T. Peternell, G. Tian, A. N. Tyurin, Transcendental Methods in Algebraic Geometry. Cetraro, 1994. Editors: F. Catanese, C. Ciliberto. VII, 257 pages. 1996.

Vol. 1647: D. Dias, P. Le Barz, Configuration Spaces over Hilbert Schemes and Applications. VII, 143 pages. 1996.

Vol. 1648: R. Dobrushin, P. Groeneboom, M. Ledoux, Lectures on Probability Theory and Statistics. Editor: P. Bernard. VIII, 300 pages. 1996.

Vol. 1649: S. Kumar, G. Laumon, U. Stuhler, Vector Bundles on Curves – New Directions. Cetraro, 1995. Editor: M. S. Narasimhan. VII, 193 pages. 1997.

Vol. 1650: J. Wildeshaus, Realizations of Polylogarithms. XI, 343 pages. 1997.

Vol. 1651: M. Drmota, R. F. Tichy, Sequences, Discrepancies and Applications. XIII, 503 pages. 1997.

Vol. 1652: S. Todorcevic, Topics in Topology. VIII, 153 pages. 1997.

Vol. 1653: R. Benedetti, C. Petronio, Branched Standard Spines of 3-manifolds. VIII, 132 pages. 1997.

Vol. 1654: R. W. Ghrist, P. J. Holmes, M. C. Sullivan, Knots and Links in Three-Dimensional Flows. X, 208 pages. 1997.

Vol. 1655: J. Azéma, M. Emery, M. Yor (Eds.), Séminaire de Probabilités XXXI. VIII, 329 pages. 1997.

Vol. 1656: B. Biais, T. Björk, J. Cvitanic, N. El Karoui, E. Jouini, J. C. Rochet, Financial Mathematics. Bressanone, 1996. Editor: W. J. Runggaldier. VII, 316 pages. 1997.

Vol. 1657: H. Reimann, The semi-simple zeta function of quaternionic Shimura varieties. IX, 143 pages. 1997.

Vol. 1658: A. Pumarino, J. A. Rodrıguez, Coexistence and Persistence of Strange Attractors. VIII, 195 pages. 1997.

Vol. 1659: V, Kozlov, V. Maz'ya, Theory of a Higher-Order Sturm-Liouville Equation. XI, 140 pages. 1997.

Vol. 1660: M. Bardi, M. G. Crandall, L. C. Evans, H. M. Soner, P. E. Souganidis, Viscosity Solutions and Applications. Montecatini Terme, 1995. Editors: I. Capuzzo Dolcetta, P. L. Lions. IX, 259 pages. 1997.

Vol. 1661: A. Tralle, J. Oprea, Symplectic Manifolds with no Kähler Structure. VIII, 207 pages. 1997.

Vol. 1662: J. W. Rutter, Spaces of Homotopy Self-Equivalences – A Survey. IX, 170 pages. 1997.

Vol. 1663: Y. E. Karpeshina; Perturbation Theory for the Schrödinger Operator with a Periodic Potential. VII, 352 pages. 1997.

Vol. 1664: M. Väth, Ideal Spaces. V, 146 pages. 1997.

Vol. 1665: E. Giné, G. R. Grimmett, L. Saloff-Coste, Lectures on Probability Theory and Statistics 1996. Editor: P. Bernard. X, 424 pages, 1997.

Vol. 1666: M. van der Put, M. F. Singer, Galois Theory of Difference Equations. VII, 179 pages. 1997.

Vol. 1667: J. M. F. Castillo, M. González, Three-space Problems in Banach Space Theory. XII, 267 pages. 1997.

Vol. 1668: D. B. Dix, Large-Time Behavior of Solutions of Linear Dispersive Equations. XIV, 203 pages. 1997.

Vol. 1669: U. Kaiser, Link Theory in Manifolds. XIV, 167 pages. 1997.

Vol. 1670: J. W. Neuberger, Sobolev Gradients and Differential Equations. VIII, 150 pages. 1997.

Vol. 1671: S. Bouc, Green Functors and *G*-sets. VII, 342 pages. 1997.

Vol. 1672: S. Mandal, Projective Modules and Complete Intersections. VIII, 114 pages. 1997.

Vol. 1673: F. D. Grosshans, Algebraic Homogeneous Spaces and Invariant Theory. VI, 148 pages. 1997.

Vol. 1674: G. Klaas, C. R. Leedham-Green, W. Plesken, Linear Pro-*p*-Groups of Finite Width. VIII, 115 pages. 1997.

Vol. 1675: J. E. Yukich, Probability Theory of Classical Euclidean Optimization Problems. X, 152 pages. 1998.

Vol. 1676: P. Cembranos, J. Mendoza, Banach Spaces of Vector-Valued Functions. VIII, 118 pages. 1997.

Vol. 1677: N. Proskurin, Cubic Metaplectic Forms and Theta Functions. VIII, 196 pages. 1998.

Vol. 1678: O. Krupková, The Geometry of Ordinary Variational Equations. X, 251 pages. 1997.

Vol. 1679: K.-G. Grosse-Erdmann, The Blocking Technique. Weighted Mean Operators and Hardy's Inequality. IX, 114 pages. 1998.

Vol. 1680: K.-Z. Li, F. Oort, Moduli of Supersingular Abelian Varieties. V, 116 pages. 1998.

Vol. 1681: G. J. Wirsching, The Dynamical System Generated by the 3n+1 Function. VII, 158 pages. 1998.

Vol. 1682: H.-D. Alber, Materials with Memory. X, 166 pages. 1998.

Vol. 1683: A. Pomp, The Boundary-Domain Integral Method for Elliptic Systems. XVI, 163 pages. 1998.

Vol. 1684: C. A. Berenstein, P. F. Ebenfelt, S. G. Gindikin, S. Helgason, A. E. Tumanov, Integral Geometry, Radon Transforms and Complex Analysis. Firenze, 1996. Editors: E. Casadio Tarabusi, M. A. Picardello, G. Zampieri. VII, 160 pages. 1998.

Vol. 1685: S. König, A. Zimmermann, Derived Equivalences for Group Rings. X, 146 pages. 1998.

Vol. 1686: J. Azéma, M. Émery, M. Ledoux, M. Yor (Eds.), Séminaire de Probabilités XXXII. VI, 440 pages. 1998.

Vol. 1687: F. Bornemann, Homogenization in Time of Singularly Perturbed Mechanical Systems. XII, 156 pages. 1998.

Vol. 1688: S. Assing, W. Schmidt, Continuous Strong Markov Processes in Dimension One. XII, 137 page. 1998.

Vol. 1689: W. Fulton, P. Pragacz, Schubert Varieties and Degeneracy Loci. XI, 148 pages. 1998.

Vol. 1690: M. T. Barlow, D. Nualart, Lectures on Probability Theory and Statistics. Editor: P. Bernard. VIII, 237 pages. 1998.

Vol. 1691: R. Bezrukavnikov, M. Finkelberg, V. Schechtman, Factorizable Sheaves and Quantum Groups. X, 282 pages. 1998.

Vol. 1692: T. M. W. Eyre, Quantum Stochastic Calculus and Representations of Lie Superalgebras. IX, 138 pages. 1998.

Vol. 1694: A. Braides, Approximation of Free-Discontinuity Problems. XI, 149 pages. 1998.

Vol. 1695: D. J. Hartfiel, Markov Set-Chains. VIII, 131 pages. 1998.

Vol. 1696: E. Bouscaren (Ed.): Model Theory and Algebraic Geometry. XV, 211 pages. 1998.

Vol. 1697: B. Cockburn, C. Johnson, C.-W. Shu, E. Tadmor, Advanced Numerical Approximation of Nonlinear Hyperbolic Equations. Cetraro, Italy, 1997. Editor: A. Quarteroni. VII, 390 pages. 1998.

Vol. 1698: M. Bhattacharjee, D. Macpherson, R. G. Möller, P. Neumann, Notes on Infinite Permutation Groups. XI, 202 pages. 1998.

Vol. 1699: A. Inoue,Tomita-Takesaki Theory in Algebras of Unbounded Operators. VIII, 241 pages. 1998.

Vol. 1700: W. A. Woyczyński, Burgers-KPZ Turbulence,XI, 318 pages. 1998.

Vol. 1701: Ti-Jun Xiao, J. Liang, The Cauchy Problem of Higher Order Abstract Differential Equations, XII, 302 pages. 1998.

Vol. 1702: J. Ma, J. Yong, Forward-Backward Stochastic Differential Equations and Their Applications. XIII, 270 pages. 1999.

Vol. 1703: R. M. Dudley, R. Norvaiša, Differentiability of Six Operators on Nonsmooth Functions and p-Variation. VIII, 272 pages. 1999.

Vol. 1704: H. Tamanoi, Elliptic Genera and Vertex Operator Super-Algebras. VI, 390 pages. 1999.

Vol. 1705: I. Nikolaev, E. Zhuzhoma, Flows in 2-dimensional Manifolds. XIX, 294 pages. 1999.

Vol. 1706: S. Yu. Pilyugin, Shadowing in Dynamical Systems. XVII, 271 pages. 1999.

Vol. 1707: R. Pytlak, Numerical Methods for Optical Control Problems with State Constraints. XV, 215 pages. 1999.

Vol. 1709: J. Azéma, M. Émery, M. Ledoux, M. Yor (Eds), Séminaire de Probabilités XXXIII. VIII, 418 pages. 1999.

Vol. 1710: M. Koecher, The Minnesota Notes on Jordan Algebras and Their Applications. IX, 173 pages. 1999.

Vol. 1711: W. Ricker, Operator Algebras Generated by Commuting Projections: A Vector Measure Approach. XVII, 159 pages. 1999.

Vol. 1712: N. Schwartz, J. J. Madden, Semi-algebraic Function Rings via Reflections of Partially Ordered Rings. XI, 279 pages. 1999.